Music Technology Workbook

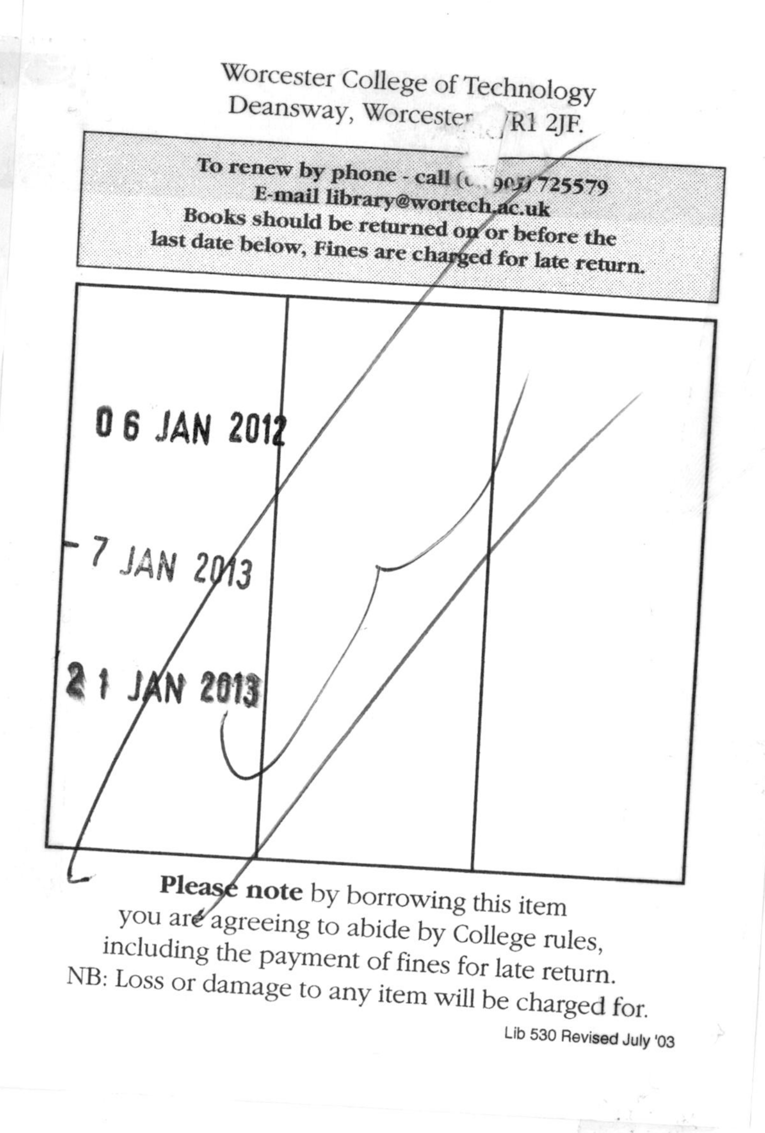

Music Technology Workbook

Key concepts and practical projects

Paul Middleton

Steven Gurevitz

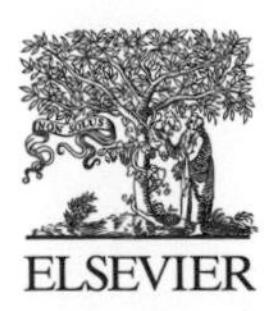

AMSTERDAM • BOSTON • HEIDELBERG • LONDON • NEW YORK • OXFORD
PARIS • SAN DIEGO • SAN FRANCISCO • SINGAPORE • SYDNEY • TOKYO
Focal Press is an imprint of Elsevier

Focal Press is an imprint of Elsevier
Linacre House, Jordan Hill, Oxford OX2 8DP, UK
30 Corporate Drive, Suite 400, Burlington, MA 01803, USA

First edition 2008

British Library Cataloguing in Publication Data
A catalogue record for this book is available from the British Library

Library of Congress Cataloging-in-Publication Data
A catalog record for this book is available from the Library of Congress

ISBN: 978-0-240-51970-8

For information on all Focal Press publications
visit our website at www.focalpress.com

Typeset by Charon Tec Ltd (A Macmillan Company), Chennai, India
www.charontec.com

Printed and bound in Great Britain

07 08 09 10 11   11 10 9 8 7 6 5 4 3 2 1

Contents

Acknowledgements

Paul Middleton would like to thank Amanda Parker for her help and support throughout this project. Thank you also to Roger Dunn, Bea Eden, Gianfranco Sciacca, Caroline Manyon and Townhouse Studios.

Thank you to all at Focal Press and Elsevier Ltd for your patience and support – Catharine Steers, Stephanie Barrett, Emma Baxter, David Bowers, Margaret Denley, Lisa Jones.

Thank you to the following software manufactures and distributors – Steinberg, Apple, Mark Of the Unicorn, Digidesign, **Propellerhead Software**, **Cakewalk**, Music Track, Roland, Tascam, Drawmer, Native Instruments, **Roxio**, Spectrasonics, Paul Kaufman – IK Multimedia, Mandy Rayment – Time + Space Distribution Ltd, Colin Meger – M Audio, Julian Coultas – Apple, Byran Borcherds – Arbiter Group.

Paul Middleton can be contacted via Focal Press.

1 Introduction

Welcome to the *Music Technology Workbook*. The aim of this book is to offer help and support to students, teachers and music enthusiasts alike. Throughout the book we will guide you through key aspects of music technology, recording techniques and basic studio skills. We will also offer hints and tips on techniques and production skills, and of course cover the basic concepts of what is included in a MIDI sequencer and digital audio recording package.

Many of you may be involved in either the teaching of or taking part in the Edexcel GCSE and A/AS challenges. This book is aimed at you. We hope to guide you through much of the coursework requirements, bringing clarity and support where there might be confusion. If you are not studying or teaching, don't worry, as all the information and techniques covered in this book are just as useful for beginners and enthusiasts alike. After all, everyone should be encouraged to use technology to write, create and perform music with confidence.

Technology

Music itself has always been changing and evolving. The people and skills involved in creating music have also changed over time. Before the advent of modern technology, if you wanted to hear music, you either had to create it yourself or listen to someone else perform it.

As time progressed it became possible to translate and express musical ideas on paper, using notation. This revolutionized the world of music as it provided a new medium for the music to exist on. A composer's music could now be read and performed by a variety of different musicians or orchestras, giving everyone a wider choice of where to listen to and perform music.

note ▶ The composer and performer concept still exists in many different forms of music today, as some artists don't write their own songs but perform songs written by other composers.

Once technology was able to capture sound and reproduce it, this provided yet another medium for music to exist on. The listener now had a choice between hearing music live or in a pre-recorded format. For the first time the connection between the performer and listener was changed.

As recording techniques developed and became more sophisticated, the actual recording process itself started to become an integral part of the artistic process, making it possible to create unique pieces of music on a recording that would be impossible to concieve on manuscript paper or perform live. The recorded medium became the music, the performance and the act of writing, all in one.

note ▶ The recording process can also be seen as an artistic process, making it possible to create unique pieces of music that could only exist in a recorded form.

So when creating your own music remember that what you hear is the most important thing, not how the dots are arranged. The sounds you choose and how they are mixed, along with many other variables, are now just as important as the actual notes themselves. A MIDI sequencer and recording system is just like manuscript paper – a medium that can be used to capture and interpret musical ideas.

producer says ▶ Throughout this book we will encourage you to compose and create your own music. We will provide you with some practical examples that hopefully can be used as a springboard for you to create your own ideas. After all, a computer or recording system, no matter how sophisticated, can only be used to express and capture your ideas – it doesn't write the music for you.

To help you learn and develop new skills, we have included some step-by-step practical exercises and examples that are common to most popular sequencing software packages and pieces of audio equipment. You should therefore be able to be complete most of the projects and exercises contained in this book using one of the MIDI sequencing and recording software packages listed below:

- Cubase
- Logic
- Pro Tools
- Digital Performer
- Sonar
- Reason
- Rewire
- Garage Band.

This book is not a manual on how to use one particular piece of software but instead a guide on how best to use them to achieve the necessary results. For a more detailed look at the different types of software mentioned, see later in this chapter.

Chapters 1–4 and 7–10 cover theory and background information and are essential reading before progressing to the relevant practical exercises and projects (Chapters 5 and 6 for the former, encompassing MIDI sequencing, and Chapters 11 and 12 for the latter, covering audio recording). Starting with simple exercises, to help you gain basic skills, we progress on to the main exercises, which are designed to reflect the curriculum and so, when completed, should result in you being able to hand in coursework.

A CD is also included so you can listen to some of the practical examples mentioned and load any of the song files.

And finally, let this book guide you in learning the basic skills of creating music using technology and achieving the best musical outcome from your technology investment.

Overview of software featured in this book

Cubase

Cubase is a versatile MIDI and audio workstation. It is generally very accessible to use for MIDI sequencing and is a good choice for beginners. The main Arrange window is clearly laid out and it allows both audio and MIDI information to be displayed and edited. There are several different edit windows that allow you to display MIDI information in a variety of different ways, such as in a list, Piano-roll-style or even as notation on a score. It also has a dedicated drum editor. The audio side features a mixer and numerous effects and several virtual instruments which include a built-in universal sound module.

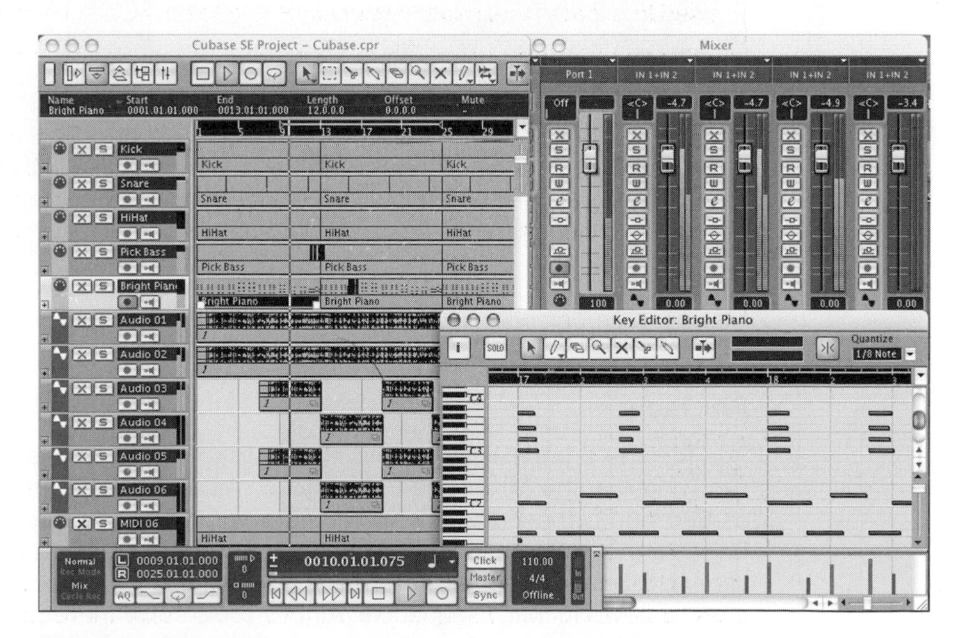

Fig. 1.1 – Cubase.

Cubase is cross-platform so it will work on both PC and Mac computers. There are also several different versions of Cubase available. The main differences between them is that the entry level versions tend to offer less tracks and 'enhanced functionality', but essentially they are the same program.

Logic

Logic software is a comprehensive MIDI and audio sequencing program that offers some excellent MIDI and audio editing features. It is based around several different windows and offers many different preferences and assignable functions that allow the program to be configured in a way that suits you. There is no shortage of sounds as there are lots of software instruments such as synthesizers and samplers built in. The audio side also features many different effects and a high-quality reverb.

There are two different versions available: Logic Express and Logic Pro. Both provide the same working environment but Express has slightly less features and is aimed mainly at students and hobbyists. Since Apple acquired Logic in 2002 this software is only available to Mac users.

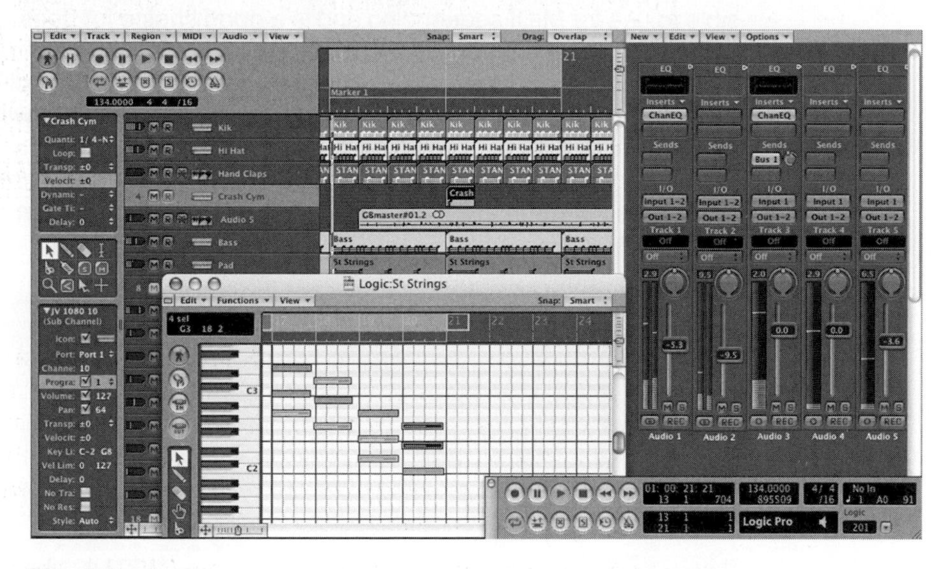

Fig. 1.2 – Logic.

Pro Tools

Pro Tools is a popular audio and MIDI editing package. It is based around two windows: the Mixer and Edit window. All editing and arranging can be carried out in the Edit window, whereas volume and pan settings can be adjusted in the Mix window. The audio side of Pro Tools is excellent and from version 6.7 a sophisticated set of MIDI features has also been added, making it into a powerful MIDI sequencer. All Pro Tools software requires some kind of Pro Tools hardware in order to work. This can range from a basic audio interface with two inputs and two outputs up to a very sophisticated system with multiple audio interfaces that provide lots of audio inputs and outputs.

There are two different versions available: Pro Tools and Pro Tools LE. Both provide the same working environment, but LE draws its processing power from the computer and has slightly less features and is limited to only 32 audio tracks. Pro Tools HD (or TDM) on the other hand provides a very powerful system, allowing you to use lots of audio and MIDI tracks, but requires additional cards to be fitted into a computer and a compatible audio interface. Pro Tools is available for both the Mac and PC.

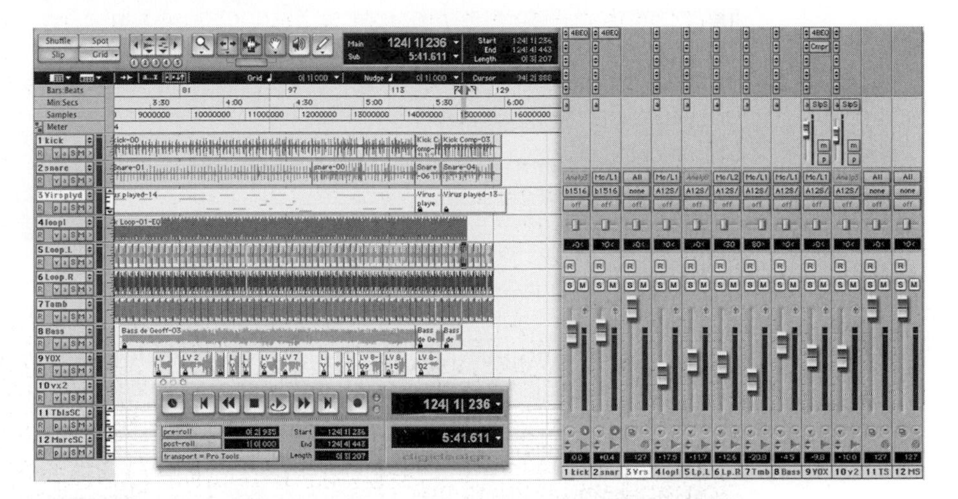

Fig. 1.3 – Pro Tools.

Digital Performer

Digital Performer is a powerful MIDI and audio sequencer that comes with plenty of plug-ins and effects for manipulating MIDI and audio. It is based around several different windows and offers a consolidated window function that allows the user to customize sections of a window to suit their own needs. Digital Performer's track layout and mixer are well laid out, which makes the program very intuitive and easy to use.

Digital Performer is made by Mark of the Unicorn (MOTU), who also make high-quality audio interfaces for recording and playing back audio with a computer. They also make an excellent software sampler called Mach 5 which can be used with Digital Performer or other popular sequencing software. All MOTU products provide excellent documentation and tutorials, which helps when getting started with them. Digital Performer software is only available to Mac users.

Sonar

Sonar is a versatile MIDI and audio workstation that originated from a program called Cakewalk Pro Audio. There are several different Edit windows that allow

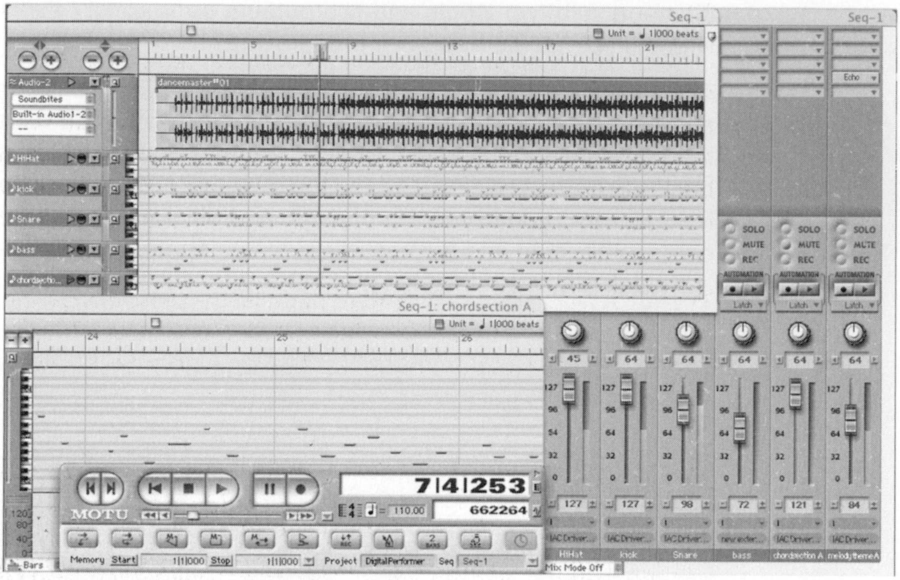

Fig. 1.4 – Digital Performer.

you to display and edit MIDI and audio information in a variety of different ways. It also features a groove clip function that allows you to easily change the tempo and pitch of an audio loop, making it ideal for loop-based composition.

Sonar is available in two different formats: Producer and Studio editions. The studio edition has slightly less features and does not include the session drummer and synth rack. This is a PC-only product.

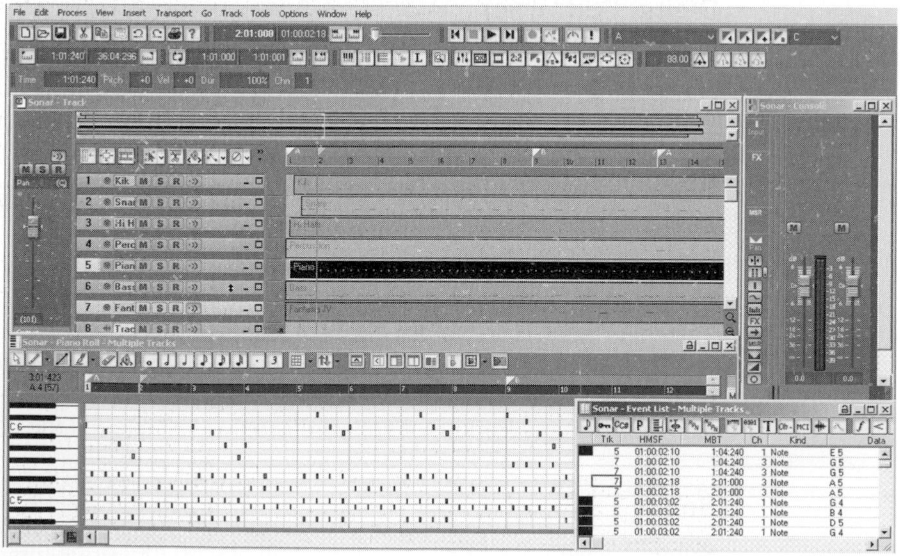

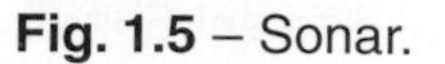

Fig. 1.5 – Sonar.

Reason

Reason has a slightly different approach to most other sequencing packages as it doesn't allow you to record audio. It does, however, contain lots of built-in sounds sources and effects and has a sampler that allows you to play back and edit existing audio files. There is even a loop player that allows you to play back special types of audio files at any speed. Each device appears as a module in a rack so you have to decide how the different devices are physically connected together.

The MIDI side offers a simple, uncomplicated way of working, allowing you to perform basic MIDI sequencing functions. It even offers a pattern-based sequencer and a drum machine. Reason works on PC or Mac and can also

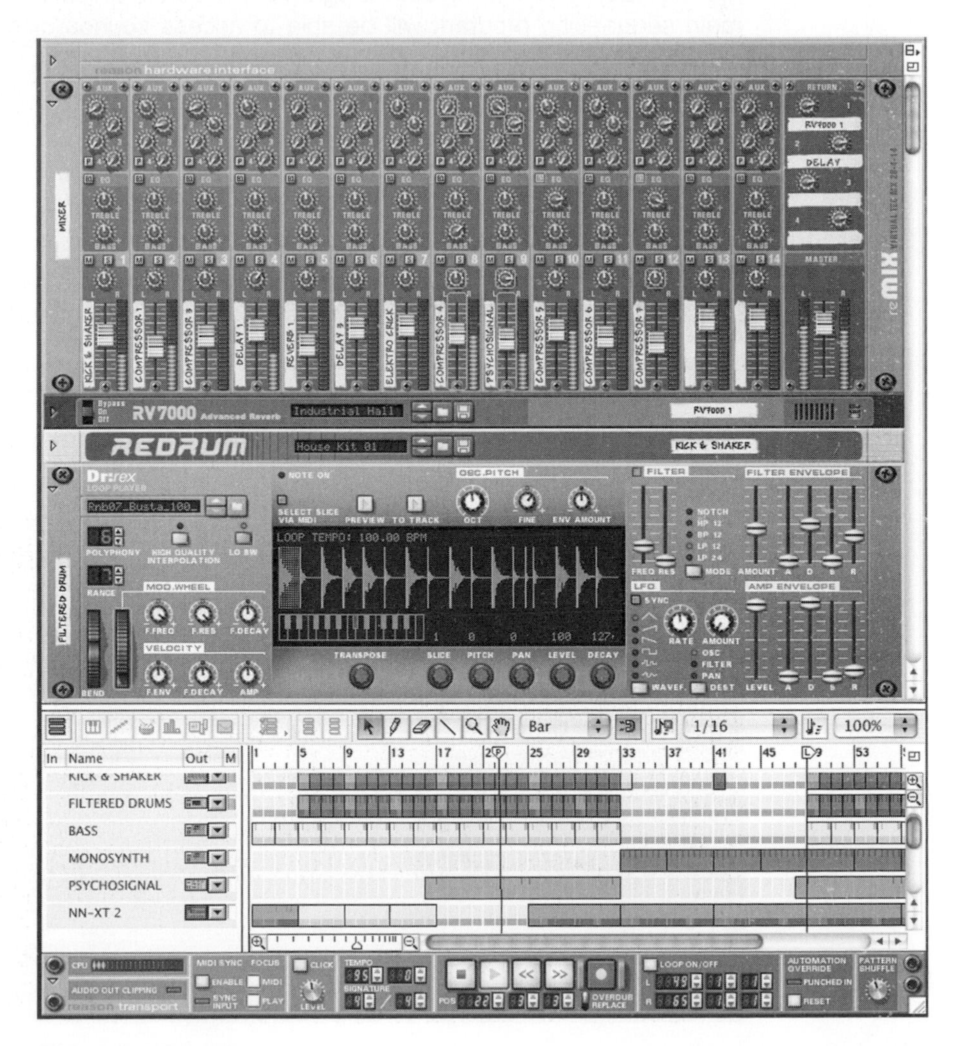

Fig. 1.6 – Reason.

be used purely as a sound source for other sequencers, such as Pro Tools, Logic or Cubase via Rewire.

Rewire

Rewire is a system that allows different pieces of audio software to communicate with each other while running together on the same computer. It works like an invisible cable that streams audio from one program to the audio mixer of another program. In order for this to work successfully you need to use two Rewire-compatible pieces of software. You then designate one program, such as Cubase, Logic or Pro Tools, as your main sequencing program and another program such as Reason as a Rewire slave. Once this has been configured the transport controls of both programs will synchronize together and the main sequencing program will be able to access sounds or audio from the Rewire slave.

Garage Band

Garage Band is a Digital Audio Workstation that is part of Apple Computers' iLife bundle. It comes with a large selection of built-in sounds and pre-recorded musical phrases called Apple Loops. This audio file format makes working with audio very straightforward, as you can just drag and drop audio clips directly into a song and easily change their speed and pitch. This makes it ideal for beinners as it instantly sounds good and is fun to use. The MIDI

Fig. 1.7 – Garage Band.

side offers a simple, uncomplicated way of working and is based around a piano-roll style edit window. This software is only available to Mac users.

producer says ▶

> It doesn't really matter which type of sequencing package you choose to work with, as they all basically allow you to achieve the same thing – to develop and capture music ideas. However, one of the key aspects in achieving this is learning how to get the best out of the system you are using.

2 Listening and Creative Skills

Approach to creating music

Music is a technical and creative subject. In order to communicate your ideas efficiently to others, you will either need to learn how to write music using traditional notation or learn how to interpret your ideas using a music technology program. It would also be an advantage to learn how to sing or play an instrument of some kind, so you can directly express your own ideas.

However, other skills you will need to learn are less tangible. You will need to learn how to listen. Hearing music is very different to listening to it. Often, people learn how to form chords or how to use a certain sequencer and sometimes they pick up those skills really quickly. However, they often struggle with the art of creating music or composing.

This is not unusual. Creating a tune or chord sequence from nothing seems almost magical. However, it is not. Almost everything we, as humans, create is based on something that has happened before. Just as in our speech – when we speak we are not inventing a new language from scratch but instead we are taking the sounds and structure of language, which we inherit from our parents and peers, and adapting it for our own use, place or situation.

Writing music is a similar process, where we can take inspiration from all the music we have heard before. We are surrounded by different forms of music and media, such as CD, vinyl, Internet, television, radio, etc., so it's difficult not to be influenced when we create our own music.

Listening skills

In order to understand the different components needed to create a piece of complex music, we need to be able to absorb music, really hear it and really listen to it. We will then find it easier to create our own music and place the instruments in an arrangement. For example, try listening to a piece of music and see if you can identify each individual instrument or part.

Examples of commonly used instruments

See if you can distinguish:

- *The bass.* A low-frequency instrument playing single notes. This could be a real bass guitar, or a synthetic bass or sampled sound produced electronically.
- *The chords.* These may be a piano or string sound playing a sustained or broken rhythmic pattern to fill the mid frequency.
- *The melody.* This is the top line and often the main focus of a piece of music. It is the main tune or theme, or the part a lead vocalist would sing.
- *The counter-melody.* Normally higher than the tune and filling in the gaps or enhancing the melody.
- *Guitar.* This can be electric or acoustic and could play a rhythm part using chords or be featured as a solo instrument.
- *Strings.* This could be a solo string instrument or an orchestra. If it's an orchestra, can you hear the first violins playing something different from the second violins? Are you able to distinguish the cellos (lower in pitch than violins) from the double-basses (even lower than the cellos)? Once you can identify which instruments are playing, listen to the rhythms being used.

Individual drum sounds

- *Kick drum.* Low-pitched percussion sound playing a regular pattern.
- *Snare drum.* Snappy mid percussion sound playing a regular pattern.
- *Hi hat.* Very thin, high-frequency percussion sound playing fast.

Creating a tune

People often find it very difficult to 'come up with' ideas or tunes. Some can do it with ease and these people have a natural talent, just as some people can play football really well, seemingly without much effort. However, composing is a skill that can be developed. All it takes is patience and perseverance. There are no set rules when creating your own music. Ideas can develop through a variety of different ways.

Here are some tips to help you create your own music:

- Listen to a wide variety of different music and absorb different styles and techniques. Don't just listen and 'sing along' but really listen to a track, over and over until you can predict what is happening and can understand how it has been designed.
- Music is not created in our fingers. Aimlessly hitting the keyboard waiting for a tune to appear will not always work. However, sometimes you may stumble across something that sounds good. If this happens, *stop* and try to recreate what you heard. Practise it over and over so it is secure in your mind.

- Ideas can develop through collaboration, perhaps in a band by jamming. *Jamming* is where everyone collectively experiments with new ideas. Out of this melting pot of music experimentation might come some interesting chords, melodies or riffs. Riffs are catchy repeating patterns. These patterns can be used as a backing track for a melody or to actually form themes in themselves.
- Try singing – instead of hitting a keyboard randomly try humming or singing a melody to yourself. Try singing a tune or single notes and, when you hear something you like, only then attempt to slowly recreate it on the keyboard. You may find it easier to sing a melody, as you are not restricted to just the notes on a keyboard.
- Experiment with changing between chords (called creating a chord sequence – see below). These create the colour and emotion in music and can trigger ideas.
- Copy an existing melody or chord sequence from a well-known tune. Use this as a template and then add new ideas on top. Once you have built up an idea, remember to delete the original material, if still in the piece, otherwise it will not be an original composition.
- Listen to a piece of music you like and create a pastiche. A pastiche is not a copy but is something that is designed to sound very similar and to be of the same genre. It should sound as though the original composer had written it themself.
- Try playing an instrument such as a piano as if it was a drum or percussion instrument. Stick to only one or two notes and play a fast, 'catchy' groove. Sometimes this help to trigger further ideas.

Experienced writers will do all the above, but do it so fast it seems that they are just magically creating music as they go along. They are doing no such thing. Not even a jazz musician who is 'improvising' is really 'making it up'. Instead, all they are doing is putting together phrases and expressions. These phrases are usually well rehearsed and practised, but it all happens so fast it seems that it is spontaneous.

producer says ▶

Inspiration

Professional songwriters often collaborate with other songwriters in order to help each other, as one person may just write music and the other the lyrics or melody. There are no rules when it comes to creating an original idea. A mood or conversation may suddenly cause inspiration totally unexpectedly.

Inspiration can strike at any time! It has also been known for songwriters to get 'writer's block', which is where they are unable to write. If this happens to you, keep trying. Ideas can develop in a variety of different ways and you might get lucky.

If you're new to composing, take each step slowly and try and listen carefully to what you are doing.

The keyboard

Figure 2.1 is an example of a keyboard. Notice how it follows a regular, repeating pattern every eight notes. Located in the centre of the keyboard is middle C. This corresponds with C3 on most electronic MIDI keyboards. A standard MIDI keyboard usually has around five octaves (61 keys); however, this can vary.

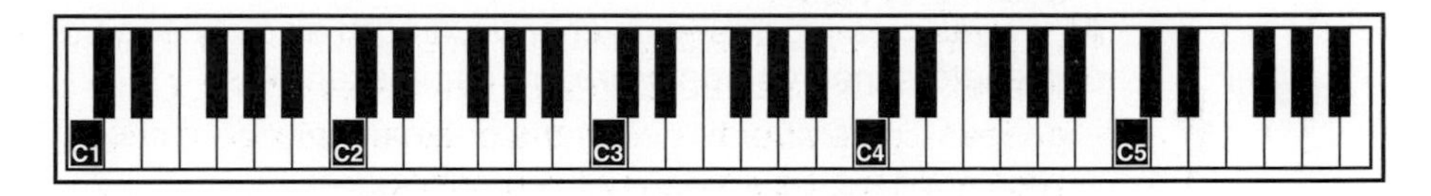

Fig. 2.1 – Octaves.

tip ▶ When using a MIDI keyboard, most sequencers will display the name of the notes you play.

Octaves

The regular interval or pattern between every seven notes is called an octave. This allows you to play the same note at higher or lower pitch.

Harmony

If you are planning to use several notes or instruments at the same time, it will help to gain a basic understanding of *harmony*, the art of mixing and matching different pitches.

note ▶ Harmony can be created by arranging the notes between separate instruments. It can also be created by the same instrument if it's capable of playing more than one note at the same time, such as a piano. When MIDI sequencing, this is usually possible with most instruments.

Chords

A chord is a selection of notes that resonate well with each other. In other words, several notes that sound good when played together.

So how do we create chords? The first technique is 'the trial and error method'. Put simply, try hitting any note and then find another note that sounds good whilst played with it. Once you have found two notes that sound good together, try adding a third note to form a chord.

There are, however, other ways of creating chords that can take less time.

Try hitting any white note. Now move five notes up from this note (when we count 5, we include the note we started with and don't include black notes). Now hold these two notes down together and find a middle note to form a *chord*.

As well as creating your own chords you can also choose from a wide variety of predetermined chords. Here are some examples.

Major chords

The C major chord contains three individual notes – C, E and G.

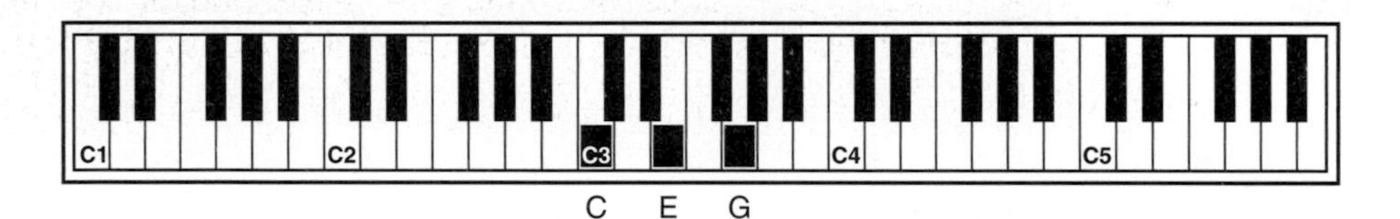

Fig. 2.2 – The C major chord.

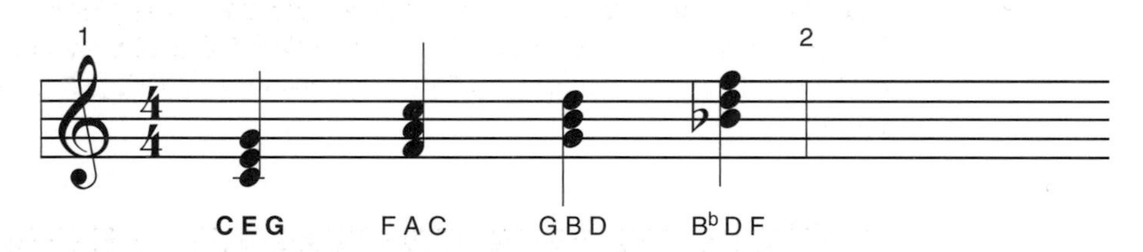

Fig. 2.3 – Examples of major chords displayed as notation.

Minor chords

The above chords are all 'happy sounding' and are called major chords. If we want something a bit more solemn, perhaps even sad, then we need to drop the middle note of each chord by what is called a half note or semitone. This is essentially a half note movement.

The C minor chord also contains three individual notes – C, E^b and G. Notice how the middle note has changed from E to E^b. (See Fig. 2.4.)

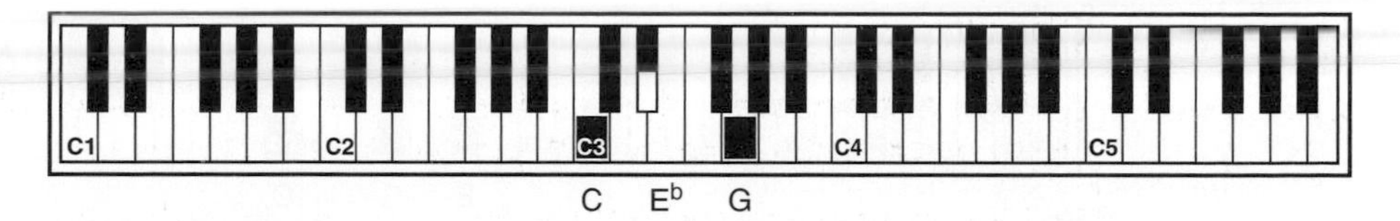

Fig. 2.4 – The C minor chord.

If we change all the previous major chords in Fig. 2.3 into minors they would look like Fig. 2.5.

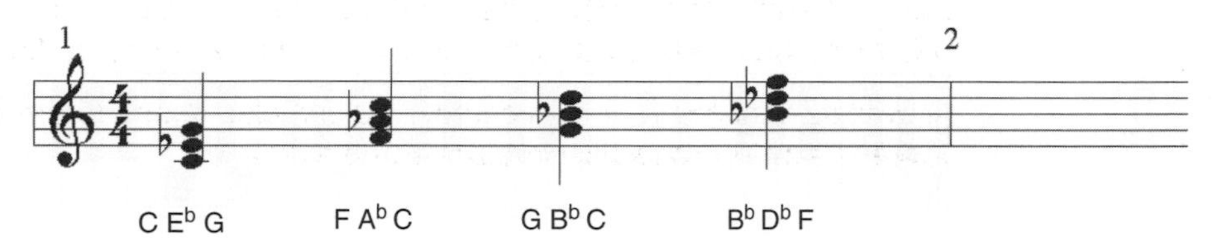

Fig. 2.5 – Examples of minor chords displayed on a music score.

tip ▶ The chords you choose will determine the mood or feel you want to create. For example, playing a major chord will sound happier than a minor chord.

Chord inversions

The same three notes within a chord can be played back in three different ways. For example, the chord of C major consists of the single notes C, E and G. When it is played back this way it is called the root position as the chord starts with its main root note, C. The other notes E and G just support and add colour.

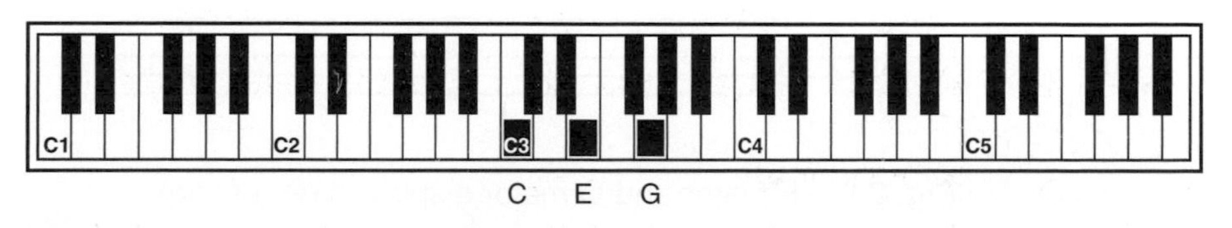

Fig. 2.6 – C major chord in root position.

Now play the chord with E as the lowest note. Move the note of C up one octave higher to the next C and you will be playing back the same chord of C, but in its *first inversion*.

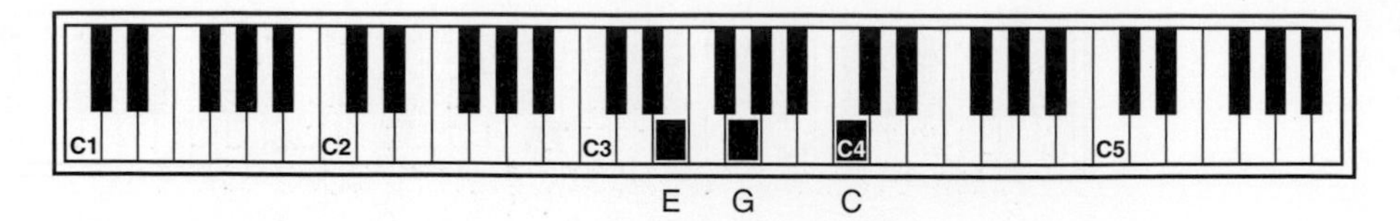

Fig. 2.7 – First inversion.

Now play the chord with G as the lowest note. Move the note of E up one octave higher to the next E and you will be playing back the same chord of C, but in its *second inversion*.

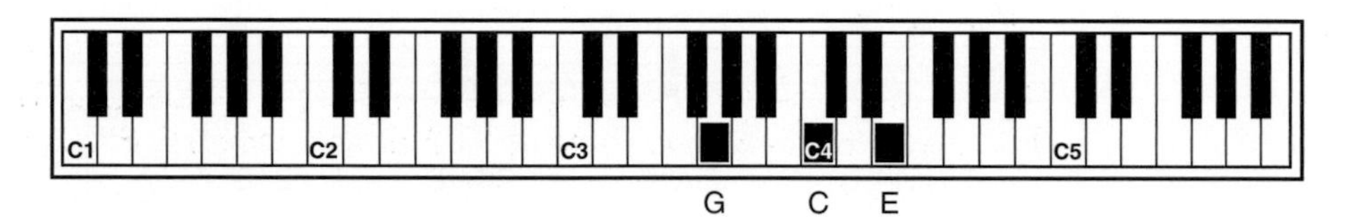

Fig. 2.8 – Second inversion.

All chords can be played back using inversions, so experiment. You may find it easier to change between different chords while using inversions, as you may only have to move one note rather than three.

Moving between different chords

An example of changing between two different chords using inversions is shown in Fig. 2.9. The F major chord contains F, A and C, and C major contains E, G and C. Notice how C is used in both chords, so you only need to change two notes rather than three.

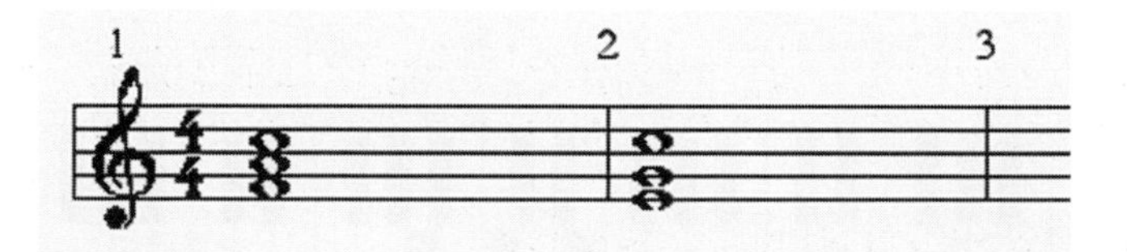

Fig. 2.9 – F major and C major displayed on a score.

An example of changing between a major and minor chord using different chord inversions – C, E, G to A, C, E – is shown in Fig. 2.10. Notice how C and E are used in both chords, so you only need to change one note rather than three.

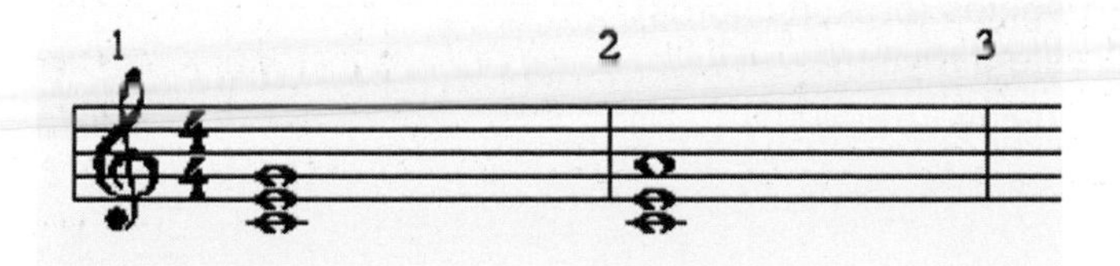

Fig. 2.10 – C major and A minor displayed on a score.

Developing a chord sequence

Once you have found a few chords you like, you can try moving between them to create a chord sequence or pattern.

Example 1

Try moving the entire chord up or down, from C major (C E G) to D major (D F# A) and back again to C major (C E G), as shown in Fig. 2.11.

Fig. 2.11 – Example chord sequence.

Example 2

Thirds are where the chord leaps a third. Some of the notes in the first chord will also be in the second, so you only need to move two notes rather than three. This also helps to make the chord sequence easier to play and results in the blend between the two chords sounding better, e.g. C E G to E G B to G B D and then maybe jump back to C E G (Fig. 2.12).

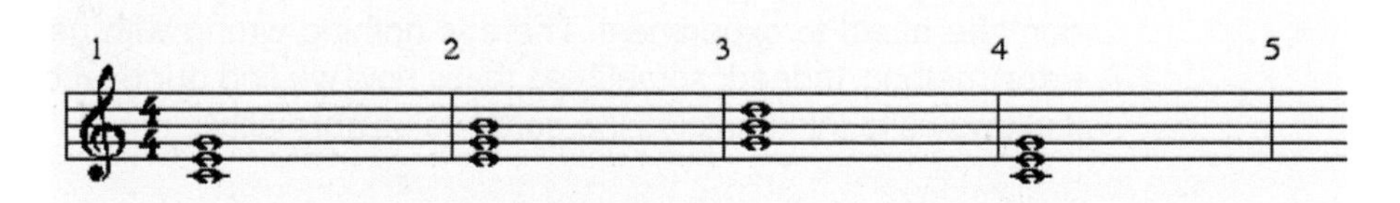

Fig. 2.12 – Example chord sequence.

Different styles of playing chords

When playing or performing a chord sequence it is possible to play the notes in a chord in a variety of ways:

- All at once
- As a stab (all at once but very quickly)
- Sustained (all at once but held on)

- Arpeggios (broken chords)
- Create a rhythm
- Changing inversions.

Bass notes

Once you have chosen a chord you will need to find a bass note or bass instrument to go with it. Try and work out what is the most dominant note in the chord. This note will probably be the most suitable to play alongside the chord. Try it and see. You need to find a note that blends with the chord so that it doesn't clash and sound discordant (unless you are deliberately trying to create discordant music). As always, listen carefully to what you create.

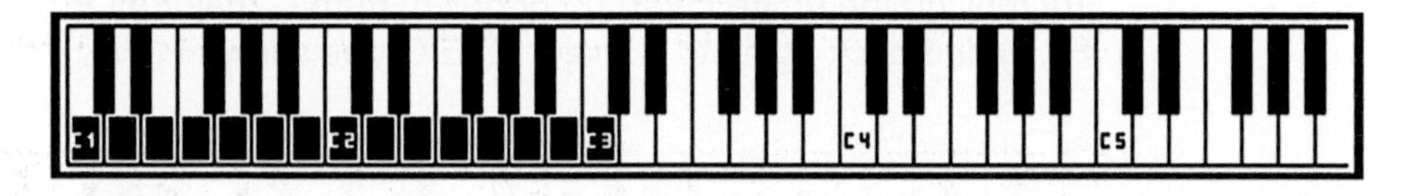

Fig. 2.13 – Bass notes are usually played on the lower section of a keyboard between C1 and C3.

producer says ▶

Start by using the root note of the chord to help form your bass note (a root note is the main note that a chord is based around). If a chord is played in its root position, try using the first note of the chord. If you have used the chord A major, for example (notes A, C, E), try playing the note A in the bass.

As each chord changes, you will often find that the bass note will too. However, don't be afraid to experiment. There is nothing wrong with using the 'trial and error' method. Indeed, sometimes that's how we find our best combinations, as it allows us to think outside the confines of 'normality'.

tip ▶

Sometimes using the same bass note while changing the chords can create an interesting harmony. This is called creating a pedal note.

producer says ▶

Creating a bass line

The bass line is usually the low-frequency element to your music, which will often work alongside the kick drum to provide a solid structure to build other instruments on. A real bass guitar usually has four strings – E, A, D and G – just like the top four strings on a six-string guitar (five-string bass guitars also have an additional low B string). The bass guitar strings are one octave lower in pitch than regular six-string guitars, so when playing a bass sound on a MIDI keyboard try and use the same note range as real bass.

Scales

Once you have a working chord sequence you will probably find that there is a scale that can be used with the music. A *scale* is a selection of notes. An example of this would be the 'C major scale', where you only play the eight white notes and where C is the dominant note. If you can 'discover' what scale works best with your chords and then use the notes in that scale, you will find making *melodies* and tunes much easier, as you will never play a 'wrong' note.

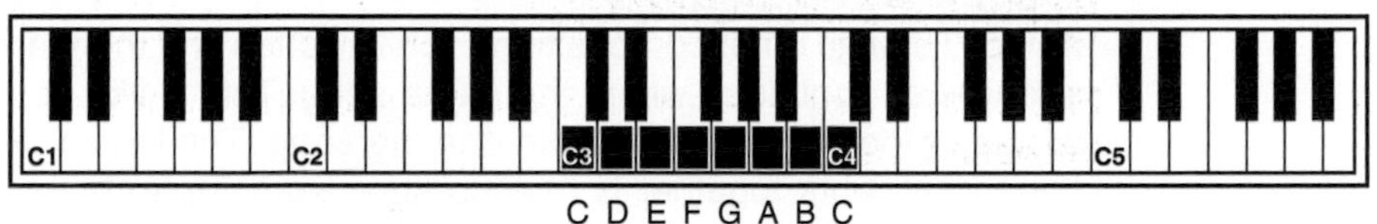

Fig. 2.14 – The C major scale.

Song structure

Music is a linear experience. This means there is always a beginning and an end. Sometimes, music can take you on a journey where there are different stages along the route. Most popular music can be broken down into smaller sections that are arranged to create the overall song. The smaller sections vary throughout a song to create a structure. When listening to some songs you can nearly predict what section is going to come next as they follow a set pattern or formula. These main sections of a song are often defined as intro, verse and chorus.

So what is the difference between a verse and a chorus? Well there are no hard and fast rules. However, there are some standard formulas, which are commonly used in pop music.

A *verse* is normally quite rhythmic, as lyrically it's where the exposition (the narrative of a story) is delivered.

A *chorus*, on the other hand, will be catchy. It should have a 'hook' or a memorable theme.

Having a chord sequence (a four- or eight-bar group of chords) followed by a separate chord sequence, which complements the first sequence, allows you to create an ABA chord structure. Adding another new unique chord sequence creates a C section. This is enough to create a song structure which reads as ABABCBBB. Here is an example of a typical song structure:

1	Intro	1| 1| 000
2	Verse	9| 1| 000
3	Bridge	17| 1| 000
4	Chorus	21| 1| 000
5	Verse	29| 1| 000
6	Bridge	37| 1| 000
7	Chorus	41| 1| 000
8	Middle 8	49| 1| 000
9	Chorus	57| 1| 000
10	Chorus	65| 1| 000
11	End	73| 1| 000

Fig. 2.15 – Typical song structure.

As you listen to other compositions you will often find there are different chord sequences contained within the same song. This is like having lots of small songs or ideas contained within one big song. The key to getting these sections to all work together is making sure each separate section blends well into the next.

Non-melodic song structures

Of course, not all music is melodic or even has a melody. African drumming has rhythmic themes but not a melody. Minimalism uses strong repetitive themes; however, it is rare that you would hear an overpowering melody in this style of music, even when it is in its purest form.

So, if you are creating pieces of music without strong melodies, how can you create the feeling of structure? Sometimes it's by arranging the instruments, rather than by changing what notes the instruments are playing; by muting certain instruments out of the music (perhaps for eight or 16 bars) to create 'space' and then slowly reintroducing them back in.

producer says ►

> Sometimes it's not only the melody or chords you need to change. Having the same bass line going across the whole tune can get a bit tiresome (unless it's creating a great groove you want to dance to), so why not vary it with a new rhythm.

The arrangement of a piece of music should take you on a journey. Imagine being on a roller-coaster ride! We have the start, which is slow and gradual; however, soon we are slowly moving up towards a peak. We feel huge suspense and tension as we approach the top. Once at the top, there is a moment of complete calm quickly followed by a 'woosh', which is when we go racing all the way down. At this moment we are feeling complete ecstasy or even fear.

However, does the ride continue on this 'high' forever? No. Once again, it slows and starts to build up, repeating the process over and over again. Eventually we come back to where we started and we find ourselves calming down and our emotions and heart rate slow. Music should be trying to create a similar experience but perhaps with some variations. This would constitute a structure.

producer says ▶

Arranging techniques

Arranging can be split into several disciplines: arranging sections of pitched music, which either have very different chord sequences or melodies, or via the discipline of arranging the order in which instruments and sounds play back together. You can also effect the arrangement by layering an instrument in terms of pitch to create an interesting sound. Additionally, be aware of the 'best' place to pitch a sound in context with the other instruments. Do you really want the bass and melody covering the same range and hence playing at the same pitch? If you are still unsure as to how to place instruments or parts, refer back to 'Listening skills' at the start of this chapter to see how other composers have arranged their music.

When creating a non-acoustic composition, getting the best out of an instrument is very important. Do not assume a sound or patch titled 'bass' must be played low down on the keyboard as a bass. Trust your ears – if you can find another use for it and *if* it sounds good, then use it.

However, you need to remember that real instruments are often limited in their ability to play certain pitches, as they have a natural range. For example, a trumpet can only play from around a low G to a high C. Hence, if you are trying to recreate an acoustic arrangement whilst using MIDI remember this, as you may be exceeding the natural range of the instrument.

Keyboard and instrument skills

It is worth learning the basics of a musical instrument. Certainly, the keyboard is a major part of today's music technology process and it would be advantageous to gain some basic skills. Whilst you don't need to become a pianist, you should try and develop some basic techniques of how to play softly, loud, fast and slow. Whilst you can edit some of your performance data after you have recorded, this can be fiddly and time-consuming and often doesn't sound as natural.

Try and observe other instruments by listening to them and watching them being played, as this will help you to perform with the keyboard. For example, when using a MIDI keyboard you may choose a harp sound. You can either thump the keyboard, or try and play it delicately by holding and releasing the keys, in a way that results in the harp sounding more natural, replicating how a 'live' harp would sound. Hence, by studying different instruments and techniques you can add more musicality to your compositions and performances.

Playing in time

Additionally, try and learn to play *in time*. This will be of great benefit if you're planning to use a MIDI sequencer and will also help you to understand how different rhythms work. When listening to music, try and identify each beat and bar by tapping your feet or try counting out loud!

Music is structured in bars, whether it is written on paper or created on a computer. The time signature determines where the bar lines fall and tells us how many beats there are in a bar. This helps determine the rhythm and fundamental pulse of the music. Here are some examples.

Fig. 2.16 – Different time signatures.

4/4	4 strong beats
3/4	3 strong beats
2/4	2 strong beats
6/8	2 strong beats

Try experimenting with different time signatures and see how they change the feel of the music.

3
MIDI Introduction

MIDI stands for Musical Instrument Digital Interface. As the name suggests, it is a way of connecting different musical devices together so they can exchange information with each other. It is important to note here that MIDI is based upon a transaction of information rather than actual sound itself.

note ► MIDI is not sound. It is simply a set of instructions that can be sent to a sound-producing MIDI device.

Understanding how MIDI works

The process of typing words on a computer is similar to transmitting MIDI data. Words are entered into a computer via a keyboard, not a musical one but a so-called QWERTY keyboard (named QWERTY due to the first six letters on the top left of the keyboard). When any one of the keys is pressed, a burst of information is sent to the computer and a character is then displayed on the screen. Transmitting MIDI data is very similar to this process in that when you press a note on a MIDI keyboard the information is transmitted to another MIDI device.

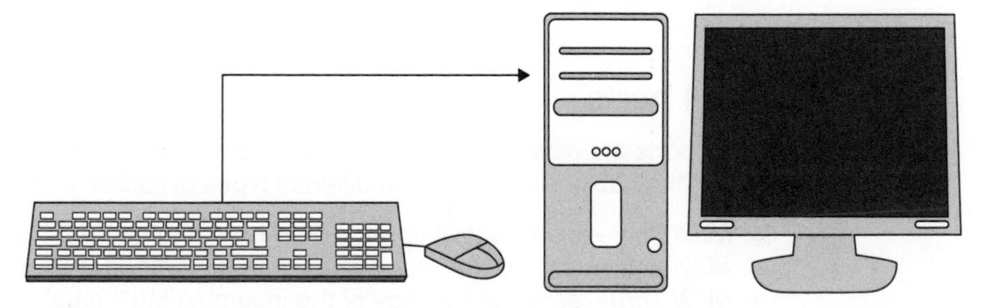

Fig. 3.1 – QWERTY keyboard connected to a computer.

MIDI connections

In order to communicate with other MIDI devices you will have to physically connect them together. A MIDI connection can be made using a standard five-pin DIN plug or sometimes directly to a computer by using USB or Firewire (see 'USB and Firewire' section below). The type of connection will vary depending on the equipment being used.

producer says ▶

Before MIDI

Before MIDI was developed it was almost impossible to connect devices made by different equipment manufacturers together, as each device often had its own style of connection that was incompatible with everyone else's. In 1983, MIDI became the agreed standard that all devices would use to communicate with. This gave everyone the freedom to make connections between a wider variety of MIDI devices made by any manufacturer.

MIDI cables have a five-pin DIN plug at each end of the cable and can be connected to any standard MIDI port. The maximum recommended length of a MIDI cable is 50 feet or 15 metres.

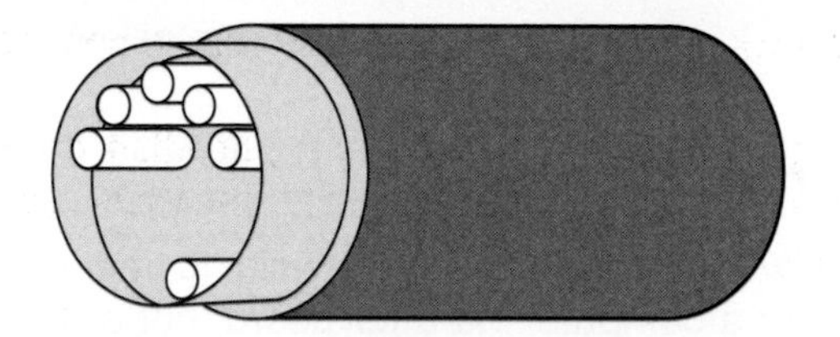

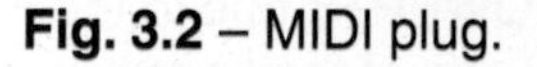

Fig. 3.2 – MIDI plug.

note ▶

DIN plugs

Five-pin DIN plugs only use the top three pins to transmit and receive MIDI data. It is therefore possible to use MIDI leads that only have three pins. Note: regular five-pin DIN hi-fi leads may not be able to transmit MIDI information as they are often configured differently.

MIDI ports

MIDI ports are usually located on the back of a device and provide connections for a MIDI cable. There are three different types of socket: in, out and thru.

- *MIDI In* – is for receiving MIDI data from another MIDI device
- *MIDI Out* – is for transmitting MIDI data to another MIDI device
- *MIDI Thru* – sends out a copy of the incoming MIDI data (this allows you to pass the MIDI data on to another MIDI device).

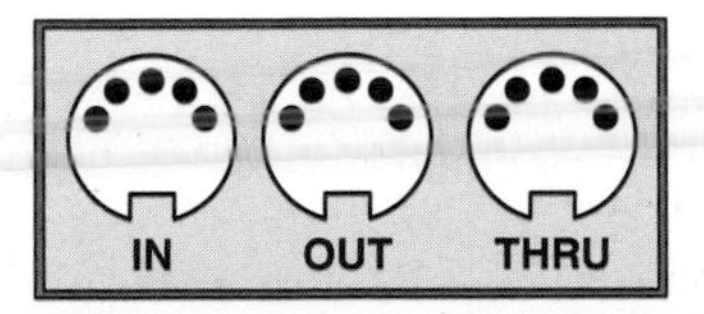

Fig. 3.3 – The three different MIDI ports.

Simple MIDI connections

The basic rules for connecting MIDI equipment together are:

- Outs go to Ins
- Ins go to Outs
- Thrus go to Ins.

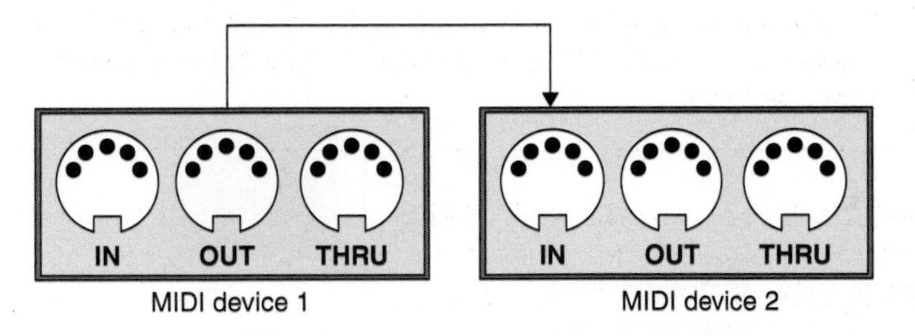

Fig. 3.4 – Example showing how two MIDI devices can be connected together.

In the example in Fig. 3.4, device 1 will send out MIDI data to device 2. Normally, the device you actually use to transmit the MIDI data is referred to as the master or controller.

note ▶ A MIDI cable allows any MIDI-equipped device to be easily connected to another MIDI device to exchange data.

Daisy chains

If you are planning to use several MIDI devices together at the same time you will need to connect them all together in a MIDI chain.

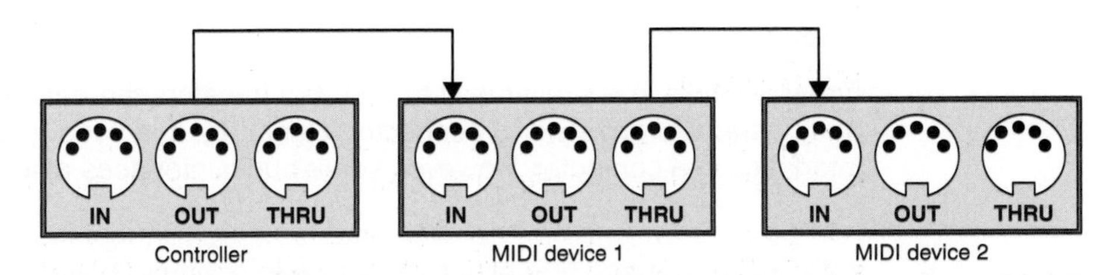

Fig. 3.5 – Example showing how three different MIDI devices can be connected together.

In the example in Fig. 3.5, the controller is able to send a MIDI signal to both devices by passing the data through the first device on to the second. This type of connection is known as a daisy chain.

tip ▶ A daisy chain provides a simple way of connecting several different MIDI devices together.

USB and Firewire

With the development of USB and Firewire on most computers, it is now possible to make a direct connection between a computer and a MIDI device without actually using a MIDI port or cable. This is only possible if the MIDI device itself has a USB or Firewire port.

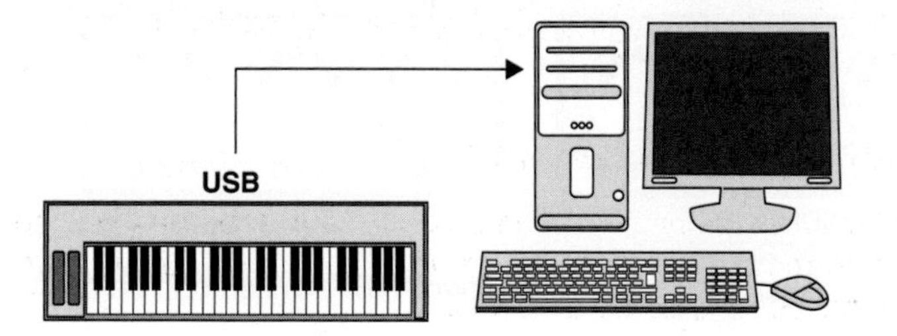

Fig. 3.6 – Keyboard connected to a computer using USB.

USB (Universal Serial Bus)

This is a computer connection widely available on Windows-based PCs and Apple Mac computers. USB allows data to be transmitted in both directions (bidirectional) with just a single connection. It is commonly used to connect printers, scanners and modems and can also be used to connect MIDI devices such as computer MIDI interfaces and MIDI controller keyboards. USB can also be used to connect audio interfaces, allowing the computer to record and play back sound.

Firewire

Firewire works in a similar way to USB but transfers the data at a faster rate and has a different style of connector. Generally, it is used to connect audio interfaces to a computer. However, some audio interfaces also include MIDI.

note ▶ Any keyboard with a MIDI output or USB connection can be used as a controller keyboard to send MIDI information and trigger sounds.

MIDI channels

If you're planning to use several different sounds at the same time or simply trying to communicate with another MIDI device, then you will need to be aware of MIDI channels. A standard MIDI system allows up to 16 separate streams of information to travel down one MIDI cable at the same time. This allows each sound or device to be assigned to a different MIDI channel number so it can be identified independently. This process is similar to posting a letter and putting the house number on it.

note ▶

So what are MIDI channels exactly?

Up to 16 different MIDI channels can travel down a single MIDI cable at the same time. This allows you to assign each separate sound or device its own MIDI channel number, therefore allowing you to communicate with it independently.

Assigning MIDI channels

In order to successfully communicate with another MIDI device, the MIDI send and receive channels need to correspond. If they don't, the MIDI data will simply be ignored by a device.

Transmit channel

The MIDI channel you transmit on will allow you to choose which device or sound you want to communicate with. This has to be determined and set up by you.

Receive channel

The device receiving the MIDI data needs to know which MIDI channels it should respond to. Don't assume this has been set up, as all MIDI devices can be set to respond to any MIDI channel.

producer says ▶

Setting up MIDI channels is an essential part of using any MIDI system. You have to decide which MIDI channels you will be transmitting and receiving MIDI data on. Generally, it's best to assign each separate device or sound to a separate MIDI channel so that you can communicate with it independently. As there are 16 MIDI channels, this allows you to have up to 16 different instruments playing at the same time.

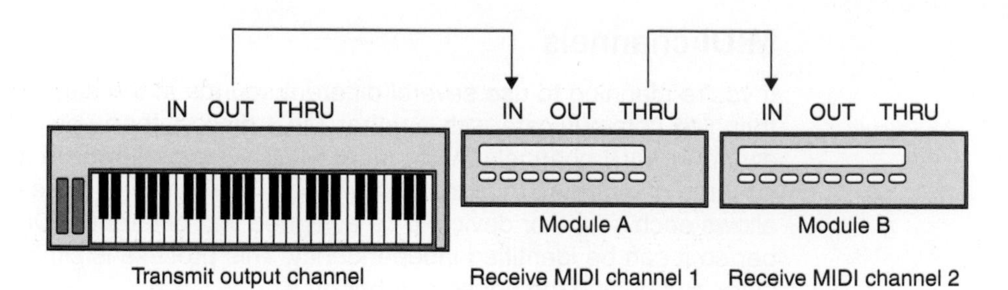

Fig. 3.7 – Controller keyboard connected to two sound modules.

In the example in Fig. 3.7, both sound modules will receive MIDI data from the keyboard as module A passes on the data via the MIDI Thru to module B. However, in order to communicate with each module separately you would need to assign a different MIDI channel to each device. Then, by changing the transmit channel on the keyboard, you could select which sound will play.

- Transmitting on MIDI channel 1, you will hear Module A
- Transmitting on MIDI channel 2, you will hear Module B.

note ▶ If both modules are assigned to the same MIDI channel, you will hear both sounds at the same time.

Even though all 16 channels may be passing through a device, the MIDI channel will let you decide if a device ignores or receives MIDI information. In the example in Fig. 3.7, transmitting data to Module B on MIDI channel 2 means Module A will ignore this.

Creating a layered sound

It is possible to combine two different sounds together. For example, if two different MIDI devices are assigned to the same MIDI channel, you will hear both sounds together.

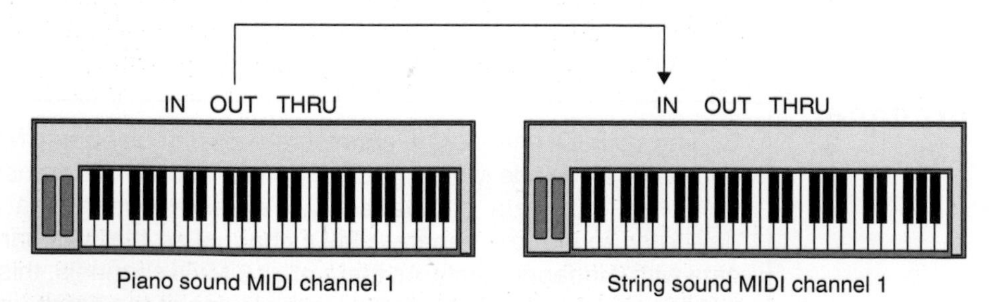

Fig. 3.8 – Two keyboards connected together using the same MIDI channel.

Omni

Some MIDI devices have an Omni on/off switch. This allows you to choose if a device will respond to one specific MIDI channel or whether it will respond to all the MIDI channels at the same time.

- *Omni on* – the device will respond to MIDI data on any MIDI channel
- *Omni off* – the device will only respond to MIDI data on the selected MIDI channel.

Multitimbral

Most sound sources are *multitimbral.* This allows a single MIDI device to generate several different sounds at the same time. Each sound is placed on a different MIDI channel and can be accessed independently. For example, if a device is 16-part multitimbral then it will be like having 16 separate synthesizers and will allow you to play up to 16 different sounds at the same time. When using a sound source with a MIDI sequencer, always choose a multitimbral mode as this will allow you to record and play back several different sounds at the same time.

producer says ▶

If a MIDI device is multitimbral it will allow you to play several different sounds at the same time. This makes it possible to build up complex arrangements using only one sound source.

Using MIDI with a computer

Most computers don't allow you to connect MIDI cables directly to them. This means that if you need to connect a MIDI cable to a computer you will need to use a MIDI interface. Most modern MIDI interfaces connect to a computer via USB.

note ▶

If you are planning to connect a MIDI device to a computer using a five-pin DIN MIDI cable, you will need to use a MIDI interface, as most computers don't allow you to connect MIDI cables directly to them.

Connecting a MIDI interface

A MIDI interface provides the link between a MIDI cable and a computer, therefore allowing you to connect MIDI devices to a computer. It is usually a small box that has five-pin DIN MIDI input and output connections and a computer connection, such as USB.

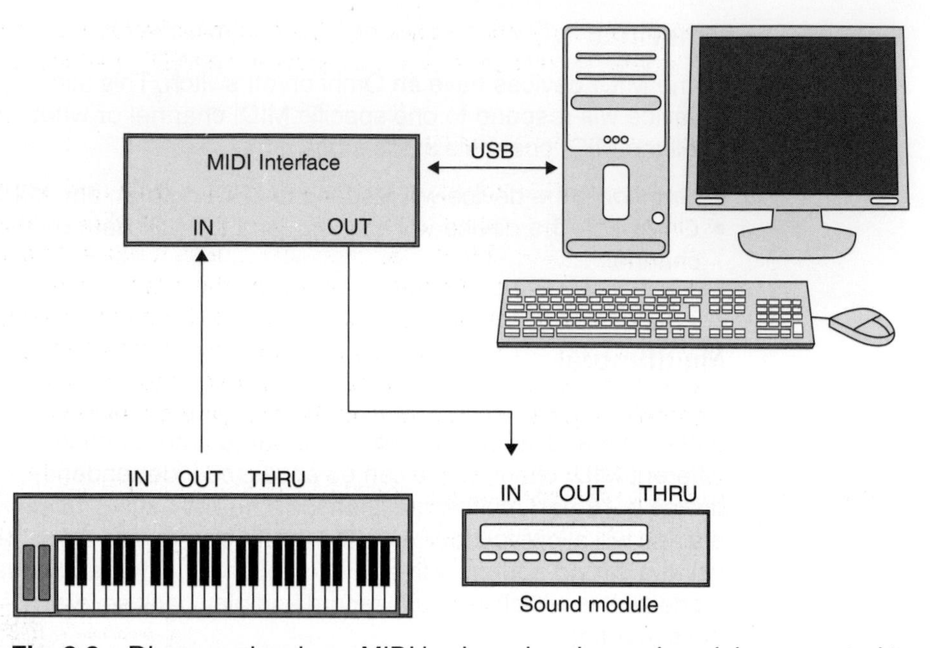

Fig. 3.9 – Diagram showing a MIDI keyboard and sound module connected to a computer, using a MIDI interface.

note ▶ You may be able to connect your MIDI devices directly to a computer using USB and avoid using a MIDI interface (see Fig. 3.6).

Changing MIDI channels in a sequencer

If you are using a MIDI sequencer you will be able to redirect the MIDI data from a controller keyboard simply by selecting a different track in the sequencer. The track you select within a sequencer will usually be assigned to a MIDI channel or sound source, therefore any incoming MIDI data will be redirected to that destination.

Soft Thru

When using MIDI with a computer, there is a special MIDI function called Soft Thru. This allows the sequencer's MIDI output to also act as a MIDI Thru, allowing any incoming MIDI data from a master keyboard to be passed directly through the computer to another MIDI device.

Visual indications

MIDI interfaces usually provide you with a visual indication of the MIDI data going in and out of the computer. This is useful for checking a MIDI system is working properly, such as seeing if the controller keyboard is actually outputting MIDI data. You should see small lights on the MIDI interface to indicate this.

Multi-port MIDI interface

A multi-port MIDI interface has several MIDI inputs and outputs, and is generally recommended for larger MIDI set-ups. Each MIDI output carries its own independent set of MIDI channels. This makes it easier to allocate MIDI data when using several different devices at the same time, as each separate device can be given its own MIDI port and therefore its own set of 16 MIDI channels. This effectively allows the use of more than 16 MIDI channels at the same time, as each port is independent. To do this you will need to assign the appropriate port and MIDI channel within your sequencer.

note ► A multi-port MIDI interface allows you to use more than 16 MIDI channels at the same time.

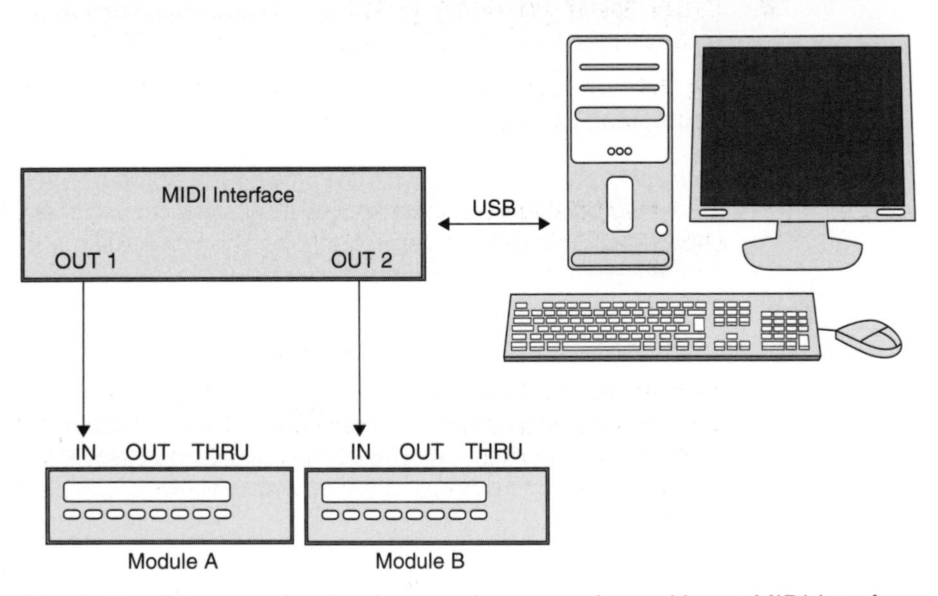

Fig. 3.10 – Diagram showing how each output of a multi-port MIDI interface can be connected to a separate MIDI device.

Following a MIDI note

Most MIDI set-ups are based around several different pieces of equipment, such as a controller keyboard and MIDI sequencer. To help you understand how information travels between these different devices we are going to explain what happens when a note is pressed down on a MIDI keyboard.

1. Pressing a note on a controller keyboard will generate a MIDI message called a note on message. This includes which key has been pressed and

other information, such as how hard the key was pressed (velocity) and the MIDI channel.

2. This note on message will travel out from the keyboard down a cable into the computer, either via a MIDI interface or directly using USB or Firewire.
3. Once the message is inside the computer, the MIDI sequencing software will either direct the message to a sound source inside the computer (see 'Virtual instruments' box below) or send it out to an external sound source.

note ▶ The sequencer may rechannel the incoming message – in other words, change which MIDI channel it gets sent out on.

4. The message will travel out of the computer via a MIDI interface and be sent down a MIDI cable to the sound module's MIDI In port.

note ▶

Virtual instruments

Alternatively, the signal could stay inside the computer and be directed to a virtual instrument. These are additional sounds built in to most sequencers that will interpret the note on information and generate a sound from within the computer.

5. The MIDI message will be received by the sound module and an electronic sound will be produced. The sound should be available from both the audio and headphone outputs of the sound module.

tip ▶ In order to get a response from the sound module, the MIDI send and receive channels will need to correspond.

6. The sound produced by the module will travel to an amplifier or mixing desk. This allows the sound to be sent on to a set of speakers.
7. The sound should now be heard from the speakers.

It is important to note here that the MIDI message does not represent sound, nor is sound. It is simply an instruction to a sound-producing device to play a sound.

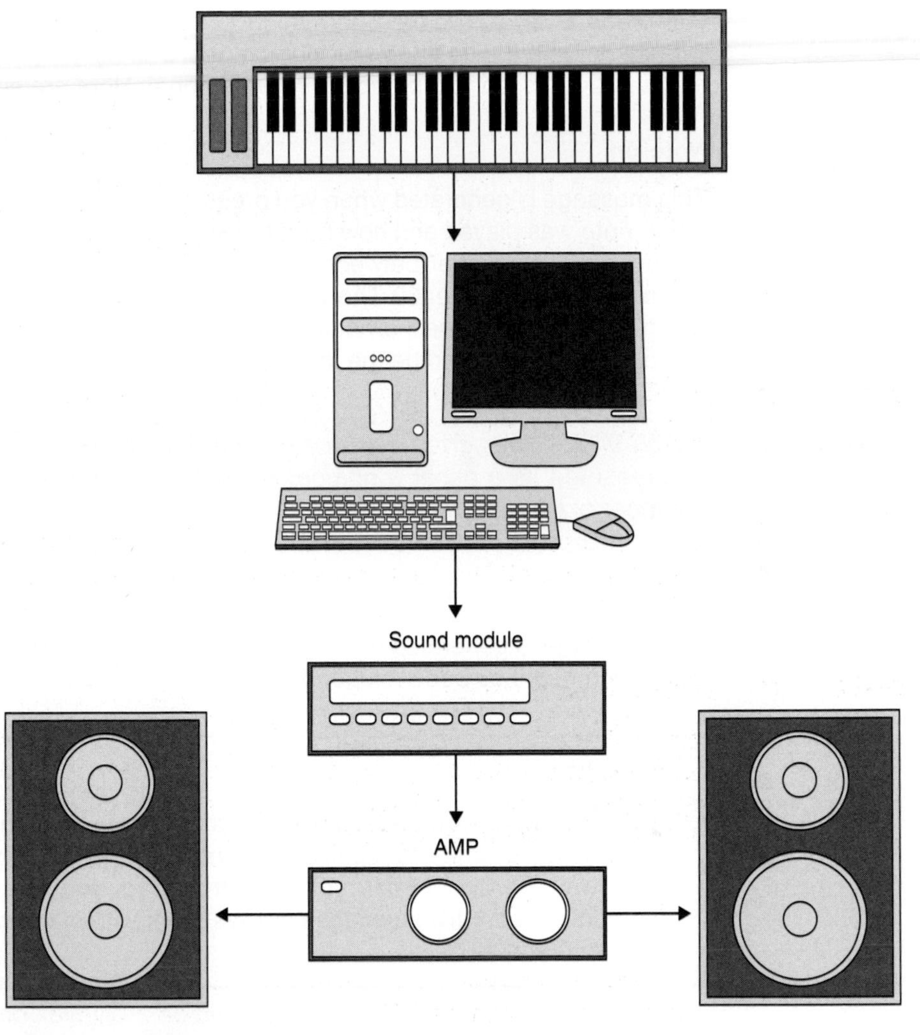

Fig. 3.11 – Typical signal flow of a MIDI set-up.

Different types of MIDI messages

MIDI messages are split into two basic types:

- *Channel messages* – these are addressed to a particular MIDI channel
- *System messages* – these are addressed to the whole system regardless of the MIDI channel.

Channel messages

These are the most commonly used messages that you will deal with. They include what notes you play and physical gestures you make, like how hard you press a key and other information you may generate, such as moving wheels or sliders. The important characteristic of all these messages is that they are addressed to a particular MIDI channel. Here are some examples.

Note on

This message is generated when you press a note on a keyboard. It transmits which note was played and how hard the key was pressed. Each MIDI note is given a unique number between 0 and 127 so it can be identified. C3, for example, is note number 60. The *velocity* refers to how hard the key is pressed, which in turn dictates the individual volume of each key. For example, 30 would be soft whilst 127 would be the loudest possible.

note ▶

Each MIDI note is given a unique number between 0 and 127 and can be represented as a either a number or letter. For example, C3 has a note number of 60.

Note off

This message is generated when you release a note on a keyboard. It transmits which note has been released.

producer says ▶

0 to 127

MIDI is based around an 8-bit binary numbering system that uses zeros and ones to transmit and receive different types of MIDI information. Due to the way the digital information is structured, you will find the most commonly used MIDI messages always work within a parameter range between 0 and 127.

Pitch bend

This can be transmitted from a controller keyboard by using the sprung wheel or joystick that is usually located on the left-hand side of the keyboard. Moving this control gives you the ability to bend the pitch of the notes played and create a similar sound that can be achieved by using a trombone slide or bending a guitar string. A pitch bend value can be adjusted to determine how sensitive a pitch bend controller becomes. For example, you could set a full movement up or down to equal moving the pitch up or down a whole octave.

Program change

This is used to change the sound of an instrument remotely using MIDI. Most sound sources contain several different sounds or presets that can be called up by pushing buttons or turning a dial on their front panel. Program change allows you to do this automatically by sending a unique number to recall a different sound from another MIDI device. Sounds can be selected between 0 and 127.

Control change messages

These allow you to control a variety of different parameters such as modulation, volume and pan. Each 'controller message' has its own unique ID number ranging from 0 to 127. Continuous controllers work like a fader moving between 0 and 127, while switch controllers such as a foot pedal simply change between two states, on and off. Control change messages can be transmitted in a variety of different ways, such as from a keyboard or computer.

System messages

These are typically used for more general functions such as synchronizing different devices together. They are subdivided into system real-time messages and system common messages and include MIDI timecode, start, stop, continue and MIDI clock. There is also system exclusive, which allows you to transmit and receive unique MIDI commands outside the standard MIDI protocol. All the devices in a system will receive these messages no matter what MIDI channel they are set to receive on.

producer says ▶

> Most MIDI devices allow you to use system exclusive to access parameters that are outside the standard MIDI protocol and that are unique to a particular device. This opens up many possibilities to use MIDI to control and automate parameters in real time and to store and recall complex parameter settings.

Panic messages

It is possible when transmitting or receiving MIDI data to get a sustaining note held on even though no keys are pressed. This is called a stuck note. There are several ways of trying to remove this:

- *MIDI interface.* Press the panic or reset button on your MIDI interface (if you have one). This should reset any sustaining notes.
- *Sequencer.* Within your sequencing software press the reset or all notes off button. This will send out a sequence of note off signals to every MIDI channel, resulting in any notes which are playing being cut off.
- *Keyboard.* Try and locate the stuck note and play it again. You may find this resets it. Make sure you use the same MIDI channel as the sustaining note.
- *Switch off.* If you are unable to stop it, as a last resort try switching off the sound source that is producing the stuck note.

Limitations of MIDI

MIDI is a *serial interface*. This means one bit of data arrives after the next, just like a row of train carriages. So, if you play several notes at the same time, technically each note will arrive one after the other. This normally happens so fast we don't notice it. However, if you use a MIDI cable over 15 metres or you transmit excessive amounts of MIDI data at the same time, you will start to hear a perceptible MIDI delay.

MIDI controllers

Devices used to transmit MIDI data are called controllers. These are the devices you actually play to trigger a sound. Generally this will be a keyboard but in fact could be any MIDI device capable of transmitting MIDI data. Here are some examples of different devices that can be used as *MIDI controllers*:

- *MIDI keyboard* – any keyboard with a MIDI capability
- *MIDI guitar* – a specially designed guitar or additional MIDI pick-up
- *MIDI drum pads* – electronic drum pads that transmit MIDI
- *MIDI wind instrument* – specially designed with a MIDI connection.

Not all MIDI devices are suitable for transmitting MIDI data, as some devices don't have a keyboard or are just not practical to use to transmit MIDI data.

note ▶ **Dedicated controller keyboards**

Some MIDI keyboards are only designed to transmit MIDI data and don't actually contain any sounds themselves. You would therefore need to connect them to another MIDI device that was capable of producing sounds, such as a sound module.

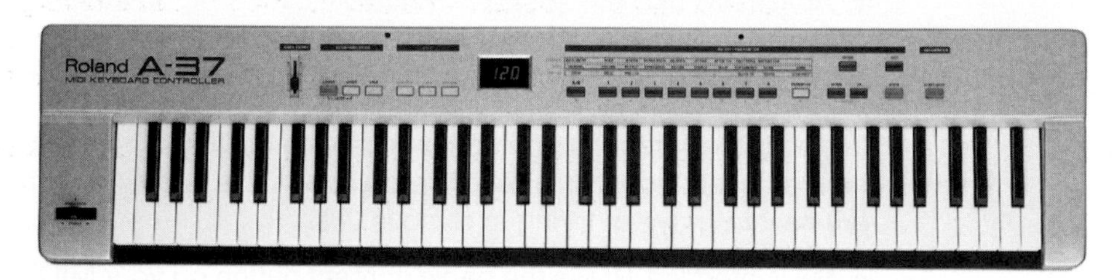

Fig. 3.12 – MIDI controller keyboard.

Any keyboard with a MIDI output or USB connection can be used as a *controller keyboard*, to send MIDI information and trigger sounds.

note ▶ Remember, MIDI only transmits information, not sound.

Sound sources

Where do all the sounds come from? There is a wide variety of different sound sources you may have access to via your computer workstation:

- Electronic keyboard or synthesizer
- External sound module
- Computer soundcard
- Virtual instruments
- Hardware or software sampler.

note ► MIDI has been implemented into a wide range of devices, such as computers, digital effects units, mixing desks, drum pads, guitars and MIDI wind instruments. It is therefore possible to connect and communicate with devices other than just musical instruments.

Using a keyboard as a sound source

There are two different types of MIDI keyboard: those with sounds and those without. Keyboards without sounds are purely designed for transmitting MIDI data and are called controllers. Keyboards with built-in sounds can be used as controllers and sound sources. A keyboard's built-in sounds can be accessed directly from the keyboard itself or from another MIDI device via the MIDI input. This allows the keyboard to be used as a sound source for another MIDI device.

Breaking the link between keys and sounds

One potential problem when using a controller keyboard with built-in sounds to play another MIDI device is that you may hear both the internal sound from the keyboard and the sound of the device you are transmitting to at the same time. One solution is to switch the local off. This is a switch inside the keyboard that makes the keyboard independent of the sound-generating section of the keyboard. The keyboard then becomes free to transmit to another MIDI device without playing its own internal sounds. Alternatively, if you don't want to use the built-in sounds on the keyboard you could just turn down the keyboard's volume.

note ► Keyboards without sounds are purely designed for transmitting MIDI data and are called controllers.

Sound modules

A sound module is a stand-alone MIDI device full of sounds. Most sound modules provide well over 100 different sounds and can be used as the main

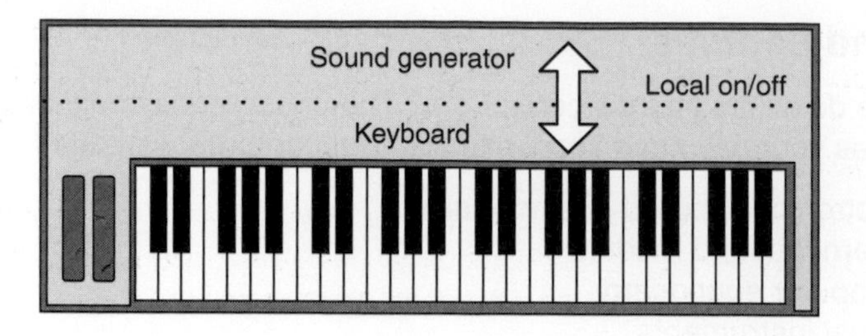

Fig. 3.13 – A *local switch* allows you to break the internal link between the keyboard and its sounds.

sound source in a MIDI system, or simply to provide additional sounds to a computer set-up. In order to access these sounds, the sound module will need to be connected to a computer or controller keyboard. Sound modules often conform to a set format called General MIDI.

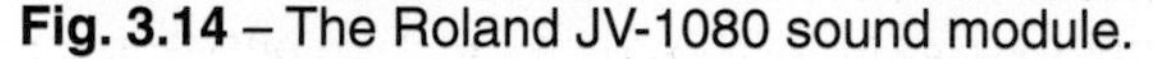

Fig. 3.14 – The Roland JV-1080 sound module.

producer says ▶

Polyphony

Polyphony determines how many notes a MIDI device can play at the same time. Be aware when using a MIDI sequencer, as you may actually run out of notes when playing back lots of notes simultaneously. Most multitambral devices allow you to play over 32 notes or more at the same time.

Computer-generated sounds

Computers are often capable of generating sounds themselves. This allows MIDI messages to stay inside a computer and be directed to a sound source actually contained within the computer. This is often more convenient, especially when you're getting started, as you don't need to use any other MIDI equipment. Internal sounds can be heard via the built-in audio outputs or via an external soundcard connected to your computer.

The quality of these sounds can vary dramatically from excellent to poor, usually depending on the complexity of the software and price. Most computers offer a basic sound source as standard, based around a GM Instrument that allows you to use up to 16 instruments at the same time.

PCs often have a soundcard, which has an onboard synthesizer. The quality of these sounds can vary dramatically, as some soundcards that come as standard with a computer are often poorly implemented.

Apple Mac computers have a basic sound capability called QuickTime. This basic GM style sound source can be accessed by any sequencing software package run on the computer.

Virtual instruments

These are sounds produced by software inside a computer. They can be anything from a synthesizer to a sampler and can turn a computer into a complex sound module. Some of these devices even emulate the functionality and actually look like real devices. You will find that most MIDI sequencing packages now come with a reasonable selection of virtual instruments built in.

Additional virtual instruments

As technology has advanced and computers have become more powerful, it is feasible to create a complete studio using software. However, as you progress you may find yourself wanting to add new instruments and sounds.

Additional virtual instruments can be added to most MIDI sequencing software and simply appear in your existing sequencing software once they have been installed. You do, however, need to be aware that there are different plug-in formats, such as VST, RTAS and Audio Units, so always check their compatibility with the sequencer you plan to use.

Latency

Often, when using a virtual instrument or sounds from a computer soundcard, there will be a large delay between hitting the keys on the keyboard and actually hearing a sound. This is a problem with most computer systems and is called latency.

This problem occurs because the computer needs time to interpret any incoming MIDI data from a keyboard before it can actually produce a sound. The time it takes to do this causes a delay between the computer's input and output. If this happens very quickly you will hardly notice it; however, a large delay caused by latency will seem unnatural and may affect the timing of the notes you play.

producer says ▶

> If you are suffering from a large delay between hitting the keys on the keyboard and actually hearing a sound being generated by a computer, then you are being plagued with *latency*.

Reducing latency

The amount of latency a computer system will produce is dependent on many different factors, such as the type of computer and software being used. Generally speaking, the faster the processing power of a computer, the less latency. However, if latency is causing a problem there are a number of different parameters that can be adjusted from within the software that may help reduce the amount of latency a computer system will produce.

The computer's *audio buffer* size can usually be adjusted to allow you to determine how much effort the computer will put into reducing latency. Generally, the lowest value is the best option. However, the lower the value, the harder the computer has to work, and this will put more strain on the computer when you try and carry out other functions.

If you are using a computer with a soundcard or audio interface, it may need a *driver* in order to communicate with the computer. A driver is a small piece of software that sits between the sequencer and the computer's operating system. If you are using a well-designed audio card which has a good driver, then the MIDI latency should be almost negligible. However, if you are unable to reduce the latency sufficiently, try contacting the soundcard manufacturer to obtain the latest update, as this could make a big difference.

Samplers

A sampler is a device that allows you to record a sound and then play it back via a MIDI keyboard or sequencer. This makes it a very flexible instrument, as you can select the sounds it contains and choose how you want to play them back. Once a sound has been recorded or loaded into the sampler it can be 'spread' across a MIDI keyboard, allowing each separate key to play the sound at a different pitch. This gives you a lot of creative flexibility to experiment with sounds at varying pitches.

note ▶

> A sampler allows you to record and play back segments of sound.

There are an infinite number of sounds available on CD-ROMs that contain ready-to-use samples. All you need to do is load them into your sampler and assign them to a MIDI channel just as you would a sound module. Samplers can be external MIDI devices or virtual instruments.

General MIDI

General MIDI (GM) is a format designed to make different MIDI sound sources more compatible with each other. A GM-compatible device is capable of playing back up to 16 different sounds at the same time and is based around 127 different sounds that are arranged in a predetermined order. By having sounds with the same designations across different equipment, it makes it easier to output your MIDI sequencing information in a format that will be compatible with another GM device.

MIDI files

MIDI files are a special type of song file that allow you to transfer MIDI data between different MIDI sequencers. They don't contain audio so they only take up a small amount of memory.

Creating a MIDI file

To save your song as a MIDI file, choose Save or Export Song as MIDI File from your sequencer. This file can then be transferred to floppy disk or CD and used by another computer or sequencer.

Playing back a MIDI file

To load a MIDI file into a MIDI sequencer choose Import MIDI File. The sounds you will hear by playing back a MIDI file will vary depending how the MIDI file has been created. You will often have to fiddle around assigning the correct instruments to each MIDI channel in order for the MIDI file to make any sense. However, if the MIDI file has been created using GM sounds and program change messages have been used to select each sound, it will automatically play back all the correct instruments when connected to a GM sound source.

4
MIDI Sequencer Basics

What is a MIDI sequencer?

A MIDI sequencer is a device that can be used to record, edit and play back MIDI data. Most MIDI sequencers are based around a software program that is designed to run on a computer. These programs offer many different ways to display and manipulate MIDI information and provide a convenient way to capture and store your musical ideas. Note: in order to use a computer and a MIDI sequencer you will need to have the relevant MIDI sequencing software installed before you start.

In this chapter we will introduce you to some of the basic functions that you would normally expect to find in a MIDI sequencer. You may also find this chapter useful if you are planning to use a Digital Audio Workstation, as many functions that can be used to manipulate MIDI information can also be used to manipulate audio.

note ▶ All the examples shown throughout this book have been created using MIDI sequencing software running on a computer.

producer says ▶ Remember, MIDI is not sound – it is information that can be used to trigger actions in devices that make sounds, such as sound modules (boxes full of sounds) or virtual instruments (programs inside the computer that make sounds).

The Arrange window

This is the main window you will see whilst making music. Most sequencers open up on this page as default. In this window you will see all the different *tracks*, a *timeline ruler* showing bar numbers, and various parameters for editing and changing MIDI data. There will also be a number of drop-down menus for accessing different functions and opening edit windows. Throughout this book we will refer to the main edit window as the *Arrange window*. This name may vary slightly depending on the sequencer used. For example, in Pro Tools it's called the Edit window.

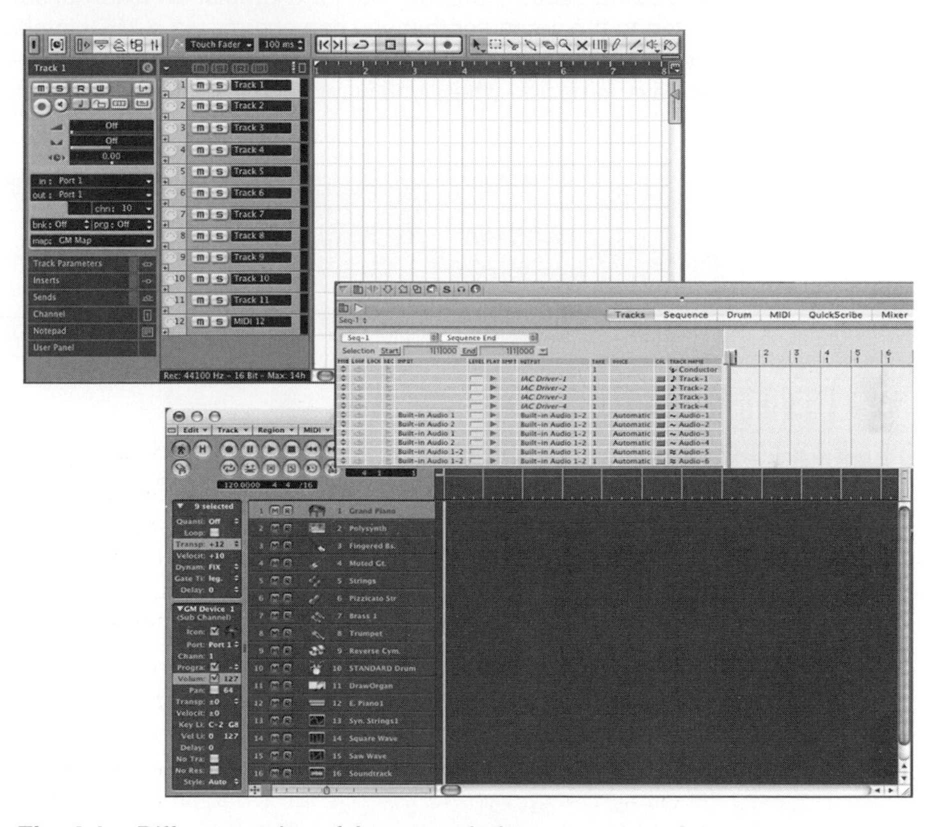

Fig. 4.1 – Different styles of Arrange window.

Tracks

Sounds and content exist on tracks. In order to produce any sound from the sequencer, at least one MIDI track will need to exist. Once *MIDI data* has been recorded or loaded into the sequencer, you will be able to see it visually on the screen displayed in blocks. These blocks of MIDI data are referred to as parts or objects.

note ▶ You will only see MIDI information displayed on the screen if it has been recorded or loaded in.

Routing MIDI data

Each individual track will have a variety of different parameters that will allow you to manipulate the MIDI data contained on a particular track. Initially, one of the most important parameters you will need to set up is the MIDI channel and sound source the track is being routed to.

The options you will have available will vary depending on the type of software being used. There may be, for example, an external sound source connected to the computer that you can access via a MIDI interface, or a variety or software instruments and sounds built in to the computer – either way, you will have to decide where you want each track to be routed to.

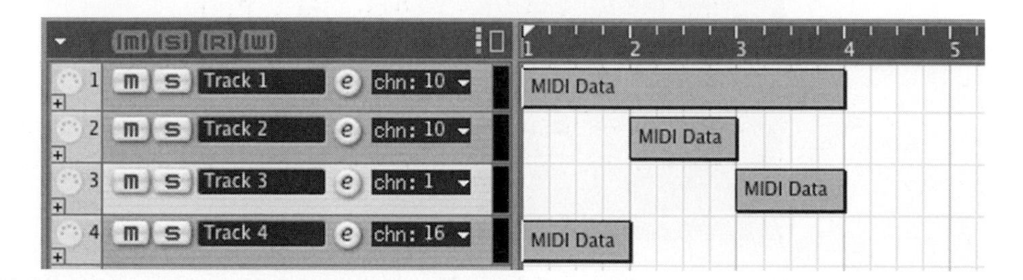

Fig. 4.2 – Example showing four different tracks routed to a selection of different MIDI channels. The data contained in each part or object will be sent to the track's selected MIDI channel.

producer says ▶

> The MIDI data contained within a track can be sent to any destination you choose by selecting a different MIDI channel or output destination. This means you can easily redirect the MIDI data on a track at any time to a different sound source.

Re-routing MIDI data

If you choose to re-route the track at any time to a different destination, the MIDI data will simply follow. The data remains the same but is simply re-routed to a different MIDI channel.

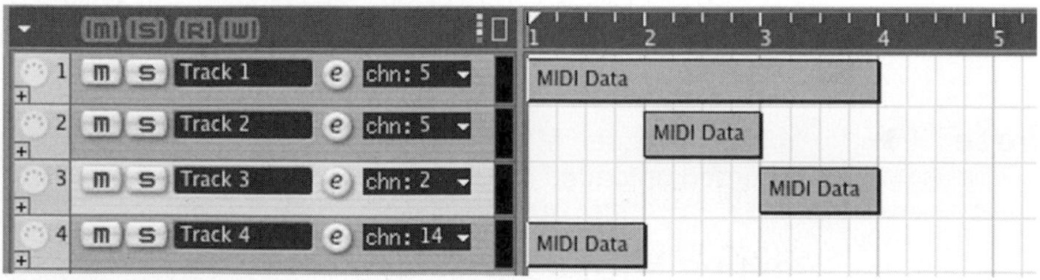

Fig. 4.3 – Example showing how the MIDI channel for each track can be changed.

tip ►

Don't confuse sequencer tracks and MIDI channels

Each sequencer track is usually given a number that is displayed on the left-hand side of the track. This number is only relevant to the sequencer and is not the MIDI channel.

Routing MIDI data to a virtual instrument

In the example shown in Fig. 4.4, a track has been routed to a virtual instrument. This is a sound that is produced by software inside a computer (see 'Virtual instruments' section in Chapter 3, p. 39). Notice how the track's output is routed to the Universal Sound Module.

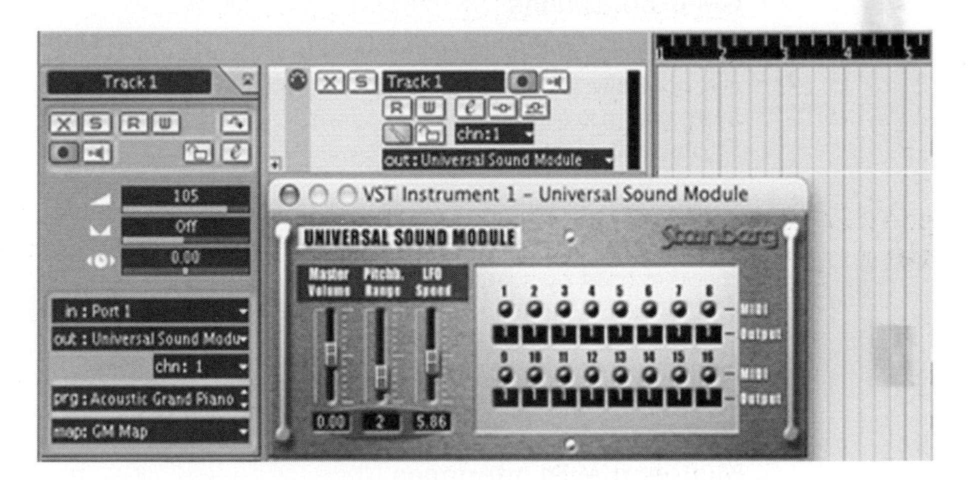

Fig. 4.4 – Example showing a MIDI track routed to a virtual instrument.

note ►

If a computer system has several different sound sources available, then you will have to decide which one you want to use for each track.

Multiple tracks to the same sound source

It is possible to route several different tracks to the same sound source, allowing tracks to share the same MIDI channel and output destination. For example, when programming a drum pattern it may be easier to assign each individual drum sound to a separate track to give you better control.

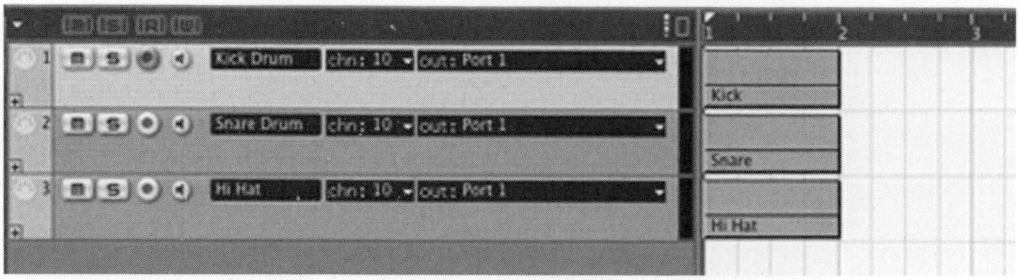

Fig. 4.5 – Example showing three drum tracks all routed to the same port and MIDI channel.

note ▶ Different tracks can share the same MIDI channel and output destination.

GM drum sounds

Drum sounds are normally assigned to MIDI channel 10, as this is where they are normally allocated on a GM-compatible sound device. Each individual drum sound is allocated to a separate key on the keyboard to form a drum kit. This means all the drum sounds will respond to the same MIDI channel.

tip ▶

MIDI Thru and changing sounds

Once several tracks have been created and routed to different sound sources, selecting a different track in your MIDI sequencer will determine which sound your controller keyboard will play.

Multi-port MIDI interfaces

If a computer system has several different external sound sources connected to it via a multi-port MIDI interface, then the user would have to decide not only which MIDI channel the data was being sent to but also which port the track will output to (see 'Multi-port MIDI interface' section in Chapter 3).

Transport controls

These are the controls for stopping and starting playback of the sequencer. They work in a very similar way to the functions on a CD or DVD player. Additionally, you may find other parameters that may be adjusted from this window, such as a metronome or cycle on/off switch. The song's tempo may also be visible from this window. This is usually displayed as a number and is normally set to 120 as default. This value can be increased or decreased to adjust the playback speed of the sequencer.

Fig. 4.6 – Transport controls.

producer says ►

Playing in time with the metronome

Before you record any MIDI information into a sequencer you will need to set up the sequencer's metronome so it will produce a click at the tempo you want to play at. It is essential that you observe this when entering MIDI information into the sequencer as it will ensure any information you play will conform to the beats and bars in the sequencer.

tip ►

Keyboard commands

Most sequencers allow you to use keys on the computer keyboard to activate various functions instead of using the mouse. Some sequencers even allow you to change these keyboard assignments to suit your own way of working. It's worth spending some time learning some of the most frequently used commands, such as play, stop and record, as this will generally speed up most operations and save you time.

The timeline

The *timeline ruler* is the section near the top of the Arrange window. This normally displays beats and bars or actual time in hours, minutes and seconds. This makes lining up parts much easier, as it provides a common reference point when copying and moving MIDI data.

Fig. 4.7 – Timeline ruler displaying bars and beats.

producer says ►

SMPTE

The timeline ruler and the transport control window will often display time in hours, minutes and seconds. This is called SMPTE time and is useful for aligning your music to visual images for video or film. It can also be used for calculating the duration of your music – for example, if you move the song play line to bar 9, the SMPTE time will also change and display the time it takes to get to bar 9 within your song.

Song position marker

This is a thin line that will move across the window when the sequencer plays so you can visually see the current song position. This value will also be displayed in the transport bar.

Locators

The locators allow you to quickly select areas of time within your sequencer. This allows you to play back a section over and over again or set up a temporary loop or cycle for practising or recording. This is achieved by setting up a *cycle start* and *end point* either by dragging the mouse across the timeline or adjusting the locator values in the transport bar. Once a cycle has been set and switched on, you will be able to see the selected area visually on the timeline.

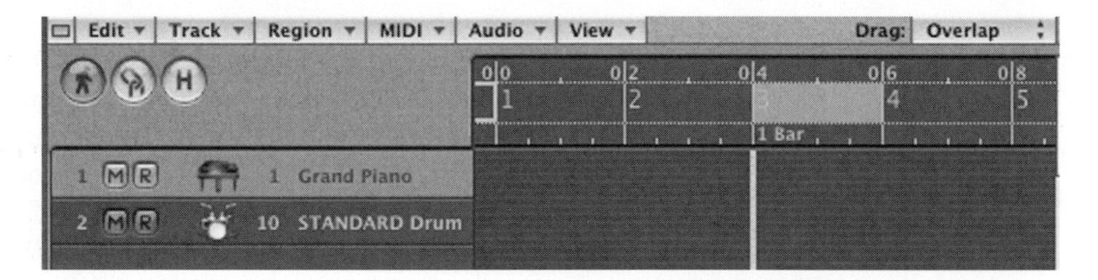

Fig. 4.8 – A one-bar cycle between bars 3 and 4 displayed visually on the timeline.

tip ▶ Setting up a cycle is ideal for practising or recording over a small section of music.

Arranging MIDI data

Once you have recorded some MIDI data into the sequencer you will see it appear in the Arrange window as a block or object. Inside these blocks are the notes you have played.

Fig. 4.9 – Blocks of MIDI data displayed on the Arrange window.

Blocks of MIDI data can be copied, moved, pasted, even cut in a similar way that words and sentences can be manipulated when using a text program such as Microsoft Word. By grouping notes into blocks or short sequences it makes music much easier to manage, especially as the majority of music is often based around small four- or eight-bar phases or a repeating theme.

tip ► Double-clicking on one of these parts will usually allow you to enter an edit window such as the Matrix, Score or List, where you can view and change the contents in more detail (see 'Different types of edit windows' section).

Tools

You will have access to a set of tools in your sequencer. These can be used to manipulate the MIDI information in different ways:

- **Pointer (Arrow)** – This is the default tool on most sequencers and can be used for selecting and moving parts.
- **Pencil** – This tool can be used to change the length of a part. It can also be used to create empty parts so notes can be drawn directly into an edit window.
- **Scissors** – These can be used to cut parts into smaller sections. This can be useful when editing MIDI information as it allows you to separate the sections you want to keep from the sections you want to delete.
- **Glue** – This tool allows you to join different parts together.
- **Rubber** – This tool allows you to delete parts.

Fig. 4.10 – Sequencer tools.

Snap

Most sequencers have a *Snap* feature. The Snap will limit how freely you can move and edit MIDI data. This means that when moving parts/regions/objects on screen they will only move to the nearest point as set by the Snap. This makes editing such as moving and cutting parts much, much easier, as one does not have to be too worried about clicking on exactly the right spot on the screen, since the data will automatically move or 'snap' to the right place. It is possible to adjust the Snap value when editing, depending on how much flexibility you need.

Commonly used Snap values:

- *Bar* – MIDI data will snap to the nearest bar
- *Beat* – MIDI data will snap to the individual beats within a bar.

note ▶ If you set the Snap value to 'bar', the sequencer will only allow you to move MIDI data to the beginning of each bar and not anywhere in between.

Additional MIDI parameters

MIDI parameters that can be used to edit parts or tracks are normally displayed down the left-hand side or via drop-down menus. These parameters are designed to affect the selected parts or the whole track. This means you have to choose carefully, as some parameters will change only the selected part whereas others will change all the parts on the selected track. These parameters usually include the following:

- *Quantize* – adjusts the timing of the MIDI data by moving it automatically to a selected value
- *Transpose* – adjusts the pitch or the selected MIDI data
- *Velocity* – allows you to increase or decrease the velocity value for the selected data
- *Gate Time* – allows you to edit the length of the notes
- *Volume* – a MIDI controller message that controls the volume of a sound over MIDI
- *Pan* – a MIDI controller message that adjusts the position of a sound within a stereo field
- *Delay* – moves the MIDI data forward or back in time.

Fig. 4.11 – MIDI parameter box.

tip ▶ When using a MIDI parameter box to edit MIDI information always ensure that the track or part you want to edit is selected.

Quantize

Quantize is similar to a spell checker in a word processor. Instead of identifying words that are spelt incorrectly it will help you identify notes that are out of time. However, the key to using Quantize effectively is selecting the correct Quantize value. For example, if the computer is told to quantize a selection of notes to the nearest 1/4, then it would typically split a bar into four equal sections, therefore limiting the number of locations within the bar where MIDI notes can fall to just four. The notes would be automatically moved forwards or backwards in time to the nearest 1/4. It's important to note that a sequencer doesn't actually know where the notes should fall – it simply moves them relative to the selected Quantize value.

tip ▶ Using Quantize can help make your music sound more in time and tighten up a performance.

Quantize can usually be activated within the Arrange window and as with most edit functions it will only affect the selected data. The Quantize value you select usually isn't permanent and can be undone or changed at any time. This allows you to experiment with different Quantize values even while the music is playing. To help you understand how Quantize can be used to change the timing of your music, try recording a basic rhythm into your sequencer and then play it back using different Quantize values.

Side-effects of using Quantize

One side-effect of using Quantize is that it can make the music sound too 'perfect' or mechanical. This is fine for dance music, but sometimes a few timing errors can make the music sound more natural and human. So, as always, experiment.

Different types of edit windows

So far, we have covered many of the main aspects of the Arrange window in a sequencer. Now let's look at some of the other windows that are common to most sequencers and which you will be working with on a regular basis.

Edit windows allow you to view and change the MIDI data you have played in more detail. Most MIDI sequencers usually have several different styles of edit window, such as Matrix, List and Score. This gives you a choice of how you want to display and edit the MIDI data. Although the presentation and functionality between edit windows will differ, you are essentially seeing the same information in different ways. This makes it possible to use a combination of different edit windows to edit the same MIDI data. Let's take a look at some of the most commonly used edit windows and how they work.

note ▶ An edit window will display the MIDI data in more detail.

Piano-roll-style edit window

The format is one of the most commonly used edit windows and is similar to a graph. Running vertically down the left-hand side is a piano keyboard, making it easy to determine the pitch of each note, whereas a horizontal timeline allows you to see the timing of each note. The notes themselves are represented by strips of colour moving horizontally towards the keyboard. The length of the strip represents the duration of each note.

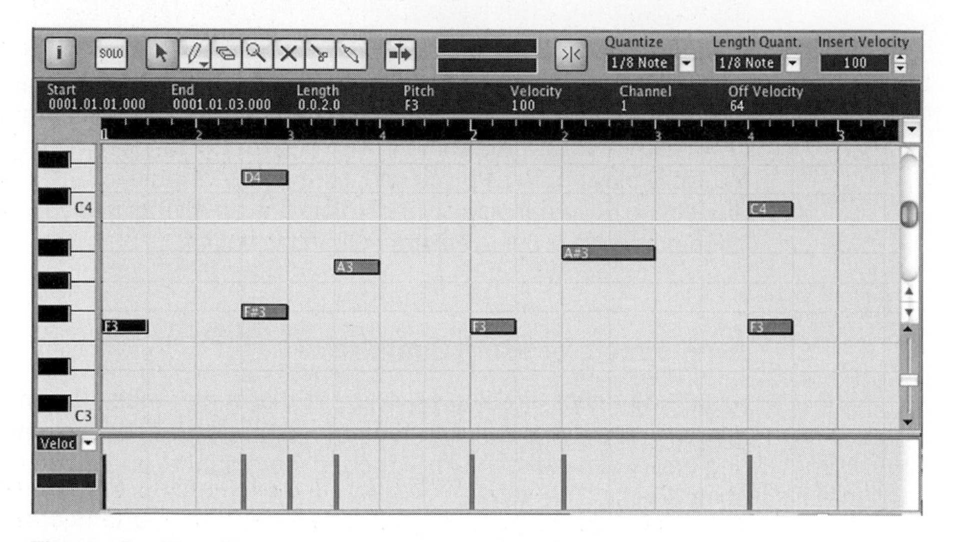

Fig. 4.12 – The Piano-roll-style edit window displays pitch vertically and time horizontally.

Editing notes in the Piano-roll-style editor

Notes can be moved by clicking and dragging them around the screen. Moving a note vertically up or down will change its pitch, whilst moving a note horizontally will change its timing. You can also edit the duration of a note by extending or shortening it. Notes can be deleted using the rubber or drawn in by using the pencil tool without having to enter into a record mode. Additionally, some software sequencers allow notes to be cut in two.

producer says ▶

> The Piano-roll-style edit window can also be used to display and edit other kinds of MIDI data, such as velocity, pitch bend or volume. This information is usually located at the bottom of the window, below the actual notes. In some sequencers you may have to choose to view this kind of data.

List editor

The List editor is neither graphical nor 'musical'. It simply lists all the MIDI events that exist in a part in a numerical form. This allows you to see each note and its associated parameters in one long list.

Edit Functions View

Position	Status	Cha	Num	Val	Length/Info
------------ Start of List ------------					
1 1 1	Note	1	C3	64	. 1 0
1 1 1	Control	1	7	127	Volume
1 2 1	Note	1	D3	40	. 1 0
1 3 1	Note	1	E3	100	. 1 0
2 1 1	Note	1	F3	75	. 1 0
2 2 1	Note	1	G3	127	. 1 0
------------ End of List ------------					

Fig. 4.13 – List edit window.

Data filters

Viewing different types of MIDI data at the same time can sometimes be confusing, so most List editors offer a filter to allow you to choose what type of MIDI data you want to see. For example, if you only wanted to see note data you could filter out all the other types of data, such as program change, volume or pitch bend.

Additional MIDI messages

Most edit windows can also be used to display and edit other kinds of MIDI data, such as velocity, pitch bend or volume. This allows you to edit the *performance data* independently of the actual notes played. For example, the *velocity data* for each note could be increased or decreased independently of the notes' pitch to help define the 'volume' of each note or the pan position could be changed to make a sound move from left to right. Table 4.1 shows a selection of some of the most commonly used MIDI messages that can be used to change the 'performance data' independently of the actual notes played. Note: controller messages such as volume and pan have their own unique ID number which is displayed in brackets.

GM mixer

A General MIDI (GM) mixer is designed to look like an audio mixer but only allows you to edit MIDI data. This makes it very visual and easy to use. Many of the additional MIDI messages, such as volume, pan, chorus and reverb, can be edited from here. It is also possible to select a patch/sound for each channel using a program change message.

Table 4.1

Name	What it does
Velocity	This value determines how hard a key was pressed and will help define the volume of each note
Pitch bend	This allows you to bend the pitch of the notes played by a set amount
Modulation (1)	This adds vibrato to the sound
Volume (7)	This will control the overall level of a sound. Individual notes are still determined by velocity
Pan (10)	This will adjust the position of a sound within a 'stereo field', allowing you to move it to the left or to the right
Sustain (64)	This emulates the effect of a sustain pedal on a real piano and allows a sound to last longer than the period the key was held down on the keyboard
Reverb send (91)	This will allow you to determine the amount of reverb that will be added to a sound when using a GM-compatible sound module

Drum editor

The Drum editor exists mainly because most drum and percussion sounds are structured in a different way to all the other instruments. Each note on a keyboard is usually set up to produce a different drum sound or sample. Pressing a different key will not play the same drum sound higher or lower in pitch, but instead play a different drum sound altogether. This makes viewing each drum by name more important than viewing its pitch.

The other change concerns the length of each note. Drum sounds tend to be short and decay quickly so having complex controls over a notes' duration is pointless.

The Drum editor looks very similar to the Matrix editor; however, the vertical axis shows the actual drum names, rather than a keyboard, and the duration of each note is simply displayed as a triangular mark. This makes editing drum and percussion sounds much clearer, as it's easy to see which sounds are being triggered.

Global Quantize 1/8 Note · Insert Velocity 100 · Insert Length Linked to · Colors Velocity

Pitch	Instrument	Channel	Quantize
C1	Bass Drum	10	1/8 Note
C#1	Side Stick	10	1/8 Note
D1	Acoustic Snare	10	1/8 Note
D#1	Hand Clap	10	1/8 Note
E1	Electric Snare	10	1/4 Note
F1	Low Floor Tom	10	1/16 Note
F#1	Closed Hi-Hat	10	1/16 Note
G1	High Floor Tom	10	1/16 Note
G#1	Pedal Hi-Hat	10	1/16 Note
A1	Low Tom	10	1/16 Note
A#1	Open Hi-Hat	10	1/16 Note
B1	Low Middle Tom	10	1/16 Note
C2	High Middle Tom	10	1/16 Note
C#2	Crash Cymbal 1	10	1/4 Note

Fig. 4.14 – Drum edit window.

Score editor

The Score edit window allows you to display MIDI note data as traditional western-style music notation. This allows musicians who can read music to edit and read MIDI note data in a familiar format. However, many of the options found in this window are to do with changing the presentation of the score, so it reflects a more traditional manuscript rather than just a MIDI interpretation of one. Here you will find many detailed and complex options for the creation of complete scores for whole orchestras.

note ▶ The Score edit window allows you to view your MIDI note information as traditional notation.

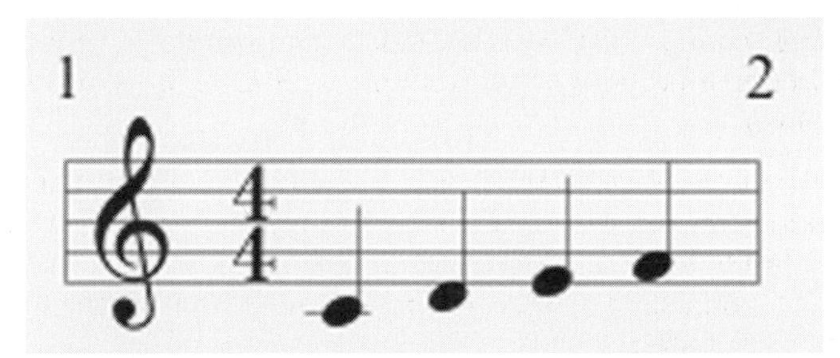

Fig. 4.15 – MIDI notes displayed on a score.

Adjusting the score

Some people may find the Score editor confusing and may prefer to use the Matrix editor, as it accurately shows you what you have played and there is a direct correlation between what you hear and what you see. The Score editor, on the other hand, can display exactly what you have played but this often looks too complicated and would make no sense to actually play. The Score editor therefore gives you the option to simplify rhythms and present them in a more traditional way. Either way, you are still viewing the same MIDI data, just in different formats.

tip ► It's possible to print out the MIDI data displayed in the Score edit window so it can be read and performed by live musicians.

Tempo editor

The Tempo editor is where you will find detailed options for editing and creating tempo changes. Often overlooked, this is the heart of any sequencer. Normally, a song's tempo is set in the transport bar; however, not all music uses the same tempo throughout. In this window it's possible to automatically create tempo changes throughout a song, allowing it to get slower or faster as the music progresses. This technique requires planning and thinking ahead (see Exercise 5.14).

5 MIDI Sequencing Exercises

This chapter contains a series of MIDI sequencing exercises that are designed to help develop the practical skills you will need to complete the main sequencing projects. Each exercise focuses one particular aspect at a time, giving you step-by-step examples and practical tips along the way.

Exercise 5.1 Initial set-up

This exercise will cover the following:

- Creating new MIDI tracks
- Receiving MIDI information
- Selecting a MIDI input
- Assigning MIDI tracks to a sound source
- Routing tracks to a sound source
- Selecting different sounds
- Choosing which sound you hear when you play the keyboard
- Sharing MIDI channels with tracks
- Assigning drum sounds to MIDI tracks
- Saving a song template.

The exercise …

What we want to do in this first exercise is to set up a working environment in a MIDI sequencer that will allow you to use several MIDI tracks and different sounds at the same time. Once this has been completed, this initial set-up file can be saved as a template for you to use over and over again. This will ensure you always have a good selection of MIDI tracks and sounds available to use instantly. When you have a great musical idea, the last thing you want to be doing is wasting time setting up tracks and locating sounds.

Before you start

In order to complete this exercise you will need a MIDI sequencer. If you are using a computer you will need to have the relevant MIDI sequencing software installed on the computer before you start. We also recommend that you have a MIDI controller keyboard connected to the sequencer in order to transmit MIDI information. Before attempting this exercise we recommend you read Chapters 3 and 4.

note ► All the examples shown throughout this book use MIDI sequencing software running on a computer.

Loading your MIDI sequencing software

What you will see once the software has loaded will vary depending on the type of software being used and whether any templates or preferences have been set up. Some sequencers will display the main Arrange window as default, while others will require you to create or load a project before you can do anything. Either way, you should now see the importance of creating your own working environment and choosing what first appears on the screen when starting a project.

tip ► A well-planned song template can save you time and provide you with a framework to get started with. It can be accessed from the File menu or even set up to open automatically by your sequencer when it loads up.

Creating a project

Most sequencers have their own special way of saving files and may require you to create a new project when you first launch the program. If this is the case, simply select New Project from the File menu, navigate to a suitable location on the computer's hard drive where you want to save your file, enter a name and press Save.

Creating tracks

Make sure that you are viewing the main Arrange window and create 12 new MIDI tracks. This is normally achieved by selecting Create or Add a new track from a menu in the main Arrange window. If the window already contains lots of tracks you will need to modify each one to suit your own working environment so, for the purpose of this exercise, it may be easier to simply delete them and create new ones (see 'Deleting tracks').

Fig. 5.1.1 – Track being created using a menu in Logic.

tip ▶ Some sequencers may allow you to create new tracks by simply double-clicking with the mouse in the space provided for a track.

Choosing the right type of track

When creating new tracks make sure they are MIDI tracks.

Fig. 5.1.2 – Creating 12 new MIDI tracks in Pro Tools.

Deleting tracks

If you accidentally create the wrong type of track or need to get rid of some of the existing tracks, you can simply delete them. This is normally done by hitting the Backspace or the Delete key on your computer keyboard or via a menu. Note that some sequencers won't allow you to delete the last track.

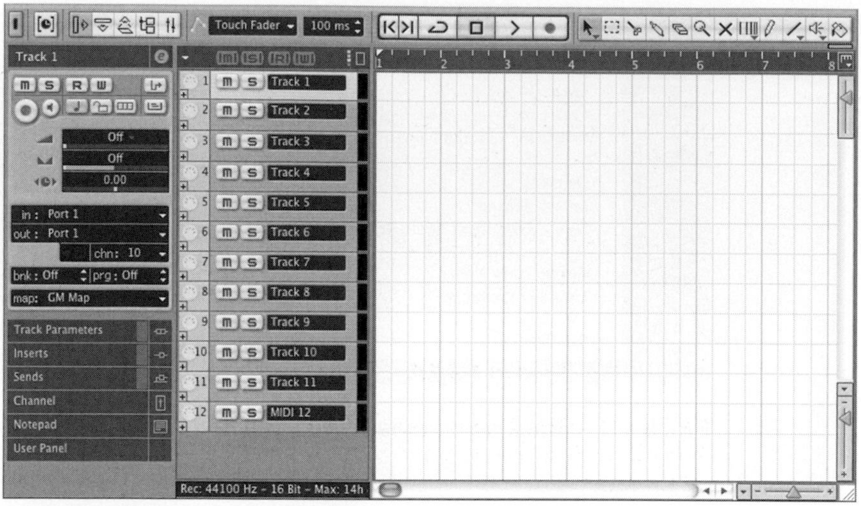

Fig. 5.1.3 – 12 MIDI tracks in Cubase.

tip ▶

Zoom setting

Try adjusting the magnification within the window by zooming in or out so you can see all of the tracks clearly at the same time.

How many tracks do you need?

Initially, when making music, you will not know the exact number of tracks you will need. Ideas will pop into your head and you will need to create tracks as you go along. It is therefore a good idea to always start with a basic selection of tracks on screen to help get you started.

Receiving MIDI information

Before we allocate any sounds to the tracks we need to check that the sequencer is actually receiving MIDI information from the controller keyboard. Select the first MIDI track and enable its record ready switch. This switch is usually red and sometimes marked with 'R'. This will enable MIDI information from the MIDI controller keyboard into the sequencer.

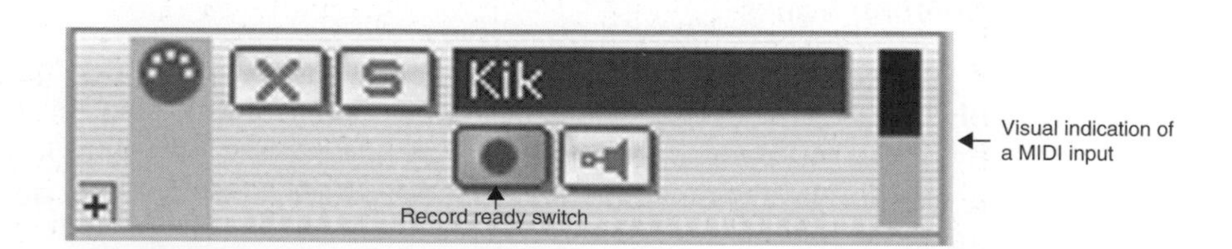

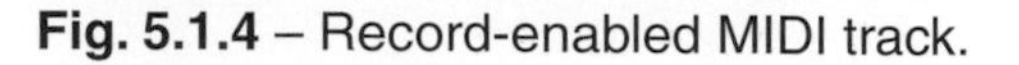

Fig. 5.1.4 – Record-enabled MIDI track.

Look for a visual indication

Press any key on your controller keyboard and look for a visual indication within the software that the sequencer is receiving MIDI data from the keyboard. This may be located on the transport bar or on the actual track itself. Note: don't worry if you are not hearing any sound at this point.

tip ▶ Always look for a visual indication that the sequencer is actually receiving MIDI information.

Selecting a MIDI input

If your controller keyboard is not being recognized by the sequencer, you may need to enable the MIDI inputs you want to use from the MIDI set up page within the software.

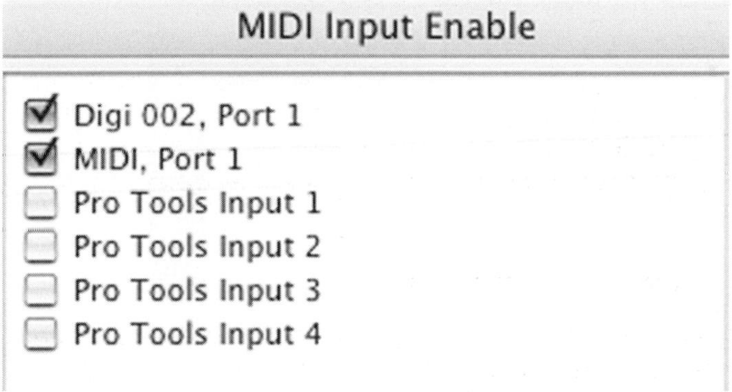

MIDI Input Enable

- [x] Digi 002, Port 1
- [x] MIDI, Port 1
- [ ] Pro Tools Input 1
- [ ] Pro Tools Input 2
- [ ] Pro Tools Input 3
- [ ] Pro Tools Input 4

Fig. 5.1.5 – Enabling MIDI inputs in Pro Tools.

You may also be required to select a MIDI input for each individual track.

Fig. 5.1.6 – Selecting a MIDI input for track 1.

MIDI problems

If your controller keyboard or MIDI interface is not being recognized at all by the software, you may have to run an installer program in order for the computer and software to recognize that an external MIDI device is actually connected to the computer. This additional piece of software is usually downloadable from the Internet.

Assigning tracks to a sound source

Once you have created 12 MIDI tracks, the next stage is to route them to a sound source. The options you will have available to produce a sound will vary tremendously, as some of you will be using sounds built into the computer, such as soundcards or virtual instruments, while others will be using external sound sources that will be connected to the computer via a MIDI interface or USB. This process is basically the same whatever sequencer or sound source you are using. Here are some examples.

Fig. 5.1.7 – The MIDI output of track 1 in Pro Tools being assigned to port 1, MIDI channel 10.

Select the first track and locate its MIDI channel and output settings. This may be located in a track's parameter box on the left-hand side of the window or via a menu.

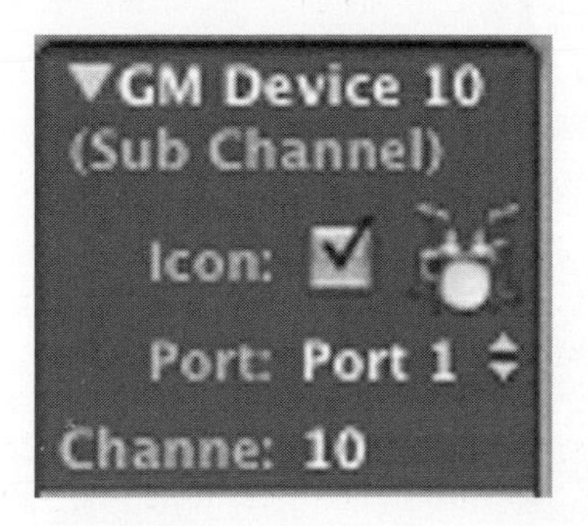

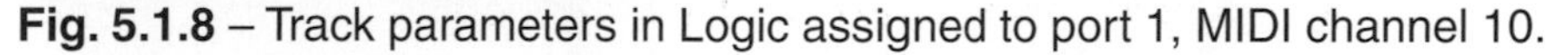

Fig. 5.1.8 – Track parameters in Logic assigned to port 1, MIDI channel 10.

note ▶ The output settings for each MIDI track are really important, as they determine where the MIDI data gets sent and the sound you will hear when the track is selected and you play the keyboard.

Routing a track to an external MIDI sound source

If you want to assign a track to an external sound source, such as a MIDI sound module, or a keyboard that has internal sounds, you will need to choose an output that is actually connected to it. External MIDI devices are normally connected to a computer via a MIDI and will appear as an output destination in your sequencer showing either the instrument's name or simply a port number. This will vary depending on how your system has been configured and the amount of sound-producing MIDI devices, if any, that are connected to the computer (see 'MIDI connections' and 'Sound sources' in Chapter 3).

note ▶ Some external sound modules may require you to select a multi or GM mode in order to produce several sounds at the same time.

Fig. 5.1.9 – Track 1 in Cubase being routed to an external sound module.

Routing a track to a sound source inside a computer

If you're planning to use a sound source that is built into the computer or software, then you won't need to use a port. Most soundcards or virtual instruments simply appear as an output destination for each MIDI track. Note: a virtual instrument will need inserting before it will appear as an output destination (see 'Virtual instruments' in Chapter 3).

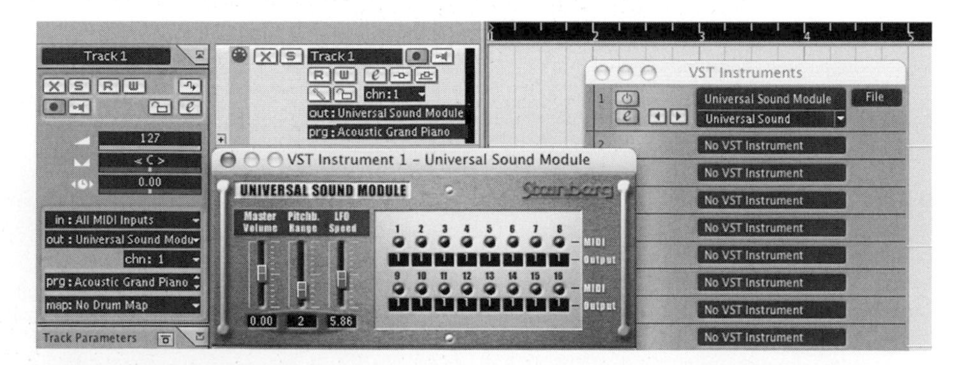

Fig. 5.1.10 – Track 1 in Cubase routed to virtual instrument.

producer says ▶

Virtual instruments

More and more people are using virtual instruments as their sound source, as they provide a convenient way of generating sounds. Most software sequencers now come with a basic selection of virtual instruments built in. It's worth taking the time to learn how to use them, as they can provide an invaluable source of sounds.

Selecting a MIDI channel

While selecting an output you will also be able to choose the MIDI channel you wish to use to communicate with a device on. If you're planning to use the same sound source to produce several different sounds at the same time, you will need to assign each track to the same sound source but to a different MIDI channel (see 'Multitimbral' section in Chapter 3).

note ▶

Aligning MIDI channels

Remember, in order to successfully communicate with another MIDI device, the send and receive channels need to correspond. To achieve this you may have to adjust the settings on the actual sound source itself (see 'MIDI channels' section in Chapter 3).

note ▶

The track's output and MIDI channel will determine the sound you hear.

Routing the first four tracks to a sound source

Once you have located a sound source you want to use, route the first four tracks to it and set the MIDI channels to correspond with the track number.

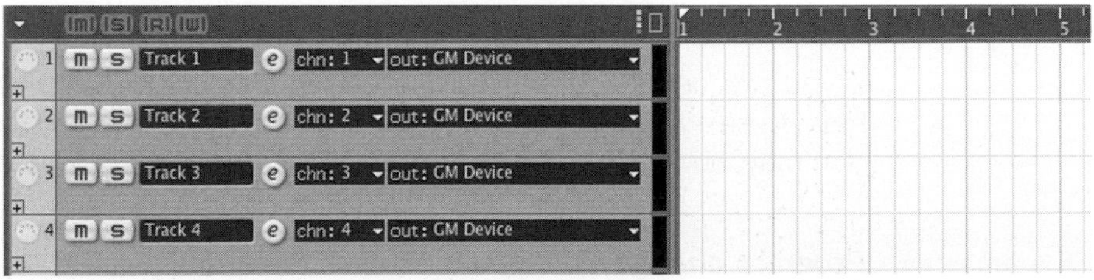

Fig. 5.1.11 – Tracks 1 to 4 routed to the same sound source using different MIDI channels.

In our example, each track is assigned to the same sound source but is using a different MIDI channel. This would allow you to allocate a different sound to each track (see 'Multitimbral' section in Chapter 3).

tip ► When using a MIDI device to produce several different sounds at the same time, each sound will need to be assigned to a different MIDI channel.

note ► The MIDI channel and port you select for a track's output are totally independent of the incoming MIDI data from the controller keyboard.

Selecting different sounds

The next stage is to choose the sounds you want to use for each track. As most sound sources are usually capable of producing lots of different sounds, it's up to you to choose which sound you want to allocate to each MIDI channel. The sounds you choose will normally be determined by the type of music you want to create. Here is an example to get you started:

- **Track 1.** MIDI channel 1 – GM 001 – acoustic grand piano
- **Track 2.** MIDI channel 2 – GM 041 – violin
- **Track 3.** MIDI channel 3 – GM 033 – acoustic bass
- **Track 4.** MIDI channel 4 – GM 057 – trumpet.

External sound sources

Sounds contained within a stand-alone sound module can usually be changed via a button or dial on the front panel of the device itself or by using a program change command from within the sequencer (see 'Program change' section in Chapter 3). You also need to be aware that some sound sources have different modes of operation, such as Multi or GM modes that will need to be selected to enable the device to play several sounds at the same time (see your manual for details).

Sounds inside the computer

If you are using a sound source generated by the computer, then you will have to select and change the sound via the mouse from within the computer. You may, however, benefit from using a program change command.

Program change

A program change message is a specific type of MIDI message that will tell a sound source which instrument to play. This allows you to select the sounds you want to use from within the sequencer. To send a program change message, select a MIDI track, locate the program change parameter, then choose a name or number from the menu. For more detailed information see Exercise 5.10 'Program change'.

note ▶ Most sequencers allow you to output program change messages to any sound sources used by the computer. This provides a quick and convenient way of a selecting new sounds.

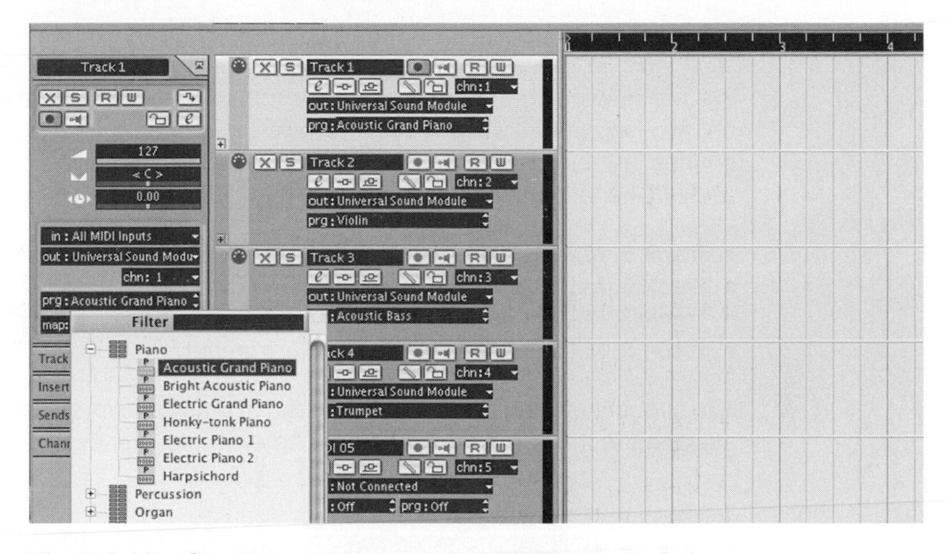

Fig. 5.1.12 – Sending a program change message in Cubase.

Located on most Arrange windows will be a program change parameter box for the selected track. This will allow you to change the sound being used by that track.

General MIDI sounds

If you are using a sound source that conforms to the GM standard, then there will be 127 sounds structured in a certain way (see 'General MIDI' section in Chapter 3).

tip ▶ Always consider using the program change command from within your MIDI sequencer to select the sounds you want to use. The advantage of doing this is that your sound selections are saved with the song file, so when you reload the song all your sound selections will be remembered.

Choosing which sound you hear when you play the keyboard

Once you have routed the first four tracks to a sound source, set a MIDI channel and chosen a sound, it now becomes very straightforward to toggle between different sounds. Simply select each track in turn while playing your controller keyboard and you should hear a different sound.

tip ►

Each time you select a different track, make sure the *record ready switch* is enabled. Some sequencers automatically switch this on when you select a track, whereas in others you have to enable it manually.

Sharing MIDI channels with tracks

We are now going to introduce you to the concept of using the same sound on two separate tracks. If you followed our earlier example of selecting sounds, try routing the next four MIDI tracks (5–8) to the same sound sources and MIDI channels used by tracks 1–4:

- **Track 5.** MIDI channel 1 – GM 001 – acoustic grand piano
- **Track 6.** MIDI channel 2 – GM 041 – violin
- **Track 7.** MIDI channel 3 – GM 033 – acoustic bass
- **Track 8.** MIDI channel 4 – GM 057 – trumpet.

You should now find you have two tracks of piano, violin, bass and trumpet all assigned to the same sound source. An example is shown in Fig. 5.1.13.

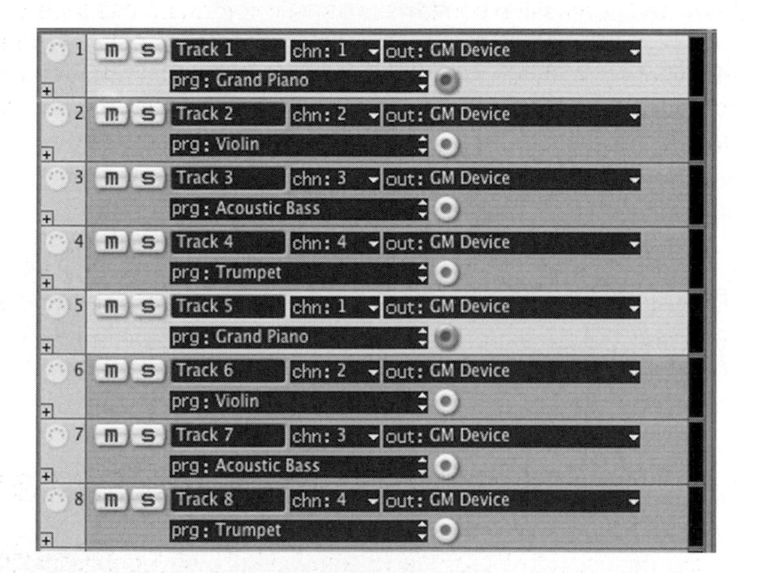

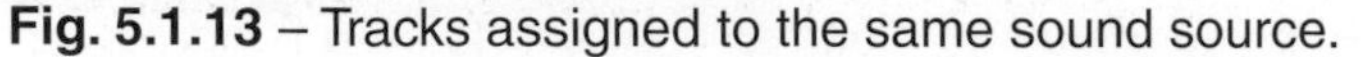
Fig. 5.1.13 – Tracks assigned to the same sound source.

producer says ►

Don't confuse MIDI channels with sequencer tracks. Each sequencer track is usually given a number that is displayed on the left-hand side of the track. This number is only relevant to the sequencer and should not be confused with the MIDI channel.

Multiple tracks to the same sound source

As you can see, it is possible to route several different tracks to the same sound source, allowing tracks to share the same MIDI channel and sound source. This can be useful when using a MIDI sequencer, as it allows you to build up complex sequences in stages across several tracks. It can also be used to place individual drum sounds on separate tracks.

Drum tracks

The remaining four tracks (9–12) will need assigning to a drum sound. If you're using a GM-compatible sound source, the drums will be assigned to MIDI channel 10. If you're using a non-GM sound source, make sure a drum or percussion set is chosen and that it is assigned to MIDI channel 10.

- **Track 09.** MIDI channel 10 – drum sound
- **Track 10.** MIDI channel 10 – drum sound
- **Track 11.** MIDI channel 10 – drum sound
- **Track 12.** MIDI channel 10 – drum sound.

Drum and percussion sounds are usually spread across several different keys on a keyboard and all respond to the same MIDI channel. When using a MIDI sequencer it's often easier to place each individual drum sound onto a separate track, as this can make recording and editing a lot easier (in the next mini-task, 'First recording', we will practise this technique).

note ▶ Drum and percussion sounds are usually spread across several different keys on a keyboard; therefore, each sound will respond to the same MIDI channel.

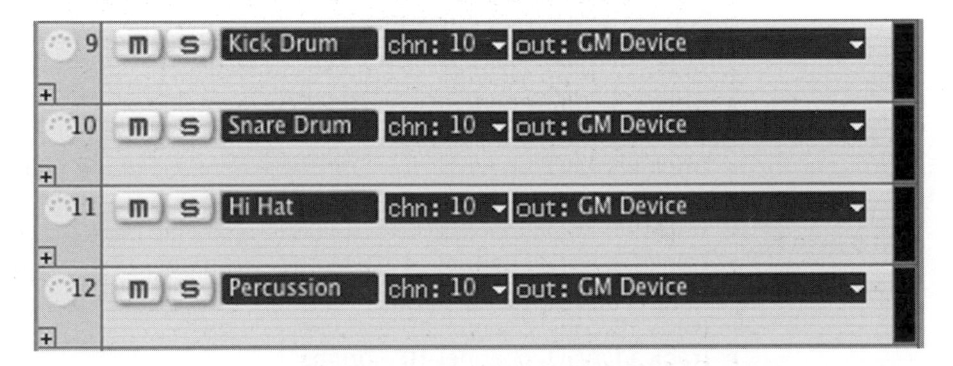

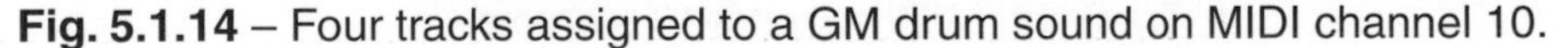

Fig. 5.1.14 – Four tracks assigned to a GM drum sound on MIDI channel 10.

In the example in Fig. 5.1.14, each track has been set up to record and play back a separate drum sound. Even though the type of drum sound has been specified for each track, all the drum sounds will be available on each drum track.

tip ▶

Naming tracks

As you start to assign each track to a sound source, you may want to consider naming each track to help you identify the sounds being used.

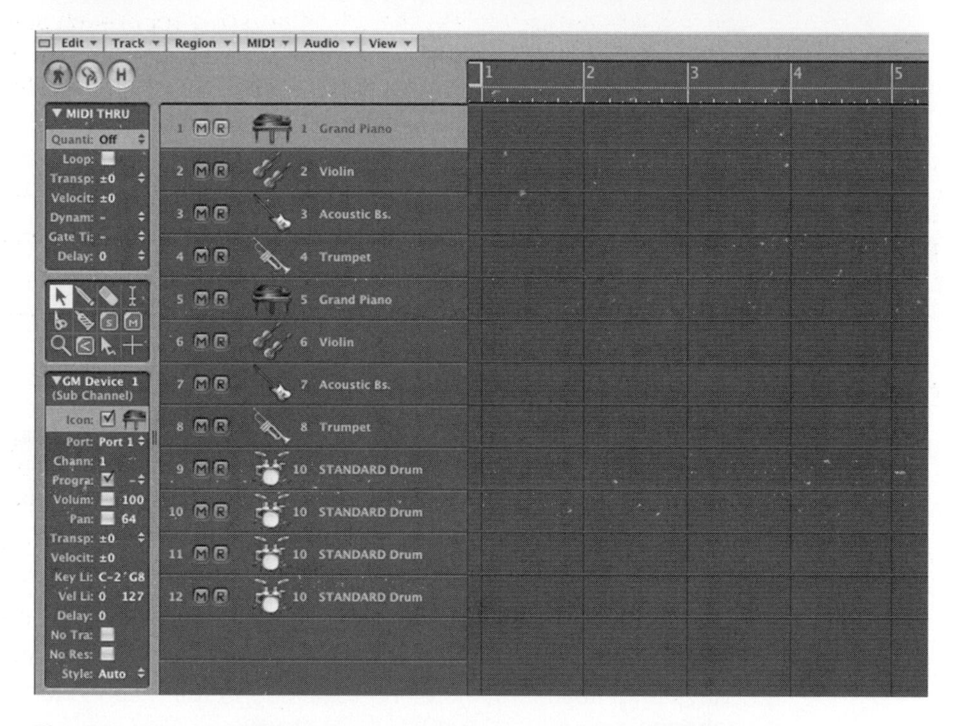

Fig. 5.1.15 – 12 MIDI tracks assigned to a sound source in Logic.

If you have followed our example you should have 12 MIDI tracks assigned to a sound source.

- **Track 01.** MIDI channel 1 – GM 001 – acoustic grand piano
- **Track 02.** MIDI channel 2 – GM 041 – violin
- **Track 03.** MIDI channel 3 – GM 033 – acoustic bass
- **Track 04.** MIDI channel 4 – GM 057 – trumpet
- **Track 05.** MIDI channel 1 – GM 001 – acoustic grand piano
- **Track 06.** MIDI channel 2 – GM 041 – violin
- **Track 07.** MIDI channel 3 – GM 033 – acoustic bass
- **Track 08.** MIDI channel 4 – GM 057 – trumpet
- **Track 09.** MIDI channel 10 – drums
- **Track 10.** MIDI channel 10 – drums
- **Track 11.** MIDI channel 10 – drums
- **Track 12.** MIDI channel 10 – drums.

As a final test, try selecting each track in turn and playing your controller keyboard. You should now hear the sounds that have been selected for each track.

tip ▶

The track you select on the sequencer will determine the sound you will hear when your play the keyboard.

You should now start to see the advantage of placing each sound on its own MIDI channel and how easy it becomes to select different sounds when using a MIDI sequencer. As most sound sources are 16-part multitimbral, you will be able to use up to 16 sounds at once.

producer says ▶

Having lots of different sounds available instantly can often inspire you when coming up with ideas.

Saving your template

Now you have completed your initial set-up, it is time to save your work to disk. Go to the File menu and choose Save As Template or Save As. Name your file and press Save. The options you will have to save the file will vary depending on the software you are using, so refer to your manual or Help file.

Most sequencers encourage you to create a template file. Logic, for example, requires you to name the template file Autoload in order for it to be opened automatically when you launch the program, whereas other programs simply offer a location where templates can be stored.

Additional things to try

Now you have learnt how to create and set up a selection of MIDI tracks, why not take the time to set up your own template file which best suits your own way of working. After all, we have just given one example of an initial set-up and everyone will work in a slightly different way.

Explore the sequencer's preferences

You may find you can make some adjustments to better suit your way of working.

Window and track sizing, and zooming in and out

You may find there are shortcuts.

Keyboard commands

Identifying the keyboard commands for some of the most commonly used operations such as Play and Stop can help speed up the operation of the sequencer.

Setting up audio tracks

If you're using a MIDI sequencing program that is also capable of playing back audio tracks, you may also want to include a selection of audio tracks in your template song.

Exercise 5.2 First recording

Before attempting this exercise we recommend you complete Exercise 5.1. This exercise will cover the following:

- Setting up tracks
- Locating a drum sound
- Setting up the metronome
- Practising playing with a click
- How to record
- Recording additional tracks
- Reviewing the recording
- Adding another part
- Setting up a cycle.

There are various different ways of recording MIDI data into a sequencer. The most obvious way is to simply drop the sequencer into Record and play your MIDI keyboard (or other MIDI controller). This method is very similar to recording onto tape and is called real-time recording. However, there are several other ways of entering MIDI data into a sequencer, such as step time recording or directly entering notes using the pencil tool in either the Matrix editor or List window. It's even possible to place notes directly onto the staves in the Score editor. For now, let us focus on a straightforward real-time recording using a MIDI keyboard.

The exercise ...

In this exercise we are going to explain how to build up a drum track by layering individual drum sounds on a separate track.

Setting up tracks

In your sequencer, create at least four MIDI tracks and assign them to a drum sound. Alternatively, use a template created in the initial set-up (Exercise 5.1).

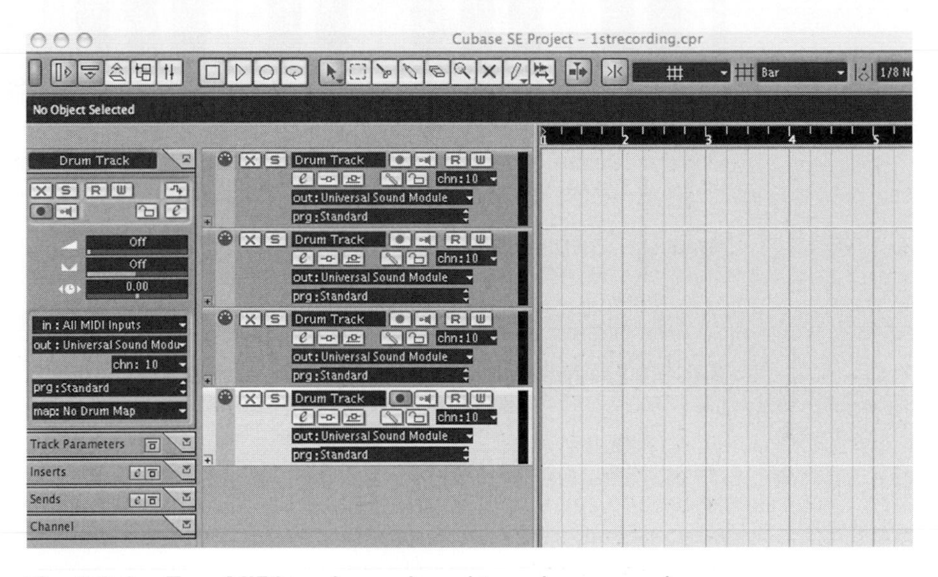

Fig. 5.2.1 – Four MIDI tracks assigned to a drum sound.

In our example we are going to use a drum sound to make our first recording. If you prefer to record a different sound, such as a piano, simply select a different MIDI channel and assign a piano sound. Alternatively, select track 1 in the initial set-up.

note ▶

Most sound sources assign drum sounds to MIDI channel 10. If your sound source works differently, just make sure a drum or percussion set is selected and it is assigned to MIDI channel 10.

Locating a drum sound

If you now play your MIDI controller, you will notice that each key triggers a different sound. In order to play a convincing drum beat you will need to play several of these keys in a suitably rhythmic fashion.

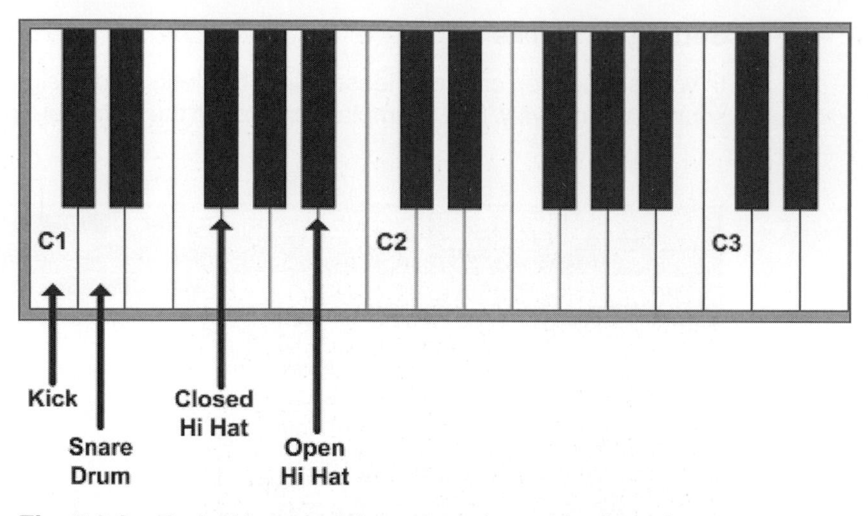

Fig. 5.2.2 – Typical layout of drum sounds on a keyboard.

tip ► Make sure the *record ready switch* is enabled on the selected track.

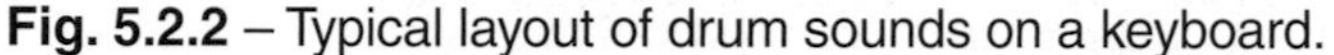

You may find it difficult to create a drum beat by playing several keys at the same time. It is therefore common practice to record each individual drum sound or percussion instrument onto a separate track in the sequencer. This allows a drum beat to be built up sound by sound by layering each sound on top of each other. For example, if you try and play several different drum sounds at once you are limiting your creative skills by linking them to your *mechanical* skills. However, if you play each part separately then you can achieve far better results than you could if you were trying to do the whole thing in a single 'take'. This technique can be confusing at first as you have to play each part separately. However, with time it will seem easy and become second nature.

Setting up the metronome

Once you're ready to make a recording it's time to set up the metronome within the sequencer. This should always be done before you make a recording, as it will provide a common reference between what you play and the beats and bars within the sequencer. Later, when you start to use some of the sequencers more advanced editing functions, you will see why this is essential.

Producing a sound from the metronome

In order to produce a sound from a sequencers metronome, you may need to set up and assign various different parameters. There is usually an option for the metronome to produce a speaker click. This allows the sound to come out of the computer's own internal speaker. Alternatively, you may be able to produce a MIDI click. A MIDI click will output MIDI data to a specified sound source, such as a sound module or sound source inside the computer. This normally needs assigning and is similar to routing a MIDI track to a sound source.

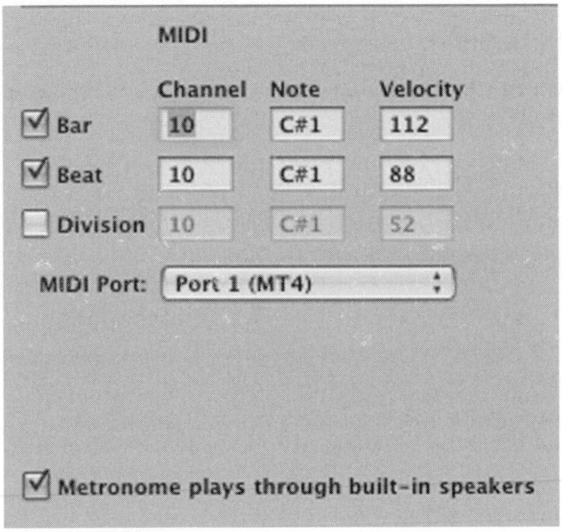

Fig. 5.2.3 – MIDI metronome settings.

Press Play and listen to the metronome

The metronome on/off switch is normally located on the transport bar or at the top of the Arrange window. This will provide a regular pulse at the selected tempo (the default tempo is usually 120 beats per minute).

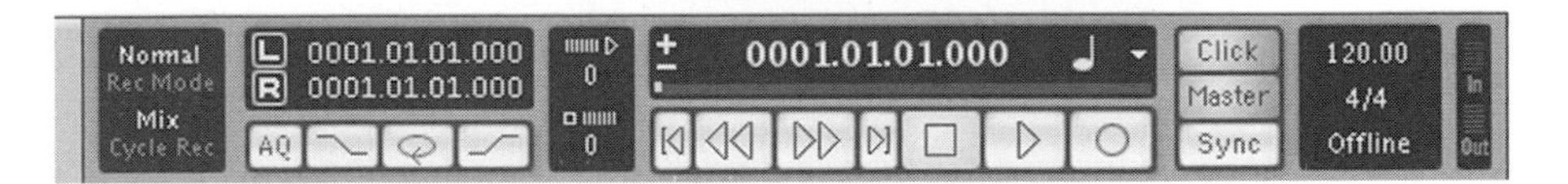

Fig. 5.2.4 – Metronome click enabled on transport bar.

Press Play on the sequencer. Notice how it makes the first click of every four slightly louder. This is emphasizing the first beat of every bar. Try watching the Song position marker, as this will help you to identify the beginning of each bar.

Song position marker

This is a vertical line that moves along the screen to display the current song position. If you press Stop, the *Song position marker* will come to a halt.

Try clicking with the mouse on the top of the Song position marker. You should be able to drag it left and right, left being back in time (in the direction of the start of the music) and right being forward in time (towards the end of the music). Note: the current song position will also be displayed in the transport bar.

Practising to the click

Once you have set up the metronome and selected a tempo, try practising and playing a rhythm along with the click. You may have to adjust the tempo to a speed you feel comfortable with. The tempo is the speed of the metronome (click). Once again, this parameter is normally in your transport bar. By increasing the BPM (beats per minute) you will speed up the click and hence your music. By using a lower number you will make it slower.

tip ▶ It is important that you always listen to the sequencer's metronome when recording MIDI information into a sequencer, as it provides a common point of reference between yourself and the sequencer.

Recording

Once you're clear in your mind what you're going to do, stop the sequencer and return to the beginning of the song again. Now press Record and wait for your count-in. Once the count-in has ended, the sequencer will go into Record mode. Record the rhythm you had previously practised.

0007.01.02.110 Click Master Sync

Fig. 5.2.5 – Transport bar with the record switch enabled.

producer says ▶

Pre-count

A count-in is the number of clicks (normally divided up into bars) which play *before* the sequencer actually enters into Record mode. For example, if the sequencer count-in is set to two bars, one would expect eight clicks before the sequencer enters into Record mode. This is providing the song has a time signature of 4/4.

Press Stop after eight full bars have been recorded. To see where eight bars end, look at the timeline.

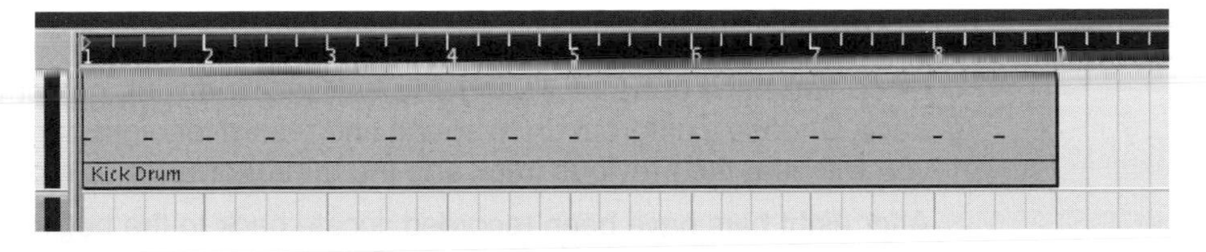

Fig. 5.2.6 – An 8 bar recording example.

Reviewing the recording

Locate back to the beginning of the song and play back what you have just recorded. Is it in time? Are you happy with the rhythm? If not, delete the part by clicking on it and pressing Backspace on the keyboard (or by using the rubber tool) and re-record the drum part.

producer says ▶

When using the controller keyboard to record information into a MIDI sequencer, it's worth remembering that what you play is going to be exactly what we hear back, so how you play it is very important. It is therefore worth spending some time developing your keyboard skills to gain more control over the instrument. For example, can you play without hesitation? Can you play using the full dynamic range or playing loud or soft as the situation demands? You can, of course, use a sequencer to edit what you have played. However, it will save hours of endless fiddling and tweaking, and will help to make your music sound more human, if you can learn to control the keyboard in the first place.

Naming

It's always a good idea to get into the habit of naming tracks, as this will help you identify the sounds being used on each track. Note: individual parts/regions can also be named independently of the tracks. This makes it possible to name individual parts in reference to the section of music. For example, you could call one part/region 'Piano chorus' and another 'Piano verse'. This would make it easier to identify different sections of a song when looking at the Arrange window and help identify the parts you need.

Saving your sequence

At this stage it would be a good idea to save what you have done to disk. If you have already created a project it will simply be a matter of updating your existing file.

Secondly, you may know exactly what it is you want to create yet it is tricky to perform. You may find yourself recording it, stopping to listen to it, deleting it if it's not up to scratch and then doing it again. However, this is a very stop start method, and you may struggle to get any momentum or feel. Often, it is better to record your idea over and over till you get a perfect recording. Then listen back and cut out the good four bars and delete the rest. For example, if you want to create a four-bar bass line or drum pattern, you could try repeating the same four bars over and over for around 32 bars. Then listen through to all 32 bars and take the best four-bar phrase or section and delete the rest.

The exercise ...

In this exercise we are going to edit out any unwanted sections from a MIDI recording; therefore, you will need to have a MIDI track within your sequencer assigned to a sound source with around 32 bars MIDI data recorded. Alternatively,

- Record 32 bars of new MIDI data
- Play something you know already
- Use the data from a previous exercise
- Load in a sequence you want to edit
- Load a MIDI file
- Load a tutorial file from our CD.

Make sure that you are viewing the main edit window (Arrange window). Most sequencers open up on this page as default. You should now have a long MIDI part on your track.

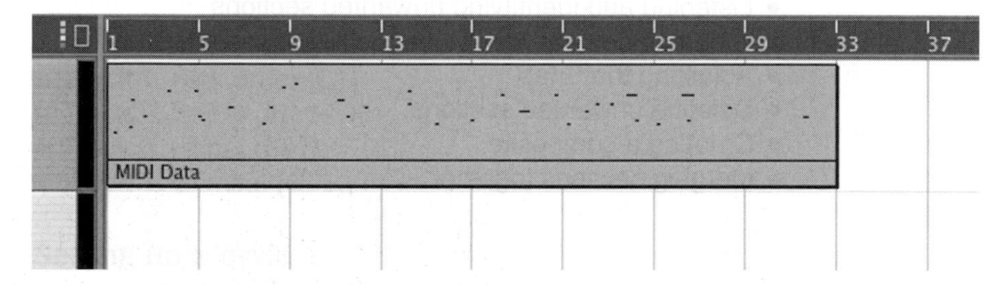

Fig. 5.3.1 – 32-bar MIDI sequence.

Listening and identifying sections

Move the Song position marker back to the beginning of the song and press Play. As you listen through the part, try and identify the sections you want to keep and the sections you want to delete. Make a note of their location using a timeline.

tip ►

Looping

You may find it easier to create a temporary loop to help you identify the sections you want to keep. Try listening to four- or eight-bar sections at a time.

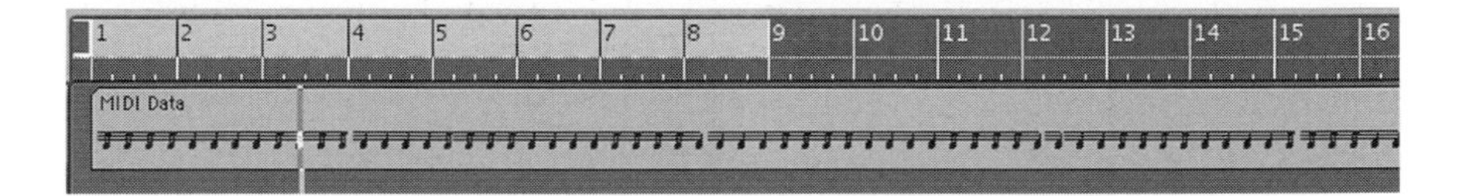

Fig. 5.3.2 – MIDI sequence with an eight-bar loop.

Where to cut and how to separate blocks of MIDI data

Stop the sequencer and separate the sections you want to keep. Use the scissors to cut where passages of notes you want keep start and end.

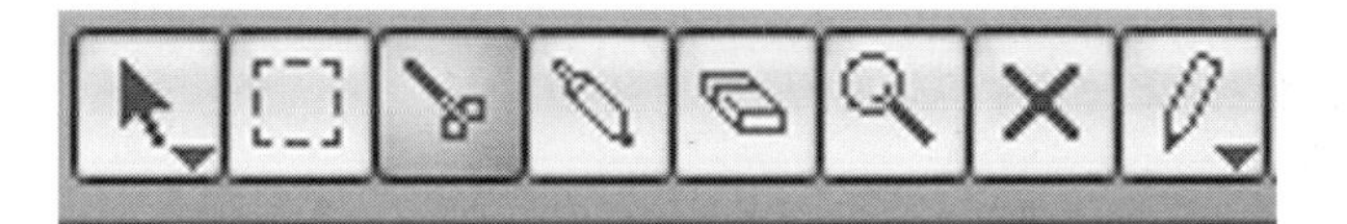

Fig. 5.3.3 – Select the scissors.

Alternatively, move the Song position marker to the position where you want to cut and select Separate or Cut from a menu.

Fig. 5.3.4 – MIDI part being separated using the Song position marker.

tip ►

In most popular music it is common to edit sections into one- or four-bar phrases. This will make ideas easier to arrange.

Adjusting the Snap

When cutting or dividing sections you need to be aware of the *Snap* value. This will restrict where you can actually make a cut, by only allowing you to cut to the nearest bar/half bar/beat or to an even smaller value. You can change the value of the Snap or even switch it off, usually via a menu on the main Arrange window.

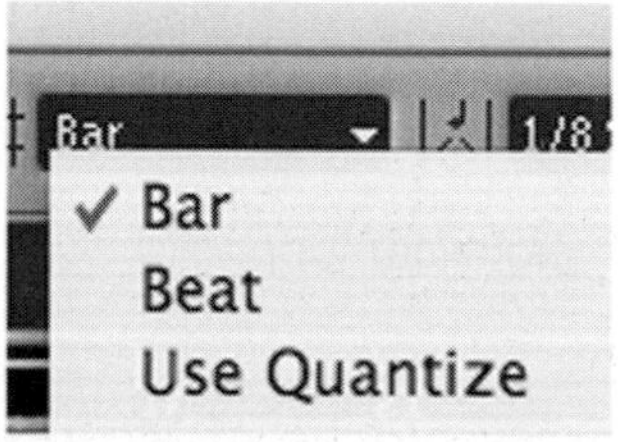

Fig. 5.3.5 – Snap value set to bars.

tip ▶ When cutting sections always be aware of the Snap value, as this will determine where you can actually make a cut.

note ▶

Splitting notes

When dividing a section, you may at times cut halfway through a note. Some sequencers automatically keep the note hanging across the cut point, while others give you the option of splitting it into two notes or shortening the note at the cut point.

Colouring & naming

As you divide the MIDI data into separate sections you may want to name or colour each individual section. This will help you to identify the different sections.

Deleting unwanted sections

Once you have separated all the sections you want to keep, it's time to delete the sections you do not require. Simply select an unwanted section and press Backspace or use the erase tool.

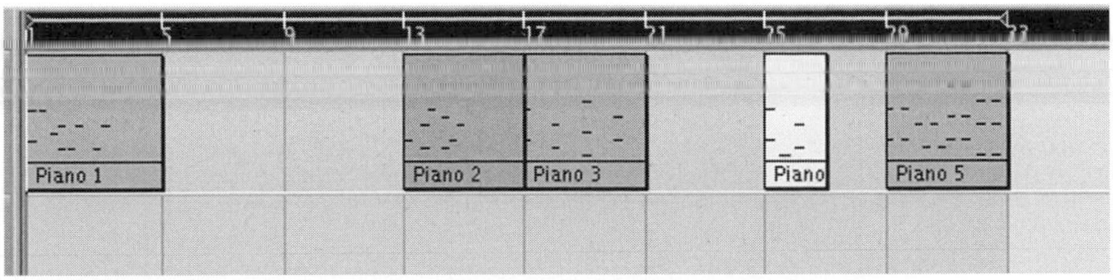

Fig. 5.3.6 – MIDI data separated into different sections.

Creating a composite

Now it's time to arrange what's left. Use the arrow tool to drag the parts/regions into an arrangement, so that as one ends another starts (see Fig. 5.3.7).

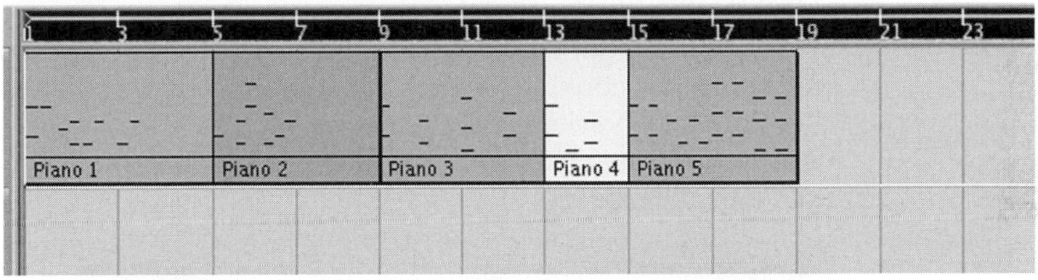

Fig. 5.3.7 – Parts arranged in sequence.

note ► Before you move the parts, make sure that each section will blend with the next. You may have to experiment by changing playback order or each part.

Joining parts together

Once you are happy with the arrangement of each part, try merging them all together using the glue gun tool.

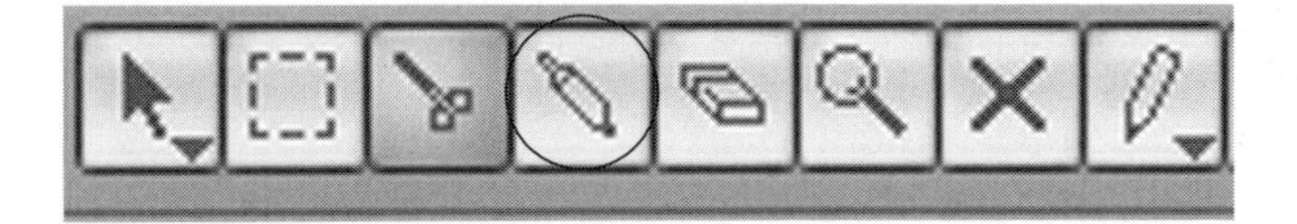

Fig. 5.3.8 – Glue gun tool.

To select all the parts you want to glue, hold Shift on your computer keyboard and click on each part with the arrow. Then select the glue gun tool and click on any of the selected parts, or alternatively select Merge from a menu. This will result in all selected parts being joined together.

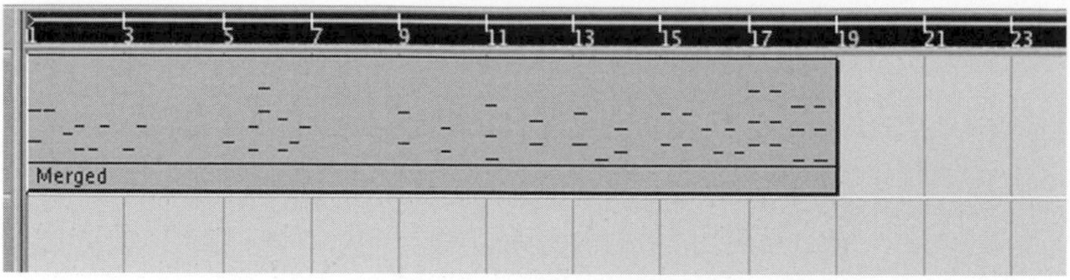

Fig. 5.3.9 – Individual parts glued together to create a composite.

Be aware, most sequencers allow you to merge data across different tracks. We don't recommend this at this stage, as it will only allow the data to be output to a single sound source. So for now only merge regions/parts on the same track.

Additional things to try

Now you have got the basic idea of how to edit a performance into separate sections, you could try attempting some more complex edits by separating sections into individual bars or even beats. This would give you greater flexibility when creating a composite, as it would allow you to choose from smaller sections of notes.

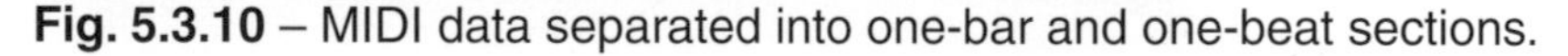

Fig. 5.3.10 – MIDI data separated into one-bar and one-beat sections.

Exercise 5.4 Looping and copying

Before attempting this exercise we recommend you complete Exercises 5.1–5.3. This exercise will cover the following:

- Duplicating MIDI data
- Cut, Copy and Paste
- Creating song arrangements.

Duplicating MIDI data

Sequencers allow us to quickly repeat sections of music, enabling us to build up complex arrangements very quickly. This saves time and avoids having to record the same section over and over again. There are usually many different ways of duplicating MIDI data within a sequencer. Let's take a look at some of the most commonly used methods.

Copy and Paste

This allows the user to take a part/region and copy it temporally into the computer's memory. Then, by using the Paste command, a new version or copy of the part/region can be inserted where the main Song position marker is placed on the timeline.

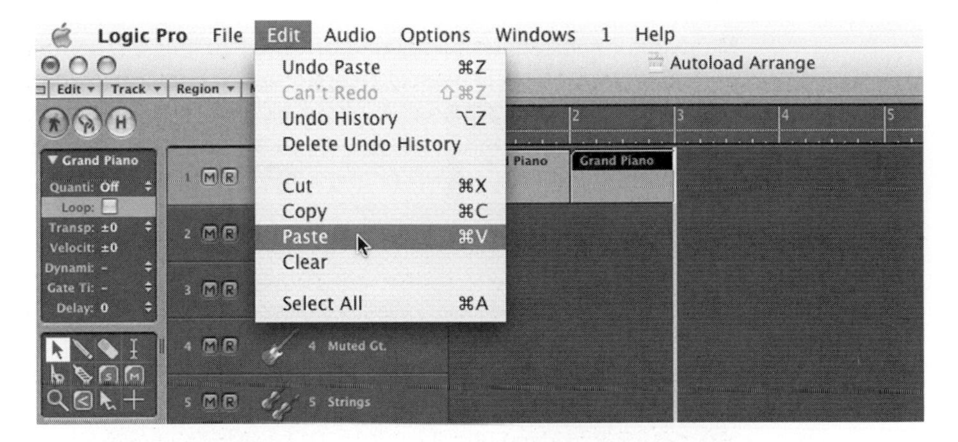

Fig. 5.4.1 – Pasting MIDI data at the song position.

note ► Before pasting an object back into an arrangement, make sure you position the Song position marker at the exact location where you want the copy to appear.

Looping

This is where you can click on a part and then repeat it endlessly. The selected part then repeats non-stop until it is stopped by another part/region.

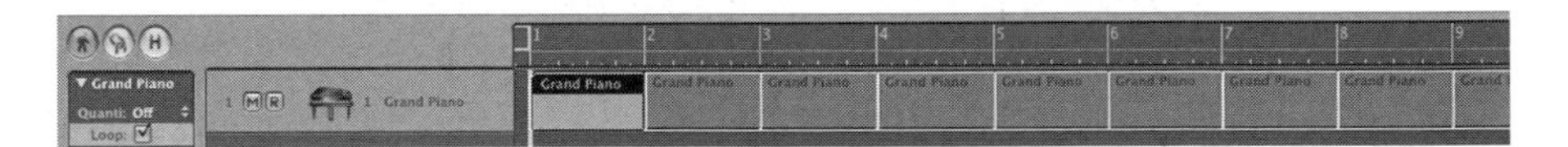

Fig. 5.4.2 – Looped one-bar object.

note ►

Cycling and looping

Looping should not to be confused with creating a cycle in the Arrange window via the timeline ruler or transport bar. A cycle is where the whole piece of music temporarily cycles between two specified points, whereas looping allows a part to repeat endlessly throughout a song.

note ▶ The main difference between looping and copying is that looping is a linear technique, which simply repeats a section over and over again, whereas, copying allows a specific section to be duplicated. Hence looping might be used to create an ongoing drum beat, whilst the Copy function might be used to duplicate a specific section, such as a chorus to a location later in the arrangement.

Ghosts/aliases

Most sequencers have some kind of 'ghost' or 'alias' system of duplicating or repeating data. There is usually an option when copying to choose between real copies or 'ghosts' or 'aliases'. A 'ghost' or 'alias' copy is a copy which is linked to the original part/region. Hence, if you then edit the original part/region, all the ghost or alias copies will change too. This also saves memory as data is only contained in the original, not the copy.

Fig. 5.4.3 – Duplicating dialogue box.

note ▶ Ghosts or aliases can be readily identified as their names are usually shown in italics. Also, at any time they can be changed into real copies.

The exercise ...

In this exercise we are going to be practising copying and looping. You will therefore need to have a MIDI track within your sequencer assigned to a sound source containing around eight bars of MIDI data. Alternatively,

- Use the data from the previous exercise (5.3)
- Use a MIDI sequence you have already recorded
- Record eight bars of new MIDI data
- Play something you know already into the sequencer

- Load in a sequence you want to edit
- Load a MIDI file
- Load a tutorial file from our CD.

Whatever method you choose to get started, you should now have a MIDI part on your track. This part should be cut so it starts and ends exactly on a bar. Use the scissors or refer to Task 5.3 to achieve this before moving on.

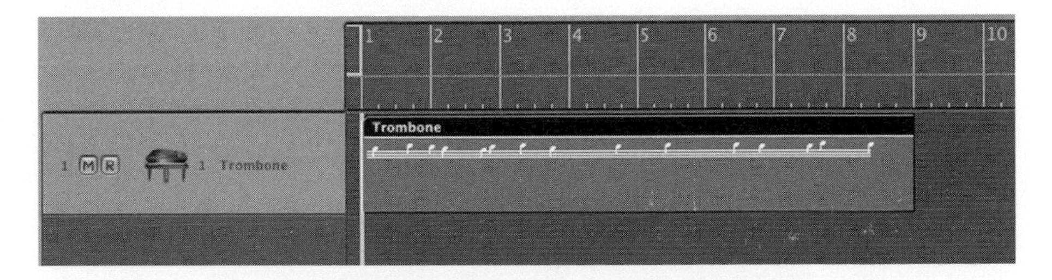

Fig. 5.4.4 – Eight bars of MIDI data.

tip ► Always try and edit parts so they start and end exactly on the beginning of a bar. This will make moving and arranging them a lot easier.

Duplicating via a menu command

Select the part/region you want to copy, then choose the repeat or duplicate option from a menu in your sequencer. We want to make this part repeat at least four times. Each sequencer will refer to this slightly differently, but the process is basically the same.

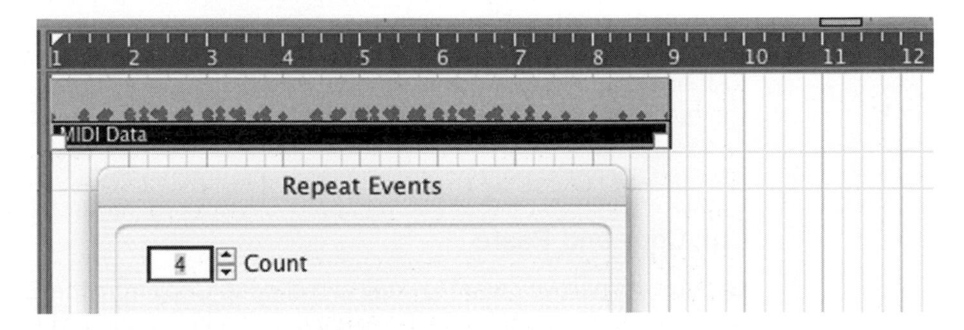

Fig. 5.4.5 – Duplicating dialogue box.

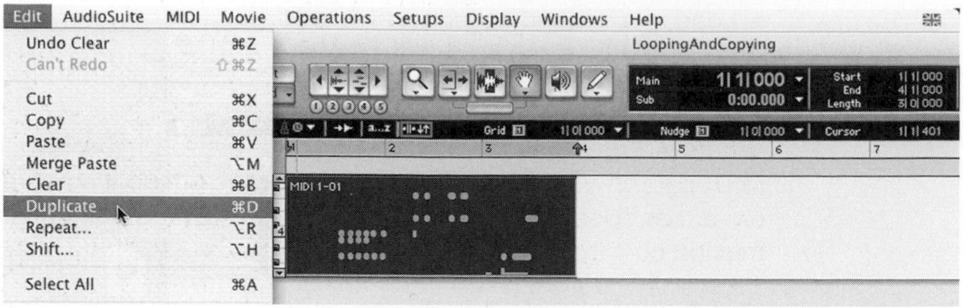

Fig. 5.4.6 – Duplicate function being accessed from a drop-down menu.

Once the original eight-bar sequence has been duplicated it should look similar to Fig. 5.4.7.

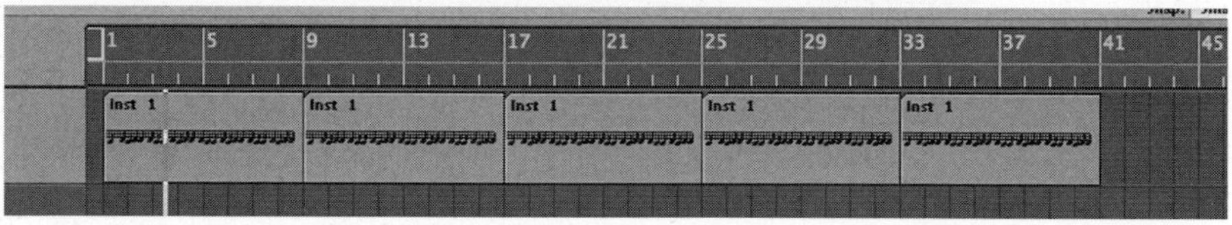

Fig. 5.4.7 – Original eight-bar MIDI sequence that has duplicated another four times.

Notice how each block of data is perfectly aligned to the previous one. This highlights why it is important to cut your parts accurately and use the snap function (see Exercise 5.3). If a part was cut unevenly and not exactly on the bar, all the subsequent repeats would be out of time and would not line up on the timeline.

tip ▶

Keyboard commands

Most sequencers allow you to assign keyboard commands to the most commonly used functions, such as copy or duplicate. Try holding down the Alt or Control key before attempting to move a part. This should allow you to quickly make a duplicate without having to acces a menu.

Cut, Copy and Paste

Click on the part you want to copy and choose Copy from the Edit menu. Move the Song position marker to the exact position in the song where you want the copy to be, then go to the Edit menu and choose Paste. This will create a copy.

Fig. 5.4.8 – Pasting MIDI data at the song position.

note ▶ When pasting an object back into an arrangement, make sure you select the track you want to paste on to.

If you want the same part to appear later in the song, simply move the sequencer's Song position marker to the desired location and press Paste. See how it's possible to build up an arrangement quickly.

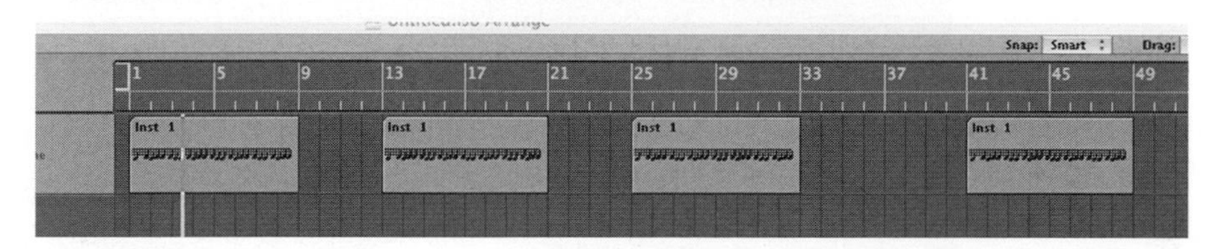

Fig. 5.4.9 – Copying the same MIDI part to different locations along the timeline.

Multiple selections

It is possible to copy multiple regions/parts at the same time even if they are on different tracks. To do this we first need to select or highlight all the data we want to copy. If you hold down the Shift key while using the arrow tool on the Arrange window you will be able to select multiple objects.

Rubber banding a selection

Alternatively, click once and hold the mouse button down (the left mouse button on a two-buttoned mouse) and drag it over the parts you want to duplicate. This should result in a wire frame box appearing that will select any parts it comes in contact with.

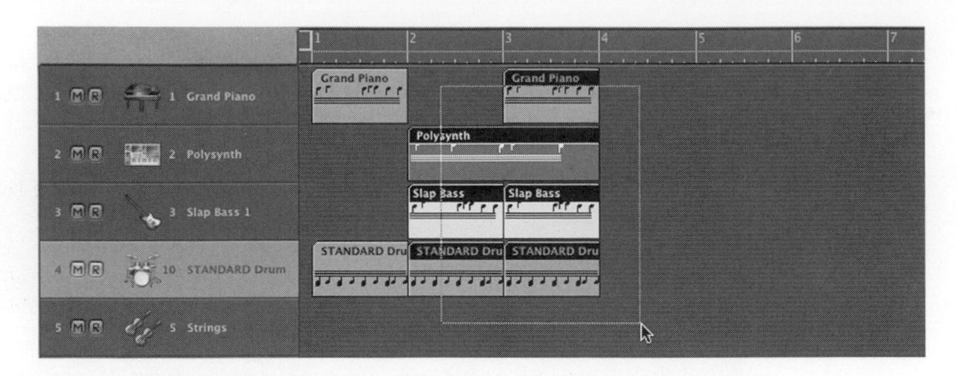

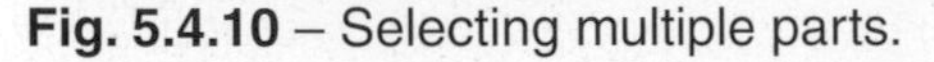
Fig. 5.4.10 – Selecting multiple parts.

Once you have selected the data you want to copy, either go to the repeat or loop option, or alternatively, go to the Edit menu and choose Copy, move the Song position marker, and then choose Paste. This will now duplicate or copy all the selected objects at the same time.

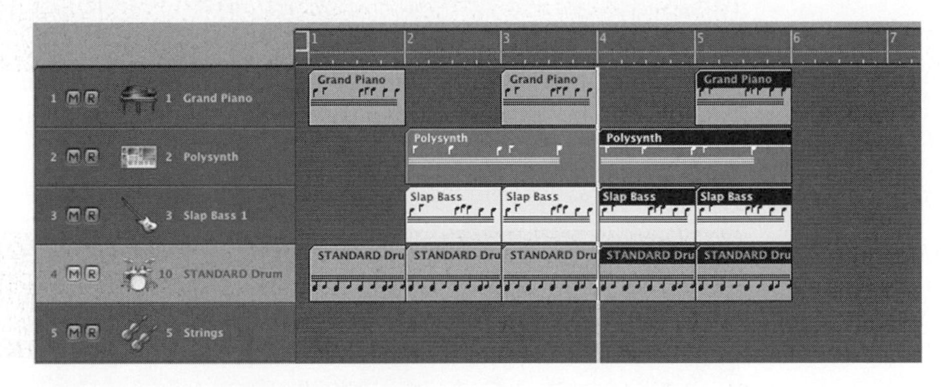

Fig. 5.4.11 – Pasting multiple parts.

Creating song arrangements

By using this technique we can quickly build up a complex song arrangement quickly and efficiently.

Fig. 5.4.12 – Complex song arrangement.

Additional things to try

Now you have got the basic idea of how to duplicate MIDI data, here are some more complex things for you to try.

Duplicating the same MIDI data onto two separate tracks to create a layered sound

In this example we have copied the grand piano data from track 1 down to track 2 and assigned track 2 to a string sound. This will produce a combined sound of piano and strings.

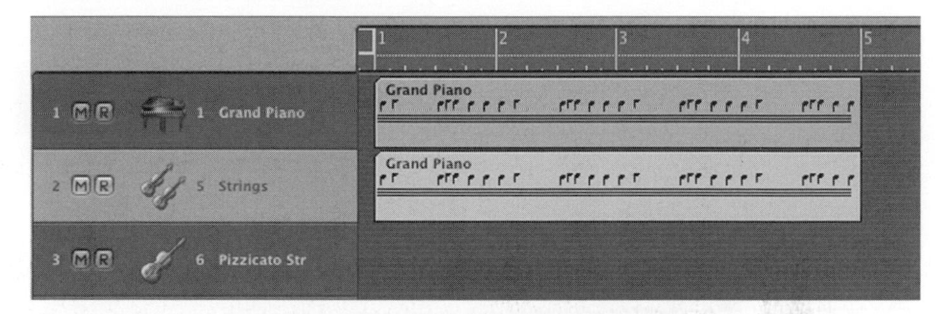

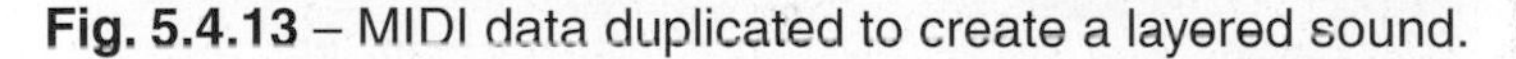

Fig. 5.4.13 – MIDI data duplicated to create a layered sound.

Use an Alias to duplicate MIDI data.

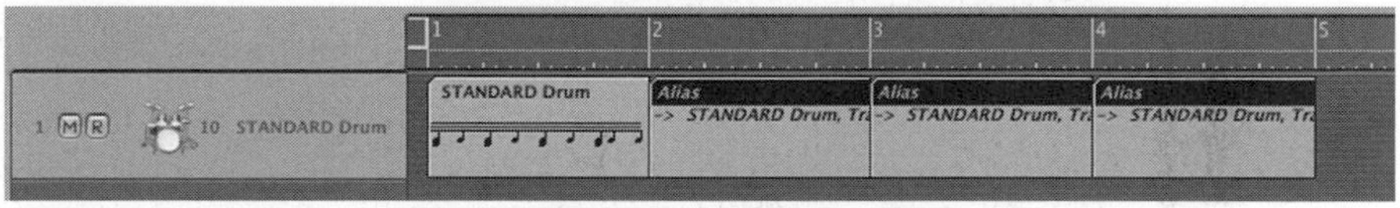

Fig. 5.4.14 – MIDI data duplicated using Alias copies.

Use the loop function to endlessly repeat a part.

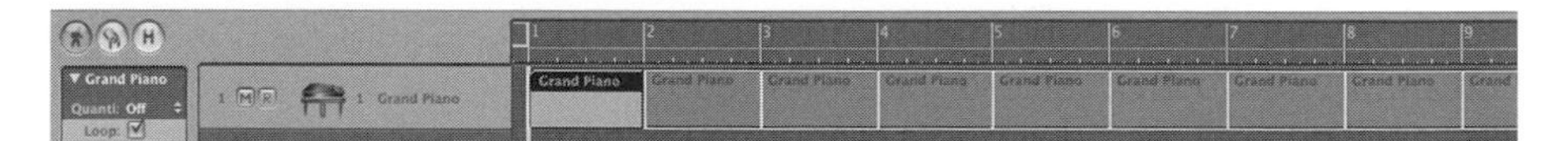

Fig. 5.4.15 – A one bar looped part.

Exercise 5.5 Using the graphic editor

Before attempting this exercise we recommend you complete Exercises 5.1 and 5.2. This exercise will cover the following:

- Opening the Piano-roll-style edit window
- Learning the common tools in the graphic editor
- Editing notes in the graphic editor.

Often, when recording MIDI data into a sequencer we make mistakes. Having made a mistake we have several options. One is to delete the part/region and try and record it again. Another is to separate the sections you want to keep and create a composite (see Exercise 5.3). However, the mistake may only be small and too difficult to edit on the main Arrange window. Therefore, it may be easier to edit and make small adjustments to the part/region using an edit window.

There are usually several different types of edit windows available on a sequencer. They all basically display the same information, just in different ways.

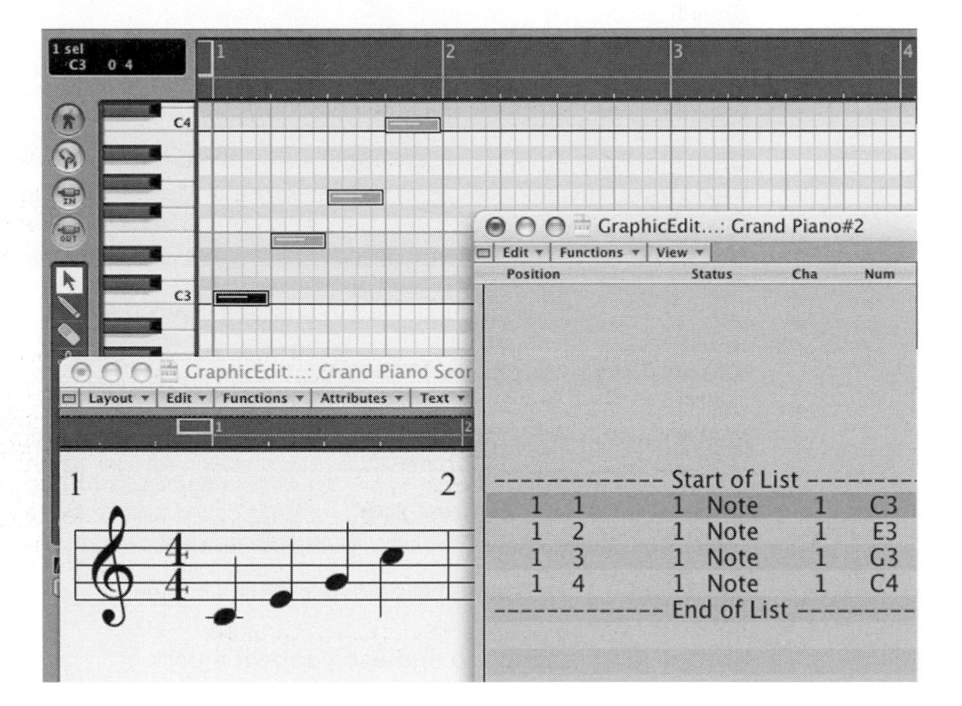

Fig. 5.5.1 – Three different edit windows all showing the same data.

- The *Piano-roll-style editor* displays notes using an old-fashioned piano-roll format
- The *Score editor* displays the notes in traditional western manuscript
- The *List editor* displays a list of events and presents them in a numerical 'list'.

The exercise ...

In this exercise we are going to explain how MIDI notes can be adjusted by using the Piano-roll-style graphic editor. You will therefore need to have at least one MIDI track within your sequencer containing some MIDI note data and assigned to a sound source. Alternatively,

- Record some new MIDI data
- Play something you know already
- Use the data from a previous exercise
- Load in a sequence you want to edit
- Load a MIDI file
- Load a tutorial file from our CD.

Whatever method you choose to get started, make sure you that you are viewing the main Arrange window. Most sequencers open up on this page as default.

Opening the Piano-roll-style edit window

Highlight the part/region you want to edit, then open the Piano-roll-style edit window via the appropriate menu. Each sequencer tends to refer to this window differently – for example, Cubase refers to it as the Key editor, whereas in Logic it is called the Matrix editor.

Fig. 5.5.2 – Opening the Key editor in Cubase.

Alternatively, try double-clicking on the part (click twice with the left mouse button quickly). It should open out into the Piano-roll-style graphic editor. If it doesn't, and instead opens into your Score editor, go to your sequencer preferences, where you can usually change what happens when you double-click on a part. Once opened, it should look similar to Fig. 5.5.3.

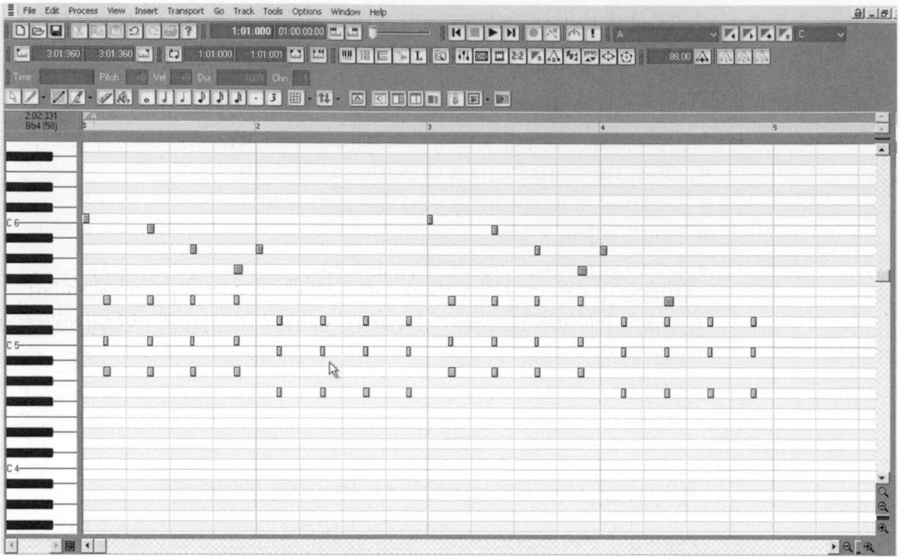

Fig. 5.5.3 – MIDI data displayed in the Piano-roll-style edit window.

This format is one of the most commonly used edit windows and is similar to a graph. Running vertically down the left-hand side is a piano keyboard that displays the pitch of each note, (the higher the note, the higher the pitch) whereas, the horizontal timeline displays time in beats and bars. The notes themselves are represented by strips of colour moving horizontally towards the keyboard, which makes it very clear to see the pitch and duration of each note.

note ▶

Pro Tools

Pro Tools does not have a dedicated Piano-roll-style edit window, so in order to achieve a more detailed view of the MIDI notes, simply increase the size of the MIDI track.

Tools in the graphic editor

The Piano-roll-style edit window usually has its own selection of tools that are similar to the Arrange window.

Fig. 5.5.4 – Graphic editing tools.

Editing notes in the graphic editor

Adjusting the pitch of a note

Use the arrow tool to click and hold a note. Drag it up and down. You should hear all the notes being triggered. This is equivalent to running your finger up and down a keyboard.

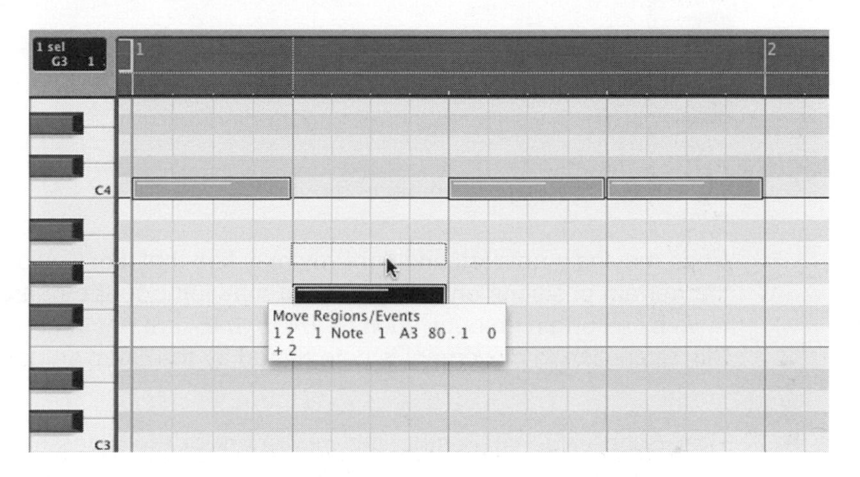

Fig. 5.5.5 – Pitch editing example.

Hearing notes in the graphic editor

Most sequencers will allow you to hear any note that you select within an edit window. If nothing is heard, make sure the track you are editing is selected in the Arrange window. You may also need to enable the MIDI output button. This is a small button that is located in the edit window, which usually has a picture of a loudspeaker or MIDI plug icon on it. This button basically allows you to choose whether the sequencer will output a MIDI signal each time you edit or move a note in the edit window.

tip ▶

Zooming in

If the note is too small to select, try adjusting the Zoom setting. This will allow you to see each note in more detail.

Changing the timing of a note

Now try moving a note left and right. Simply click and hold the mouse in the middle of a note and move it. This determines its location within the song and will allow you to make fine adjustments to its timing.

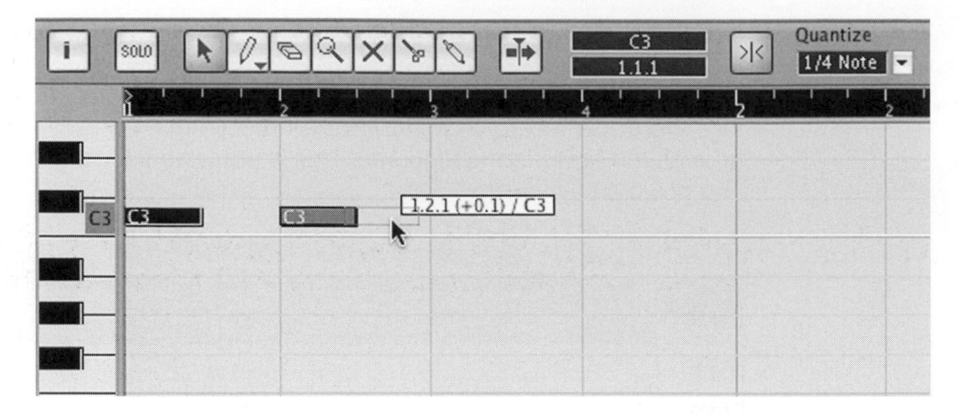

Fig. 5.5.6 – Moving the timing of a note.

When moving a note left and right you may find that it only allows you to place it at certain positions on the timeline. It may jump to the nearest line or only move by a certain amount, such as 1/4 or 1/16, but not to an actual line itself. This restriction in movement is determined by the Snap value.

note ▶

The Snap value will restrict the movement of the notes by a set amount and will determine how freely you are able to move notes horizontally across the timeline.

Adjusting the Snap value

Look for a Snap value in the edit window. By switching it on we can make sure that any notes we create or edit will snap into position and only move to the nearest grid line. For example, setting it to 1/4 will only allow you to place or move notes to each quarter (1/4) in each bar. For more detailed movement you will need to increase the Snap value. For totally free movement switch the Snap option off. This will allow you to ignore the grid lines and place notes anywhere.

tip ▶

Some sequencers allow you to hold down a key on the computer keyboard to bypass or adjust the Snap value. This provides a convenient way to toggle between the Snap value and free movement.

Adjusting the display resolution

The Piano-roll-style editor is based around a grid that displays bars, beats and other subdivisions of the bar. You can change how many subdivisions are displayed by experimenting with the resolution options inside the editor. This means changing the number, which normally looks like 1/4, 1/8 or 1/16. The higher the number, the more subdivisions of the bar will be displayed.

Figure 5.5.7 shows an example of four notes placed *on* the main beats of a bar. Notice that altering the display resolution or the Snap value affects the amount of grid lines on the screen. By making the Quantize 1/4 we see the bar broken up evenly by four lines. By making it 1/16 we see it evenly broken up by 16 lines.

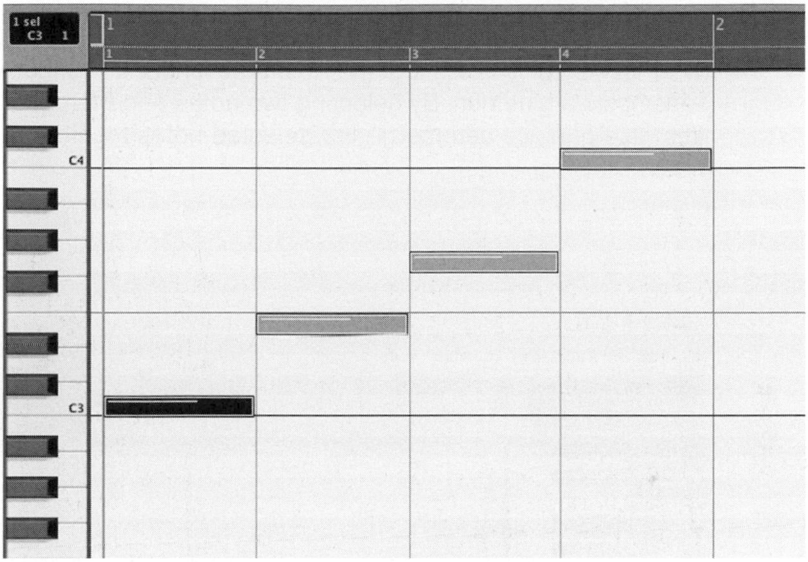

Fig. 5.5.7 – Edit window displaying a 1/4 grid.

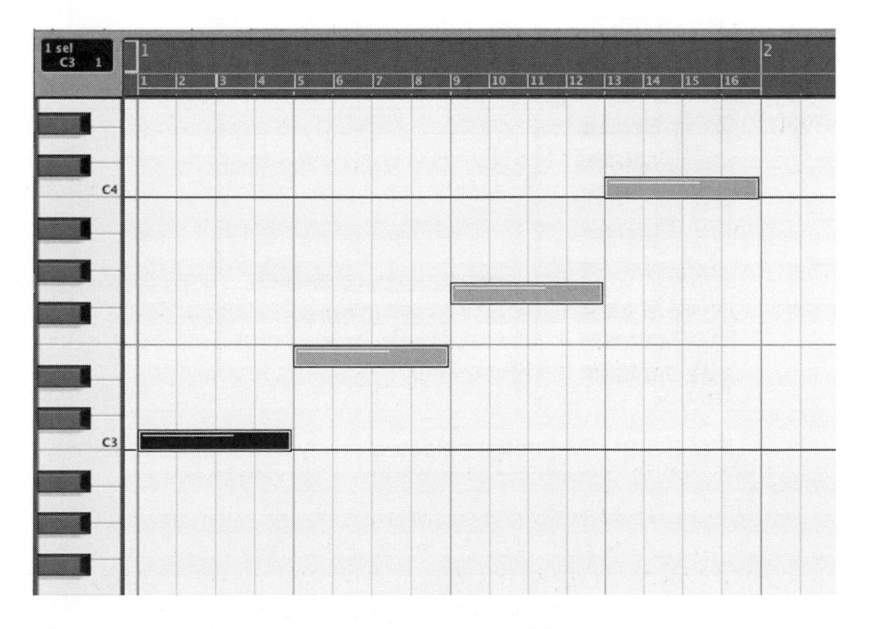

Fig. 5.5.8 – Edit window displaying 1/16 grid.

Changing the duration of a note

Use the Pencil to click on the *end* of notes. This allows us to change the duration of a note. The duration is how long each note actually lasts. As with moving notes, the Snap value may affect this.

Joining notes

We can also *glue* notes together to create longer notes. Look for a tool which resembles a glue gun. By selecting two notes and then clicking on them with this glue gun, we can merge the selected notes together to create a longer note.

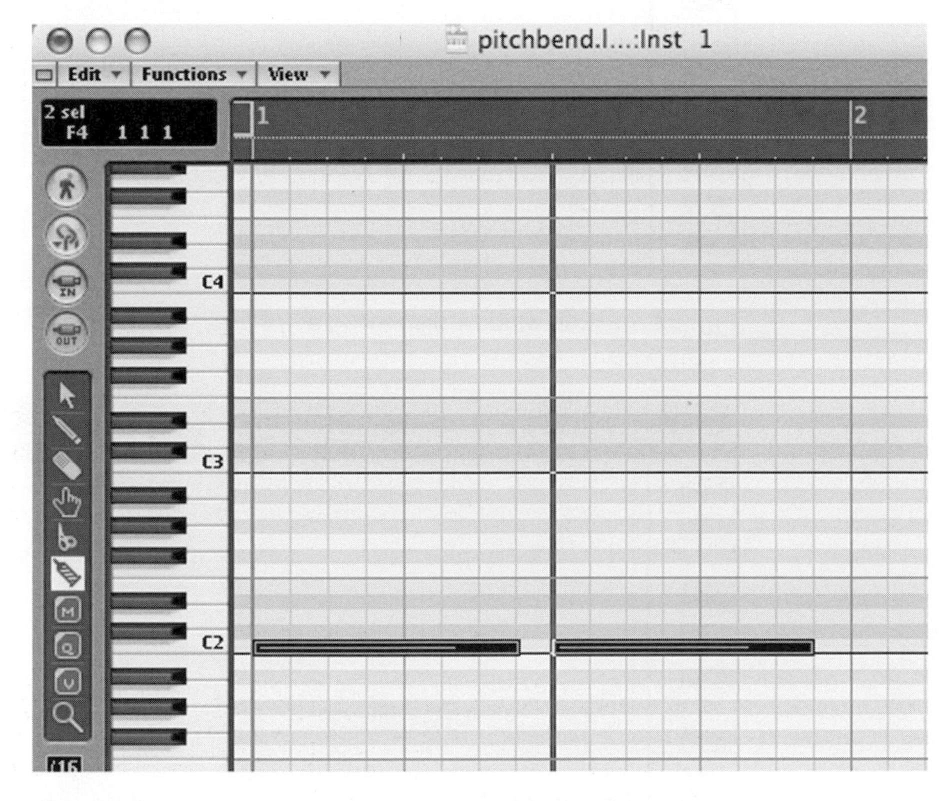

Fig. 5.5.9 – Two notes selected ready to be glued together.

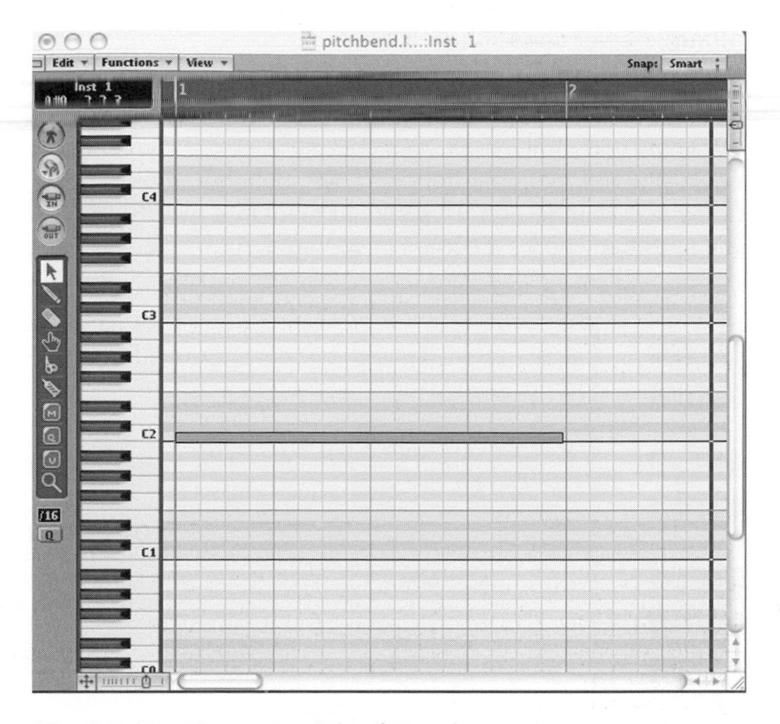

Fig. 5.5.10 – Two notes joined together.

Cutting notes

You may want to try other editing tools such as the scissors. These will allow you to split notes. Hence if you play a note that lasts two bars and you decide you would rather have two notes at the same pitch instead of one long note, you can use the scissors to split them. Once again the Snap value may restrict where you are able to make a cut.

tip ▶ When editing notes within the Piano-roll-style editor, it's easy to get carried away without listening to your adjustments in context with the overall arrangement. We therefore recommend you always listen back at regular intervals.

Multiple selections

It is possible to edit more than one note at the same time by making a multiple selection. This can be achieved by holding down the shift key and selecting the notes you want to edit. Alternatively, click once and hold the mouse button down (the left mouse button on a two-buttoned mouse) and drag it over the notes you want to edit. This should resut in a wire frame box appearing that will select any notes it comes in contact with.

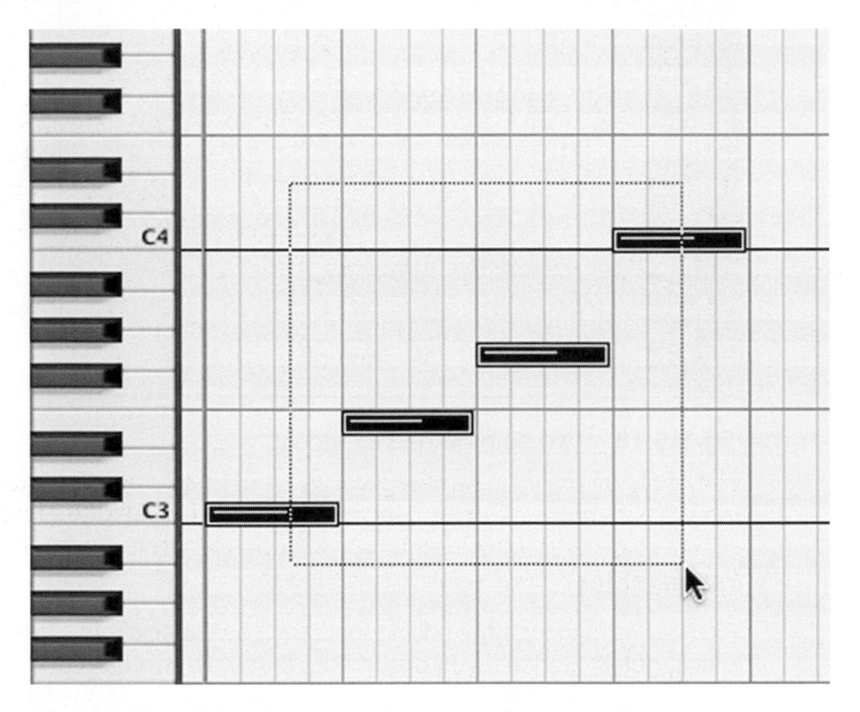

Fig. 5.5.11 – Selecting several notes at the same time.

Editing your selection

Once several notes are selected, try moving the notes around the edit window. Notice how they all move yet keep their relative distance from each other.

Advanced editing and selection options

Often, you will find more advanced editing and selection options. These will help you make a more detailed selection. For example:

- *Select all notes of similar pitch.* By selecting a note in the editor and then using this function from the menu, it will automatically select all identical notes (by pitch) in the editor.

- *Select all notes inside locators.* This function allows you to specify which notes get selected by entering a value in the timeline ruler or transport bar. All the notes within the locator's range will be selected.
- *Select notes of similar velocity.* Select a note in the editor and then use the menu function to automatically select all identical notes that have an equal velocity in the editor.

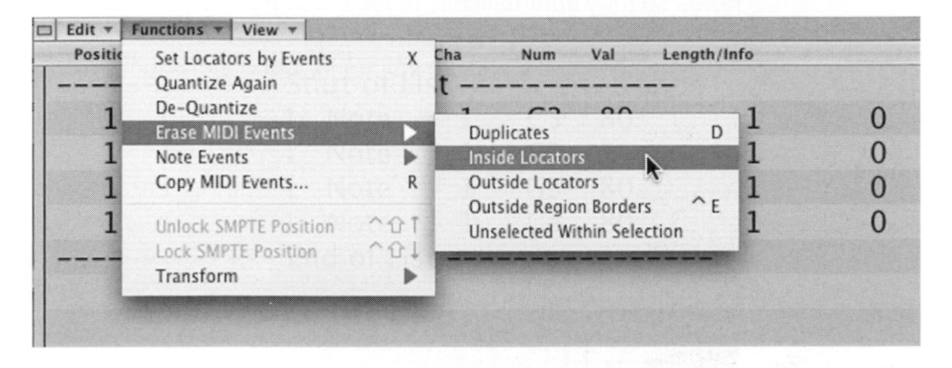

Fig. 5.5.12 – Advanced selection options in Logic.

Hopefully by now you have a good grasp of how to use the graphic editor and what it is capable of. From now on, you should be able to make small adjustments and edit a MIDI performance to perfection without having to become the greatest keyboard player in the world.

Additional things to try

Try opening the List editor

This will display each note in a numerical 'list'.

Edit | Functions | View

Position			Status	Cha	Num	Val	Length/Info		
------------ Start of List ------------									
1	1	1	Note	1	C3	80	.	1	0
1	2	1	Note	1	E3	80	.	1	0
1	3	1	Note	1	G3	80	.	1	0
1	4	1	Note	1	C4	80	.	1	0
------------ End of List ---------------									

Fig. 5.5.13 – MIDI notes displayed in a list.

Drawing notes

The pencil tool can also be used to draw notes onto the edit window and hence into the part/region which is open. Learn more about this in Exercise 5.11.

Quantizing notes in the graphic editor

After making a selection it is usually possible to quantize a note or group of notes so they automatically move left or right to the nearest subdivision line of a bar. This allows a very detailed approach to quantizing, as you can often choose individual Quantize values for each individual note (learn more about Quantize in Exercise 5.6).

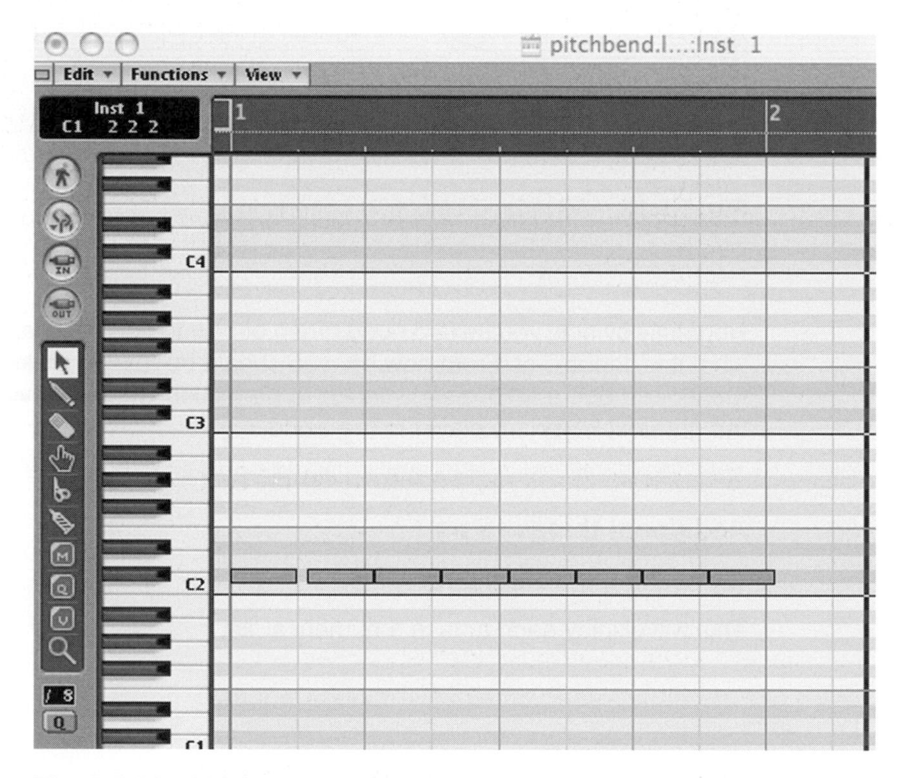

Fig. 5.5.14 – Unquantized MIDI data.

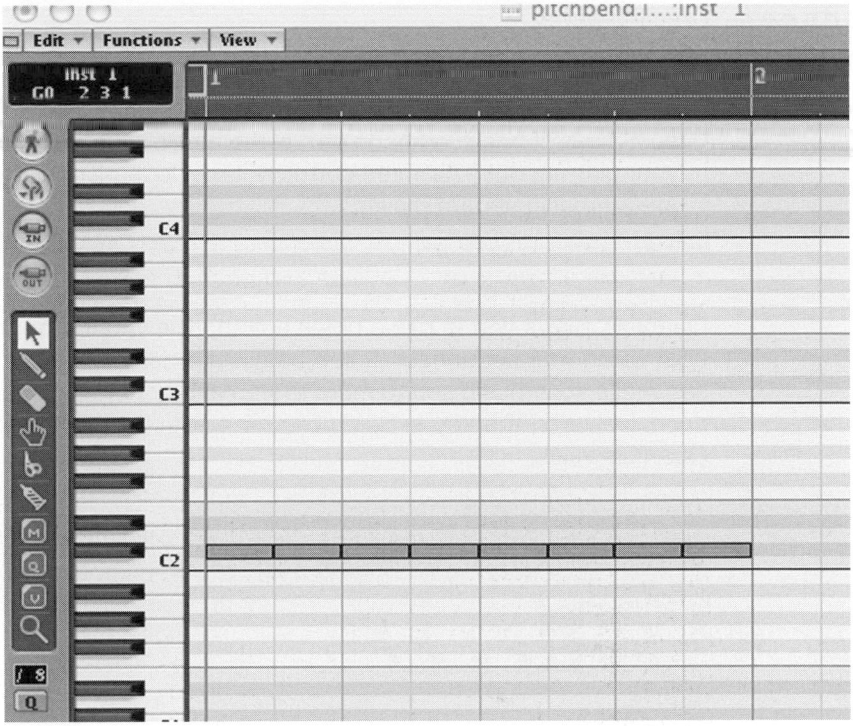

Fig. 5.5.15 – Quantized MIDI data.

Exercise 5.6 Quantizing MIDI information

This exercise will cover the following:

- How to correct the timing of notes
- Displaying Quantize
- Quantizing different rhythms
- Using advanced Quantize functions.

Using Quantize with your MIDI information

When listening to music these days, have you noticed how no one makes any mistakes any more? Everything is perfectly in time. You never hear a drum beat out of place or a piano chord slightly late. This is not because musicians have become amazingly accurate and consistent, but it is essentially due to the introduction of a sequencer function called Quantize.

producer says ▶

> When recording MIDI we often make rhythmic or timing mistakes. It's natural and not something we should worry about. Traditionally, music has been full of mistakes and generally this has been accepted, and sometimes it has even been part of the music. Quantize doesn't improve your playing, but if it's used correctly it can fool anyone into thinking you are a great keyboard player!

Quantize allows you to move notes automatically to the nearest beat or sub-beat within your sequencer (a sub-beat is a division between each beat). This can make your music sound more in time and generally tighten up a performance. However, the key to using Quantize effectively is selecting the correct Quantize value.

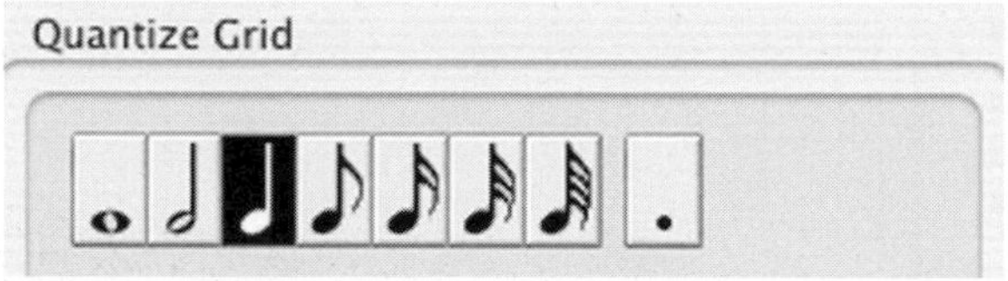

Quantize Grid

Fig. 5.6.1 – Quantize values displayed as notation.

Quantize values are sometimes displayed as notes or in a numerical form.

Quantize values: 1/1, 1/2, 1/4, 1/8, 1/16, etc. tend to create a regular pulse and are used to make rhythms sound 'on the beat'. For example, sustaining chords which play only once per bar might be set to a Quantize value of 1/1. This means each note of the chord will be automatically moved to the beginning of each bar.

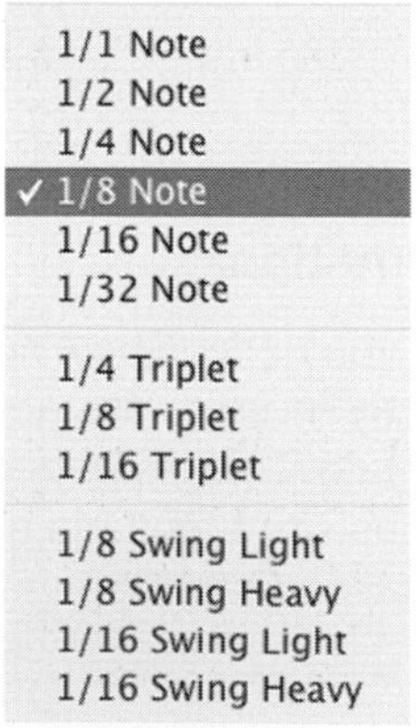

1/1 Note
1/2 Note
1/4 Note
✓ 1/8 Note
1/16 Note
1/32 Note
1/4 Triplet
1/8 Triplet
1/16 Triplet
1/8 Swing Light
1/8 Swing Heavy
1/16 Swing Light
1/16 Swing Heavy

Fig. 5.6.2 – Quantize values displayed in numerical form.

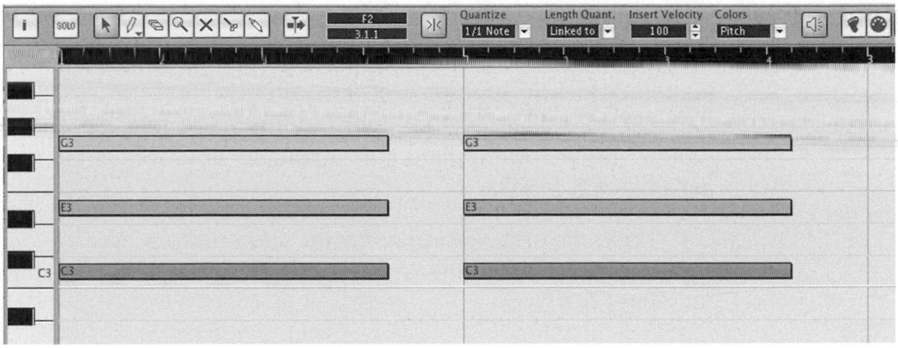

Fig. 5.6.3 – Quantized chords.

A hi hat pattern is usually busier than a chord sequence and may need to be set to higher quantize value such as 1/8 or 1/16. Notice how it needs more divisions in order to place notes between each beat of the bar.

Fig. 5.6.4 – A Quantized hi hat pattern.

tip ► An understanding of different rhythms will also help you choose the most appropriate Quantize settings.

How Quantize works

By selecting a Quantize value you limit the number of locations within a bar where MIDI notes can fall. Each note is then automatically moved to the nearest selected location. However, in order to use Quantize effectively, you need to be able to play roughly in time with the metronome. Quantize needs a reference point within the sequencer so you should always try and play in time with the metronome, otherwise the Quantize function may move your notes to where it wants and not where you want them to go.

The exercise ...

In this exercise we are going to be recording a short section of music to experiment with Quantize. If you already have an existing MIDI sequence containing a basic rhythm, jump to the stage headed 'Correcting the timing' below. Before attempting this exercise you will need to have completed Exercises 5.1–5.3.

1. *Load the sequencing software and create a new project.* Make sure that you are viewing the main sequencing window for recording and arranging MIDI data.
2. *Set the tempo to 120 and time signature 4/4.*
3. *Switch the metronome on.* Make sure it is set up to produce a sound.
4. *Select a MIDI track and assign it to a sound source.* Create a new MIDI track or use an existing one. Assign it to a MIDI channel and select a drum sound as your sound source.
5. *Enable the record ready switch and try playing your keyboard.* You should hear a sound. If you do not, refer to Exercise 5.1.
6. *Locate a sound you want to use.* You should find a different drum sound on each key of the keyboard.
7. *Create a simple rhythm.* Try and create a regular rhythm. In our example we have copied the same rhythm as the sequencer's metronome (the metronome is usually set to play four clicks in each bar). If you want to try this, make sure the metronome is switched on, then try playing the keyboard as if you are a robot, exactly in time with the metronome click.

Fig. 5.6.5 – Regular four-beat rhythm.

note ▶ If you were to try recording the same 'rhythm' as the metronome, then you would be playing four notes in one bar (if we assume that the time signature is 4/4). These four notes should in theory be equally placed across the bar.

8. *Press Record.* Make sure you locate the song position back to the beginning of bar 1 before you drop into Record.

9. *Record your rhythm into the sequencer.* Press Stop when you have recorded your idea and locate back to the beginning of bar 1.

10. *Listen back to what you have just recorded.* Is it in time? Are you happy with the rhythm? If you're not, delete the part and record it again. If it sounds OK, move on to the next stage.

Correcting the timing

1. *Select or highlight the MIDI data you want to edit.* What we want to do now is put each note in time.

2. *Choose a 1/4 note Quantize value.* Quantize menus are usually located on the main Arrange window. In our example we have recorded a kick drum, and selected a 1/4 note Quantize value.

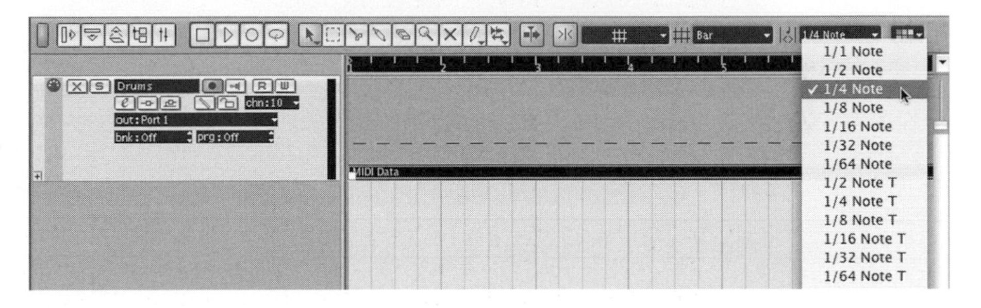

Fig. 5.6.6 – Selecting a 1/4 note Quantize value.

tip ▶ If you have recorded a simple rhythm, then try quantizing it to 1/4 or 1/8.

3. *Now listen back and see how the timing has changed.* By selecting a quantize value of 1/4 you have limited the number of locations within one bar where a MIDI note can fall to just 4. This means each note can only fall on each 1/4 and will be automatically moved to its nearest 1/4.

Displaying Quantize

If you open up one of the graphic edit windows, such as the Piano-roll-style editor, you will be able to see how Quantize adjusts the timing of each note visually. Before the notes were quantized (Fig. 5.6.7), you may have played almost perfectly in time with the metronome, but notice how some notes may be placed slightly before or after the bar division lines.

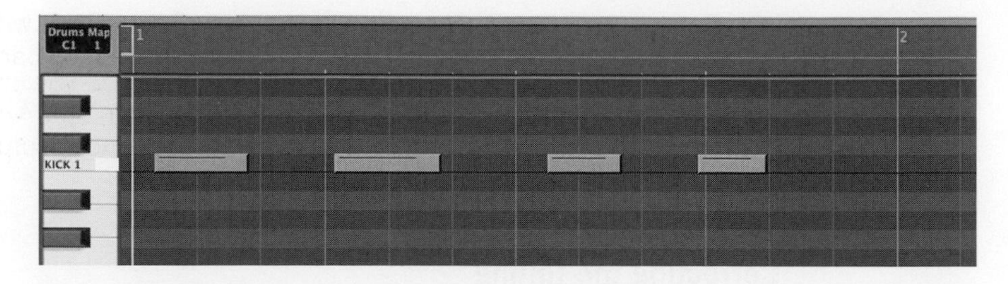

Fig. 5.6.7 – Unquantized kick drum pattern.

After the notes were quantized (Fig. 5.6.8), notice how each note has moved so it sits exactly on the bar division line.

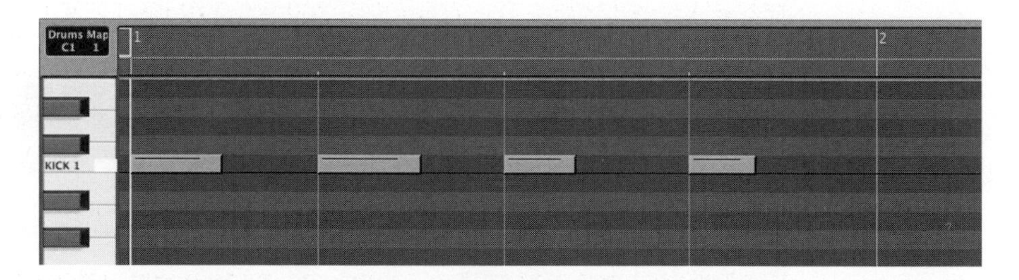

Fig. 5.6.8 – Quantized kick drum pattern.

note ► Quantize doesn't automatically put your music in time, it just moves the notes you've played to the nearest beat.

Quantizing a different rhythm

Now try recording a different rhythm into the sequencer, perhaps using the hi hats. Make it a fast rhythm, using more than four notes in each bar. Once this has been recorded, quantize it using a different Quantize value, such as 1/8 or 1/16.

note ► Usually, Quantize is non-destructive. This means the Quantize value you select is not permanent and can easily be changed or switched off at any time. This allows you to experiment and flick through all the different Quantize values in the list to find the most suitable one to use. You can even do this while the sequencer is playing.

When Quantize doesn't work

Sometimes Quantize cannot put your music 'in time' and you may find it fixes some of the notes but others seem to move more out of time. This may be because you are choosing the wrong Quantize value or the original recording was not close enough to the beat.

Choosing the wrong Quantize value

This can make some of the notes jump to the wrong location, as each note will simply move to the nearest selected value not in time. Experiment by trying different Quantize values.

The original recording was not close enough to the beat

If you are too far away from a beat or sub-beat (such as 1/8 or 1/4) when you record the notes into the sequencer, you will give the computer more options of where to position the note when you use Quantize. The computer may shift your note to the wrong division line because it is simply the nearest. (Remember, the computer doesn't know where each note should be, it just chooses the nearest.)

If this happens, you have two options:

1. Delete your part and then record again.
2. Use the graphic editor to manually move any notes that have not quantized correctly.

Remember, what matters is what the music sounds like, not what it looks like in the edit window. Trust your ears.

note ► Quantize can be used on any type of MIDI data, not just drums.

Using advanced Quantizing functions

Once you have learnt the basic principle of how Quantize works, it's time to explore some of the more advanced features that are usually available.

Creating a swing feel

Quantize values such as 1/1, 1/2, 1/4, 1/8, 1/16, 1/32, etc. create a regular pulse and are used to make rhythms sound 'on the beat'. However, values such as 1/6, 1/12 and 1/24 create more of a syncopated triplet-type feel, which can be useful if you are trying to create a piece of music which should 'swing'. Experiment by choosing one of these values to hear how it changes to rhythm.

Another way to achieve a similar result is to apply a percentage of swing to an existing Quantize value. This option allows you to select a regular 1/8, 1/16, 1/32 on the beat Quantize value and then decide in percentage how

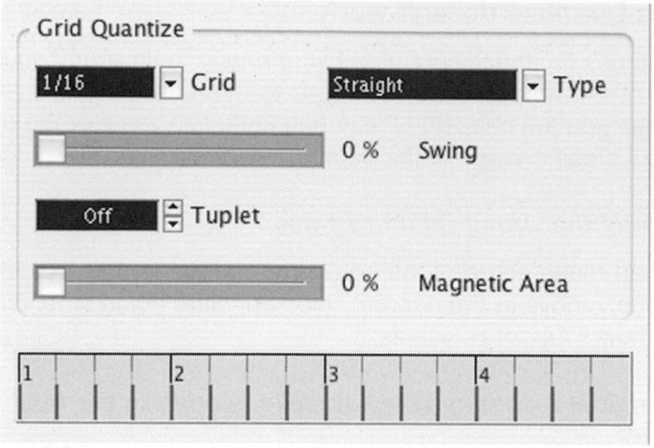

Fig. 5.6.9 – Quantize window with no swing.

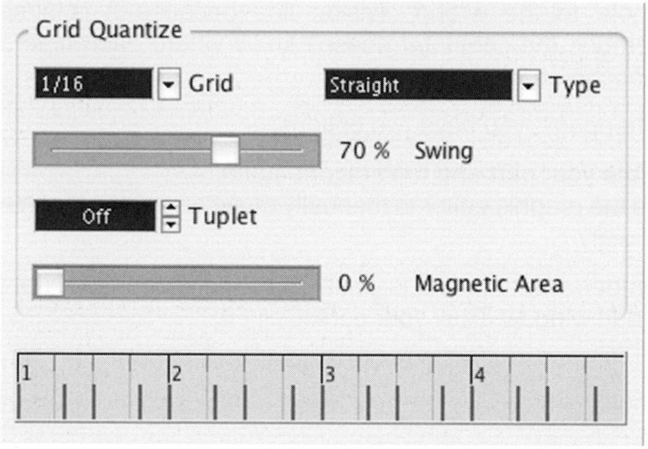

Fig. 5.6.10 – Quantize window with 70% swing.

much swing you want to add. In our example in Figure 5.6.10 we have selected a regular 16th Quantize value and then added a 70% swing. Notice how it moves every second 16th value slightly to the right.

note ► Most sequencers default to a Quantize value of 1/16.

Using different Quantize values on the same track

In this exercise we have explained how to quantize all the notes in the selected part to one particular value. Sometimes you may find different sections of your song may need to be quantized to different values. This may be because a rhythm changes at a certain point or the chosen Quantize value only corrects some of the notes and not all of them. One way to solve this problem is to cut your part into the relevant sections and then quantize each individual section separately (see Exercise 5.3). Alternatively, you may be able to quantize notes separately whilst in the graphic editor (refer to Exercise 5.5).

Exercise 5.7 Editing velocity

This exercise will cover the following:

- Recording velocity information into a MIDI sequencer
- Displaying velocity information
- Editing velocity information.

Velocity is part of a 'note on' message and is sent each time you press a key on a MIDI keyboard. It measures how hard or the speed at which the key is pressed. This in turn determines the individual volume of each key. Hence, velocity can have a fundamental effect on the dynamics of a sound.

Measuring velocity

Velocity has a parameter range between 0 and 127, with 0 being the quietest and 127 the loudest. For example, a note with a velocity of 30 would be quiet and a velocity of 120 would be very loud.

The exercise ...

1. *Initial set-up.* Load your sequencing software and create a new project. Make sure you that you are viewing the main sequencing window for recording and arranging MIDI data, and you have a least one MIDI track that it is assigned to a sound source. Try playing your controller keyboard to check that MIDI signals are being received by the sequencer and that you can hear the sound source.
2. *Selecting a sound.* Try selecting a sound that will respond well to changes in velocity. In our example we have selected GM057, a trombone sound.

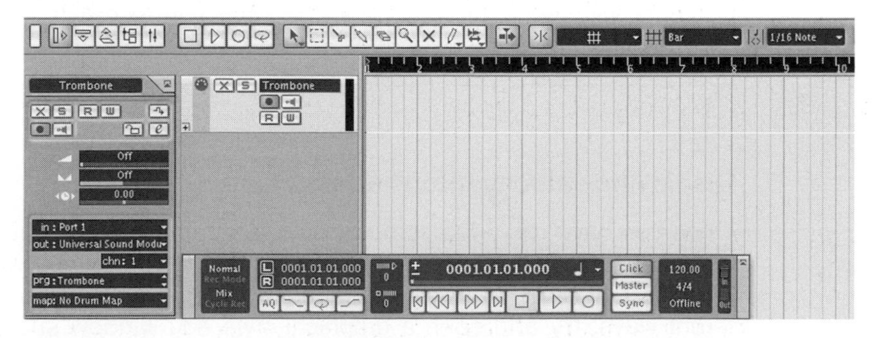

Fig. 5.7.1 – Cubase track assigned to a trombone sound.

Recording velocity information into a MIDI sequencer

Put the sequencer into Record mode. You may get a one- or two-bar pre-count, which is the time before the sequencer starts recording. After this, you will see the marker on the screen start to move and any notes you play will be recorded.

Whilst playing the controller keyboard, experiment by striking the keys (i) softly and (ii) hard and listen to how the sound changes.

After playing for a while, press Stop on the transport bar to make the sequencer stop recording. You should have now created some note and velocity data on your track.

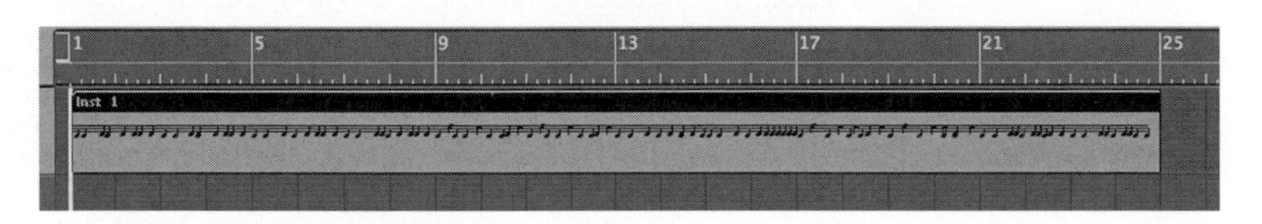

Fig. 5.7.2 – MIDI information.

producer says ▶

Different types of keys

The actual keys on a MIDI controller keyboard can vary tremendously. Some keyboards are very light and their keys need hardly any effort to press down. Other keyboards are semi- or fully weighted. This means that the keys have a weight built into them to emulate the feeling of playing on a real piano. Remember, real piano keys have a tension due to the string and the weight of the actual materials. The lesson here is to get to know the keyboard you are playing on. Experiment playing loud and then soft to see how responsive the keys are. You will find that some keyboards make it easier to be expressive while others will seem unresponsive and force you to press the keys down harder.

Displaying velocity information

Once you have recorded some velocity information into a MIDI sequencer it can be displayed and edited graphically. Select the part you want to edit and open an edit window. Different sequencers display velocity in a variety of different ways. Try and open a graphical-style edit window such as the Hyper Edit or Piano-roll.

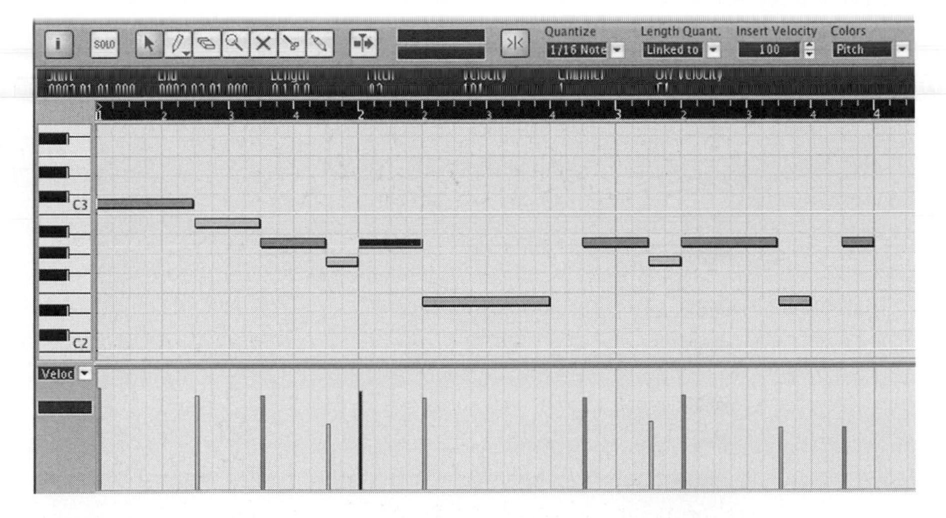

Fig. 5.7.3 – Note and velocity information displayed graphically.

You will see a number of markings underneath each note which look a bit like a line graph. These display the relative strength of each note's velocity, i.e. the higher the line the higher the velocity. They may even vary in colour, with red normally representing a high value near to 127, and blues and greens representing lower values.

note ▶ Some edit windows will display velocity only in the bottom half of the window. In Logic, for example, this part of the window has to be switched on and you have to choose what kind of parameter you want to see.

What tools to use

Using the pencil or arrow tool you will be able to select individual velocities and change them. This will allow you to make each note sound louder or softer. The pencil tool also gives you a 'free-hand' approach to changing the levels, allowing you to sweep across the velocity lines and create gradual crescendos and diminuendos.

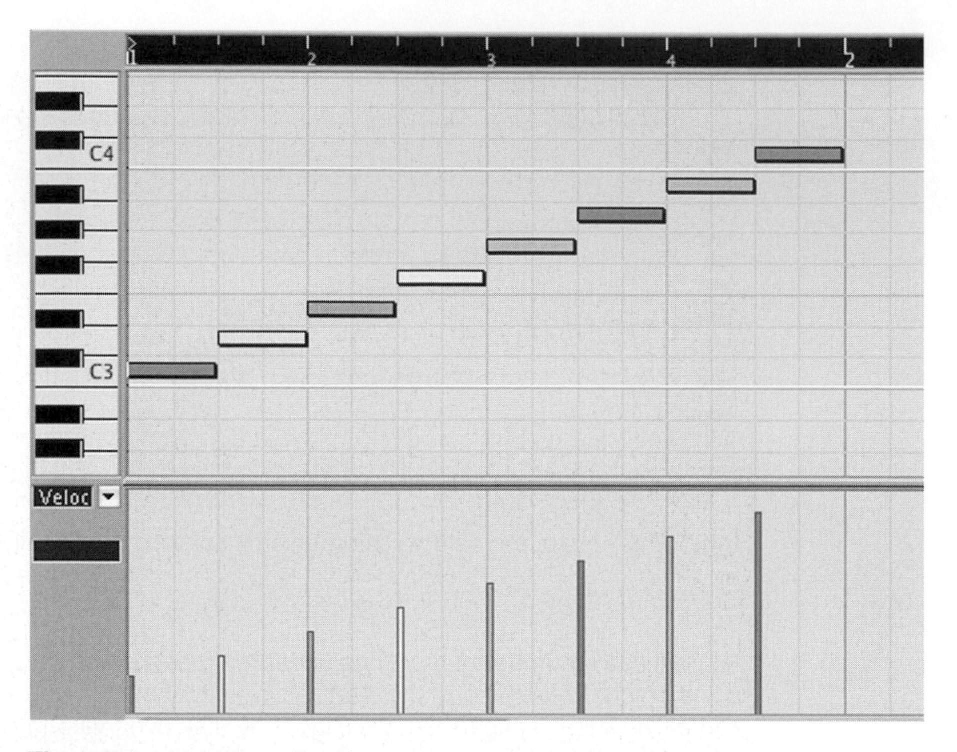

Fig. 5.7.4 – Velocity edited to create a crescendo.

Creating a velocity crescendo

Try starting with a low velocity value of around 40 and slowly increase the level throughout the bar to 127. This will automatically increase the volume of the notes and create a build throughout the bar. For another example of how velocity can be used to create a crescendo, see 'Creating a velocity snare roll,' p. 206.

note ► Editing the velocity allows you to increase or decrease the volume of each note.

Editing velocity

Now open the List editor and see how the same data would look in a numerical format. The first few columns usually list the position of the notes within the song, the note number and the MIDI channel. In the example in Fig. 5.7.5, the 'val' column lists all the velocity values. Notice how all velocities from song position 3.2.1.65 have been edited to a value of 98.

Untitled.lso:Inst 1

Edit | Functions | View

Position				Status	Cha	Num	Val	Length/Info			
Start of List											
0	4	4	206	Note	1	C2	98	.	.	.	132
1	1	4	222	Note	1	C2	67	.	.	2	12
1	3	3	9	Note	1	D#2	56	.	.	.	127
1	4	1	29	Note	1	D2	103	.	.	.	118
1	4	2	210	Note	1	C2	87	.	.	.	157
2	1	2	209	Note	1	D2	64	.	.	.	178
2	2	2	210	Note	1	D2	46	.	.	.	128
2	2	4	209	Note	1	D#2	46	.	.	.	235
2	3	4	190	Note	1	D2	84	.	.	.	196
3	1	1	16	Note	1	C2	64	.	.	.	139
3	2	1	65	Note	1	C2	98	.	.	2	196
3	3	2	237	Note	1	D#2	98	.	.	.	168
3	4	1	6	Note	1	D2	98	.	.	.	82
3	4	2	208	Note	1	C2	98	.	.	.	207
4	1	3	102	Note	1	D2	98	.	.	.	139
4	2	2	198	Note	1	D2	98	.	.	.	138
4	2	4	188	Note	1	D#2	98	.	.	.	157
4	3	4	226	Note	1	D2	98	.	.	.	176
4	4	4	178	Note	1	C2	98	.	.	.	168
5	1	4	225	Note	1	C2	98	.	.	2	112
5	3	3	4	Note	1	D#2	98	.	.	.	139
5	3	4	225	Note	1	D2	98	.	.	.	139
5	4	2	222	Note	1	C2	98	.	.	1	27
6	1	3	1	Note	1	D2	98	.	.	.	139
6	2	2	165	Note	1	D2	98	.	.	.	184
6	2	4	178	Note	1	D#2	98	.	.	.	146
6	3	4	164	Note	1	D2	98	.	.	.	216
6	4	4	221	Note	1	C2	98	.	.	.	196

Fig. 5.7.5 – Velocity information displayed in a list.

producer says ▶

Random velocities

Velocity can be the key to making a MIDI instrument sound more interesting. For example, try playing in a straight 1/8 or 1/16 drum rhythm. Then go into the editor and randomly change the velocity settings so you accent or emphasize certain beats in a bar. This is a fast and simple way to create a pattern which will have more dynamics. By changing the individual volume of each velocity we are changing the accents, which is similar to what a real drummer does in order to create a 'groove'.

Making all the velocities the same

If you have a rapidly changing velocity pattern, such as a snare hit, which you would like to change so it plays back at a consistent level, you wll need to make all the velocities the same. If you change them to 127, for example, you get the loudest hit possible.

tip ► To provide a constant level, try making all the velocities the same.

Adjusting the volume of individual drum sounds

Drum sounds usually form part of a drum kit, where each individual sound is placed on a separate key of the keyboard. This makes it very difficult to control the volume of each individual drum, as they all share the same MIDI channel and will respond to the same MIDI messages. For example, adjusting the volume for one drum will change the volume for the whole drum kit, even if each drum is recorded on a separate sequencer track. One solution is to use velocity to change the level of each individual drum sound.

In the example in Fig. 5.7.6 we have increased the velocity level of the kick drum to a value of 120 and set all the other drums to 80.

Position			Status	Cha	Num	Val
------------ Start of List ------------						
1	1	1	♩ KICK 1			120
1	1	1	♩ Closed HH			80
1	1	481	♩ Closed HH			80
1	2	1	♩ KICK 1			120
1	2	1	♩ SD 1			80
1	2	1	♩ Closed HH			80
1	2	481	♩ Closed HH			80
1	3	1	♩ KICK 1			120
1	3	1	♩ Closed HH			80
1	3	481	♩ Closed HH			80
1	4	1	♩ KICK 1			120
1	4	1	♩ SD 1			80
1	4	1	♩ Closed HH			80
1	4	481	♩ Closed HH			80
------------ End of List ------------						

Fig. 5.7.6 – Individual drum velocities.

Exercise 5.8 Pitch bend

This exercise will cover the following:

- Locating the pitch bend controller
- Recording pitch bend information into a sequencer
- Recording notes and pitch bend information at the same time
- Displaying pitch bend information
- Editing pitch bend.

As well as transmitting MIDI notes by sending 'note on' and 'note off' messages, there are other types of information which can be sent via a MIDI keyboard. *Pitch bend* is one example of this.

A pitch bend message can usually be transmitted by using the sprung wheel or joystick that is usually located on the left-hand side of a controller keyboard. Moving this control gives you the ability to bend a note. By bend we mean making a note slightly lower or higher in pitch, which therefore makes a note flat or sharp. A similar sound can be achieved by using a trombone slide or bending a guitar string.

Pitch bend operates between a parameter range of 0 and 127. It allocates half of this value to increase the pitch and the other half to decrease it. When displayed graphically, information placed on the centre line indicates that no bend has been applied to the note.

- A value above the centre line will increase the pitch
- A value below the centre line (64) will decrease the pitch
- A value placed on the centre line will have no effect.

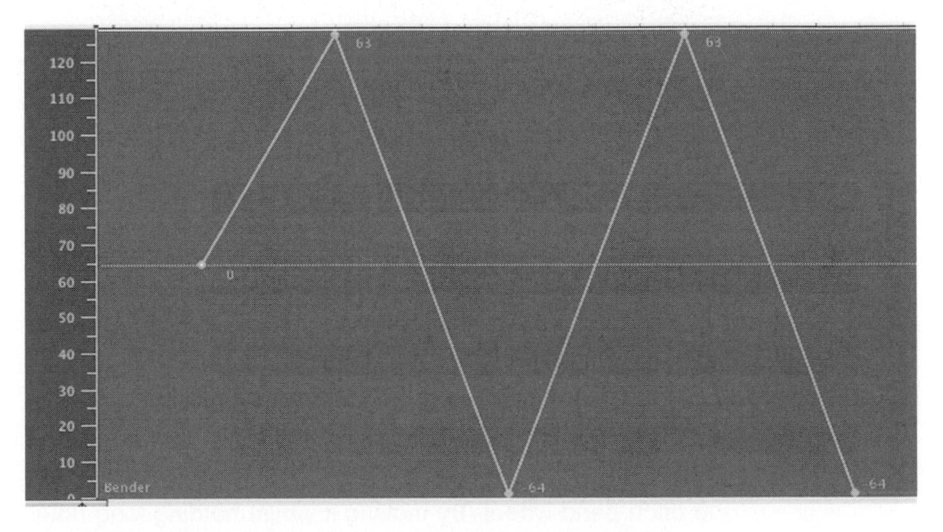

Fig. 5.8.1 – Pitch bend information displayed graphically.

The exercise ...

In this exercise we are going to be recording a short section of music to experiment with pitch bend. If you already have an existing MIDI sequence that you simply want to add pitch bend to, jump to the stage 'Recording pitch bend on a separate track' on p. 122.

1. *Initial set-up.* Load your sequencing software and create a new project. Make sure that you are viewing the main sequencing window for recording and arranging MIDI data, and you have a least one MIDI track that it is assigned to a sound source (see Exercise 5.1).
2. *Selecting a sound.* Try selecting a sound that will respond well to pitch bend, such as organ or a brass sound. In our example we used a GM sound no. 57, trombone.

note ▶ You can use pitch bend with any sound, such as piano, trombone, bass, etc. However, avoid using pitch bend on drum sounds, as this will usually change the pitch of the whole drum kit and not just one individual sound.

Locating the pitch bend controller

The pitch bend wheel is the sprung slider or dial on the left-hand side of the keyboard. It may swivel left and right horizontally or it may glide up and down vertically. Some keyboards have more elaborate controls such as touch strips (similar to those on laptops used to control the mouse) as ways of transmitting performance information such as pitch bend.

Fig. 5.8.2 – A typical pitch bend controller.

Try playing your keyboard

Start playing some notes on the keyboard. Whilst playing, experiment moving the pitch bend wheel. Try moving it whilst holding long notes. See how it can be used to increase or decrease the pitch of a note.

note ▶

Spring back to normal

Notice how the pitch bend controller always springs back to a central point when you let go. This is really a safety feature so the pitch always returns back to normal.

Recording pitch bend information

Now drop the sequencer into Record and start playing some notes on the keyboard whilst moving the pitch bend wheel at the same time. Any notes you play, along with the pitch bend information, will be recorded.

tip ▶

Sliding with the beat

Try recording with the sequencer's metronome switched on. This will provide a regular timing reference and allow you to experiment using the pitch bend rhythmically by moving the pitch bender in time with the metronome so the notes slide with the beat.

Listening back

Once you have recorded a selection of ideas, press Stop on the transport bar. You should now have a MIDI part that contains note and pitch bend information.

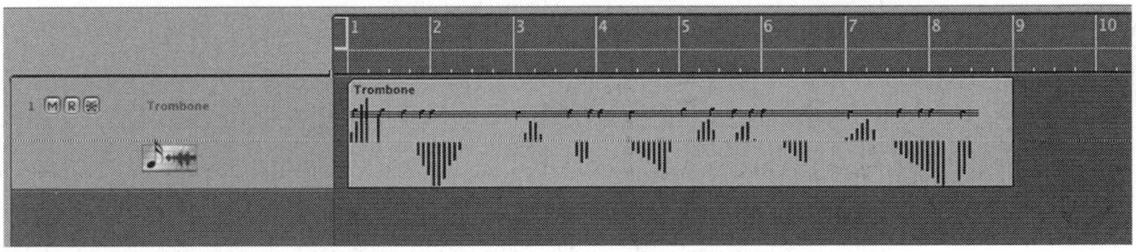

Fig. 5.8.3 – MIDI note and pitch bend information.

Press Play and listen back to what you have recorded. You should hear the pitch of the notes being adjusted by the pitch bend information.

Recording notes and pitch bend at the same time

Moving the pitch bender and playing notes on the keyboard at the same time will transmit two different types of MIDI data:

- Note data (which notes are being played)
- Pitch bend data (how much pitch bend is being used).

How much pitch bend to use

The amount of pitch bend a wheel or dial is capable of is determined by the sound source, not the keyboard itself (unless it's a keyboard containing sounds, of course). This allows the user to adjust how sensitive the pitch bend wheel is and determine the bend range. For example, you could set it so that a full movement up or down equals moving the pitch up or down a whole octave or just one note. You may have to review the manual which came with your sound source to do this.

Displaying pitch bend information

Now select the part you have recorded and open a graphic edit window such as the Piano-roll or List editor. In the example in Fig. 5.8.4, the pitch bend information is displayed in a list alongside the note data.

Position			Status	Cha	Num	Val
-------------			Start of List -------------			
1	1	1	Note	1	C3	108
1	1	175	PitchBd	1	68	68
1	1	203	PitchBd	1	71	71
1	1	232	PitchBd	1	72	72
1	1	288	PitchBd	1	73	73
1	1	317	PitchBd	1	75	75
1	1	346	PitchBd	1	79	79
1	1	375	PitchBd	1	82	82
1	1	401	PitchBd	1	85	85
1	1	430	PitchBd	1	86	86
1	1	459	PitchBd	1	87	87
1	1	488	PitchBd	1	88	88

Fig. 5.8.4 – Pitch bend information displayed in a list.

Display filters

Inside the edit window you will often find a display filter. This will allow you to choose the type of MIDI data you want to see.

note ►

> Pitch bend data can be displayed in a variety of ways depending on the type of sequencer you are using. Most of the time it is displayed alongside the actual note data.

Other types of edit window, such as the Piano-roll-style editor, may display the pitch bend separately, in the bottom half of the window for example. In some sequencers this part of the window has to be switched on and you actually have to specify the type of data you want to see.

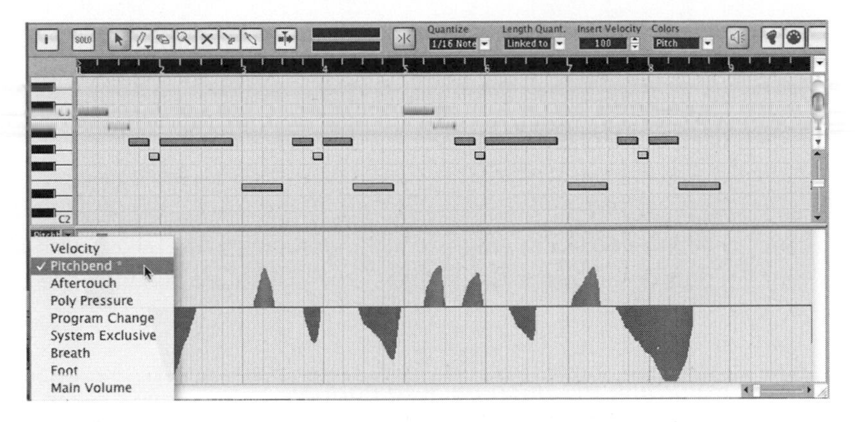

Fig. 5.8.5 – Pitch bend information being displayed below the MIDI notes.

When displayed graphically pitch bend looks a bit like a line graph. The middle of the graph is the normal unchanged position – the same as the centre position for the pitch bend controller. Any information above the line shows an increase in pitch; below the line shows a decrease in pitch.

Editing pitch bend

Make sure you can clearly see the pitch bend information that has been recorded. In our example we are viewing it graphically in the bottom section of the Piano-roll-style editor.

Use the pencil tool to edit and change the pitch bend line or use the rubber to delete the information:

- Moving the line lower makes the music drop in pitch
- Moving the line higher makes the music rise in pitch.

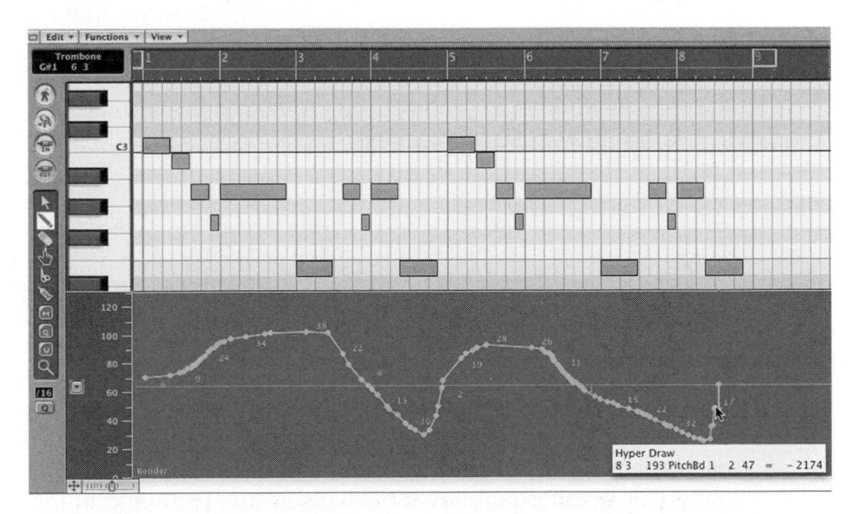

Fig. 5.8.6 – Note and pitch bend information displayed graphically.

Listen back to the changes you've made

Once you have made some changes to the pitch bend data, press Play and listen back. Listen to how the notes change as they follow the curves of the pitch bend display.

note ▶

When making changes to pitch bend data the MIDI notes in the edit window do not move or change. They just sound like they are being bent.

Returning the pitch back to normal

When editing pitch bend always make sure you return the last piece of pitch bend data back to the centre position. If you don't, the pitch of all the notes for that sound and MIDI channel will remain stuck at a lower or higher pitch. This can really confuse people, so please take note of this!

The reason this happens is because a sound source will remember the last piece of information it was sent and stay at this value until it receives some new information. So if you tell a device to increase its pitch, it will stay there indefinitely until you draw the line back to the centre position or move the pitch bend controller. Note: you may also encounter this problem by stopping the sequencer halfway through a pitch bend change.

producer says ▶

MIDI choke

Remember that by creating MIDI 'events' we are creating a situation where more MIDI information has to be sent down our MIDI cables. Normally, this is not a problem. However, be aware that if there is a lot of information you could end with slight timing errors. This is because there might be a slight delay from the time the information is sent by the computer to when it is received by the sound source. Hence, try not to go overboard with the pitch bend!

Additional things to try using pitch bend

Recording pitch bend on a separate track

It is possible to record pitch bend data onto a separate track from note data. You may find this method useful if you're finding it difficult to coordinate playing the notes on a keyboard and using the pitch bend wheel at the same time, or if you want to add pitch bend to an existing MIDI sequence. Separating gives you better control over each discipline. Remember, the two tracks *must* share the same MIDI channel.

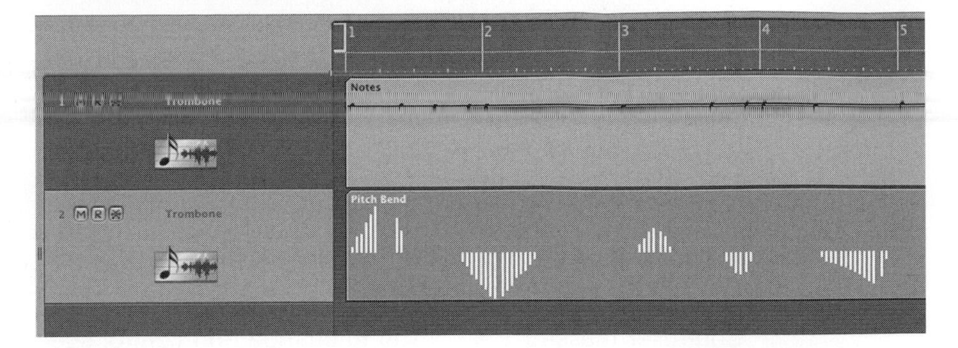

Fig. 5.8.7 – Note data and pitch bend on separate tracks: track 1 contains note data, track 2 pitch bend data.

note ▶ By placing the pitch bend data on a separate track it can be displayed and edited separately.

Sliding up to a note

Pitch bend can be used to accurately slide between different notes (see 'How much pitch bend to use' section above). For example, set the bend range to +2 via the controls on your sound source. Now when you play the keyboard it will allow you to pitch bend up a tone. Try playing C and sliding up to D:

- Hit C once and you hear C
- Push the pitch bender up and hit C and you will hear D
- Now press C and slide up to D.

Octave jumps

By changing the bend range to +12 to −12 we can make the pitch bend jump up or down a whole octave. Try this with single notes and chords.

Pitch bend can be used to enhance different styles of music

Pitch bend can be used in various different styles of music to create quite a cool effect. Try using this later in this book for styles such as:

- *Dance* – maybe for a fast and crazy lead theme
- *Minimalism* – to create a relaxing chord sequence but using the pitch bend to move between the chords
- *Blues* – try using pitch bend to create a 'blues-type effect', to 'bend' the notes and make them sound slightly flat and 'blue'
- *Indian raga* – to create small tuning variations to simulate authentic Indian instruments.

Exercise 5.9 Creating and editing controller data

Before attempting this exercise we recommend you complete Exercises 5.1, 5.2 and 5.5. This exercise covers the following:

- Displaying controller data
- Drawing and editing controller data
- Different ways of transmitting controller data
- Recording controller data in real time.

Controller messages allow you to change and control specific aspects of a sound, such as volume and pan, remotely using MIDI. Each controller message has its own number between 0 and 127, so it can be identified separately from other controller messages. Here are some examples:

- *Controller number 7 is volume.* This allows you to adjust the overall volume of an instrument on a particular MIDI channel. Note that it will also adjust the volume of any other instruments assigned to the same MIDI channel.
- *Controller number 10 is pan.* This allows you to adjust the placement of a mono signal in the stereo field. This means sending a sound to either the left or right speaker or placing it somewhere in between the two. Note: this will only be possible if you are monitoring the sound in stereo.
- *Controller number 91 is reverb.* This allows you to determine how much reverb is added to a sound on a particular MIDI channel (reverb adds ambience to a sound). Note: this is only possible if the sound source itself is capable of producing reverb.

note ▶ When using controller messages, be aware that they get sent to all the sounds and instruments on a particular MIDI channel. Controller messages cannot be used to affect individual notes or instruments that share the same MIDI channel.

The exercise ...

In this exercise we are going to be experimenting with controller data. You will therefore need a MIDI sequence to add controller data to, so:

- Record some new MIDI data
- Play something you know already
- Use the data from a previous exercise
- Load in a sequence you want to edit
- Load a MIDI file
- Load a file from our CD.

Whatever method of starting you choose, make sure you have at least one MIDI track with around eight bars worth of music assigned to a sound source.

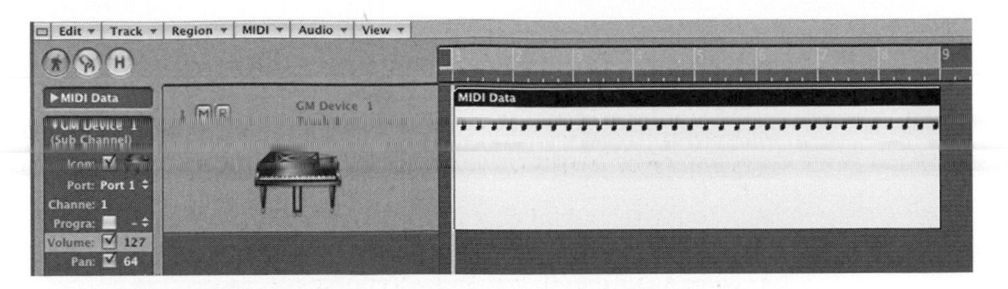

Fig. 5.9.1 – MIDI data displayed on the Arrange window.

Displaying controller data

Select the MIDI part/region you want to add controller data to and open up the Piano-roll-style graphic editor (also known as the Matrix or Key editor). Make sure that you can see the MIDI notes and that you have enabled the section that displays the controller data. In some sequencers this part of the window has to be switched on and you also need to specify the type of data you want to see.

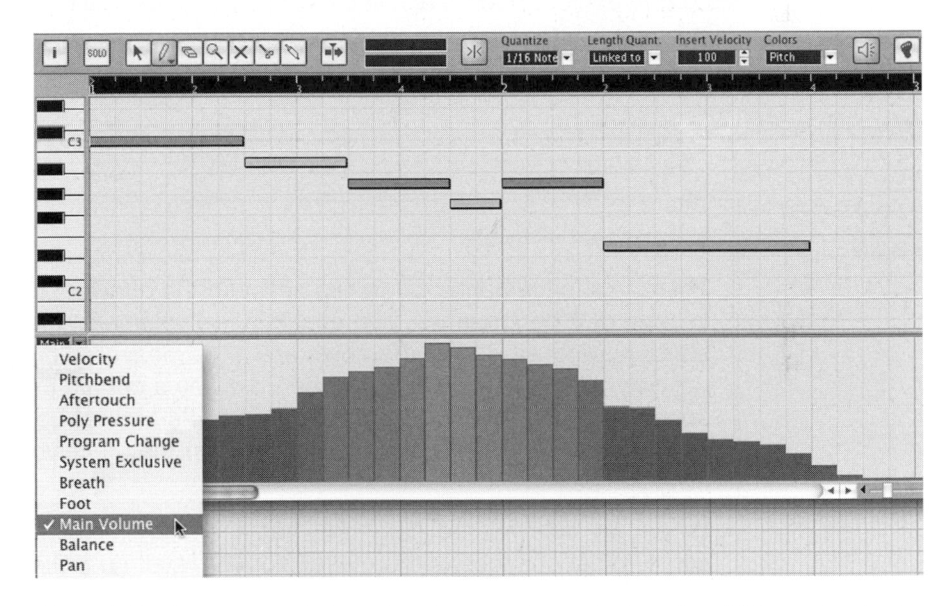

Fig. 5.9.2 – Volume data that has been entered below the MIDI notes in the bottom half of the edit window.

Drawing and editing controller data

Tools

Look at the tools you have available. The pencil will allow you to draw in volume changes onto the screen. The higher the line, the louder the channel; the lower the line, the quieter the channel (see Fig. 5.9.2). If you make a mistake, simply draw over what you have done or use the rubber to delete the controller information. You may find other tools, such as the cross-hair or line tool. These allow you to create very smooth and linear increases or decreases in volume accurately.

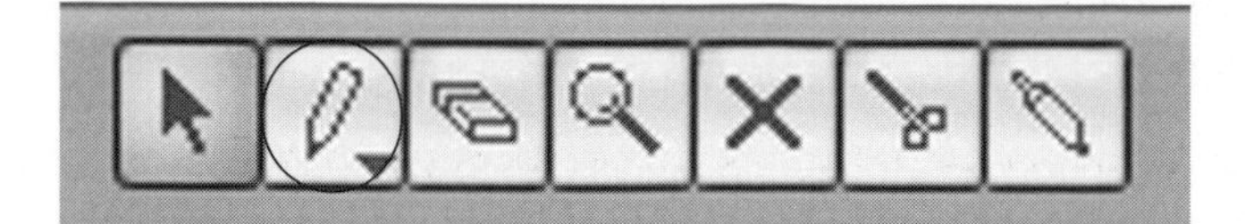

Fig. 5.9.3 – Editing tools.

note ▶

Pro Tools

Pro Tools does not have a dedicated Piano-roll-style edit window, so in order to achieve a more detailed view of the notes and controller information, simply view the track as notes, increase the size of the MIDI track to large, then choose controllers from the track view menu.

producer says ▶

Volume vs velocity

There is a fundamental difference between the attack of a note and the volume of a performance. The attack is the volume at the point a note starts playing. How hard the notes are hit is measured by velocity. This not only affects the volume, but often the way a certain instrument may sound too. This is because some instruments are set up to produce a slightly different texture to their sound when they receive different velocity values.

When we edit volume data, the notes and their velocities stay the same. What changes is the overall volume level (just like turning up a volume dial or fader), not the performance or aesthetic of the sound or instrument.

A practical example of where you may choose to adjust the volume rather than the velocity would be if you wanted to have a sustained note rise in volume. With velocity you can only change the attack (the point at which the note is hit), not the volume for the duration. With volume you could make a sound gradually rise in volume whilst the note is being held down.

Drawing in pan

Pan data can be added in exactly the same way as volume. Simply change the controller number to 10. However, there is one change. As we are dealing with pan, you have a central position were the sound is equal in both speakers. This is going to be exactly in the centre of the display, as increasing the value towards 127 will move the sound towards the right, while moving down towards 0 will move the sound towards the left. Apart from that, using the editor is very similar.

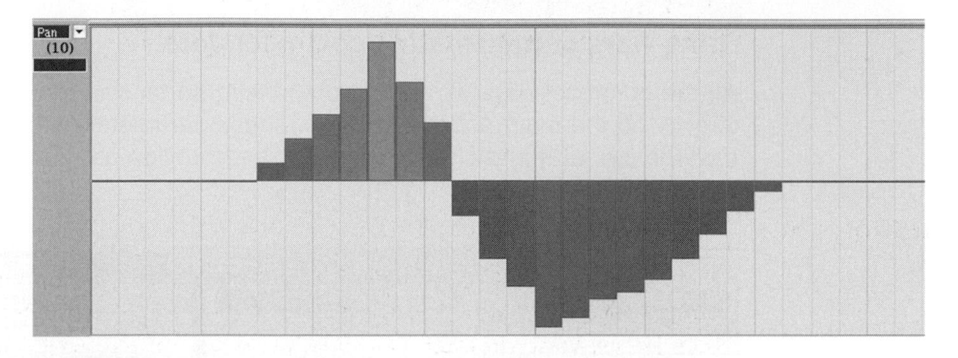

Fig. 5.9.4 – In this example the pan data moves the sound over towards the right speaker, then to the left speaker and then back in the centre.

The List editor

Try opening the List edit window. This will display controller data in a numerical 'list'.

Edit | Functions | View

Position			Status	Cha	Num	Val	Length/Info
------------ Start of List ------------							
1	1	1	Control	1	7	1	Volume
1	1	91	Control	1	7	1	Volume
1	1	121	Control	1	7	2	Volume
1	1	151	Control	1	7	3	Volume
1	1	181	Control	1	7	4	Volume
1	1	211	Control	1	7	5	Volume
1	1	241	Control	1	7	6	Volume
1	1	481	Note	1	A1	88	. .
1	1	481	Note	1	C2	88	. .
1	1	481	Note	1	E2	97	. .
1	1	481	Control	1	7	7	Volume
1	1	511	Control	1	7	8	Volume
1	1	541	Control	1	7	9	Volume
1	1	571	Control	1	7	10	Volume
1	1	601	Control	1	7	11	Volume
1	1	631	Control	1	7	12	Volume

Fig. 5.9.5 – Volume and note data displayed in a list.

Display filters

Inside the List edit window, you will often find display filters that allow you to choose the type of MIDI data you want to see. Make sure you can view controller data, not just notes.

Fig. 5.9.6 – Display filter.

Other ways of transmitting controller data

All the controller data we have worked with so far can usually be changed directly on the main Arrange window. These parameters are usually located towards the left-hand side of the track and provide a quick and convenient way to make changes.

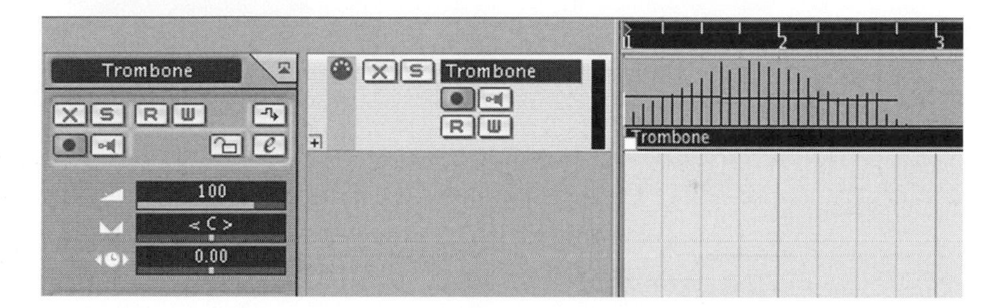

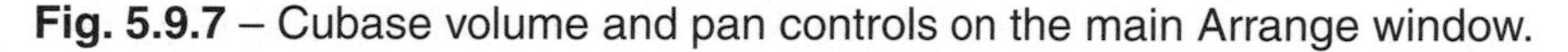
Fig. 5.9.7 – Cubase volume and pan controls on the main Arrange window.

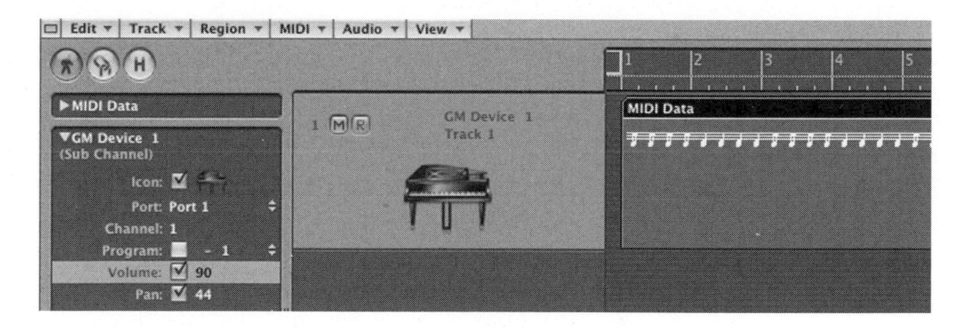

Fig. 5.9.8 – Logic volume and pan controls on the main Arrange window.

GM and MIDI mixers

MIDI mixing desks usually look like an audio mixing desk, but only control MIDI parameters. Some sequencers include both audio and MIDI tracks together in a mixer that mirrors the tracks on the Arrange window, while others have dedicated GM mixers.

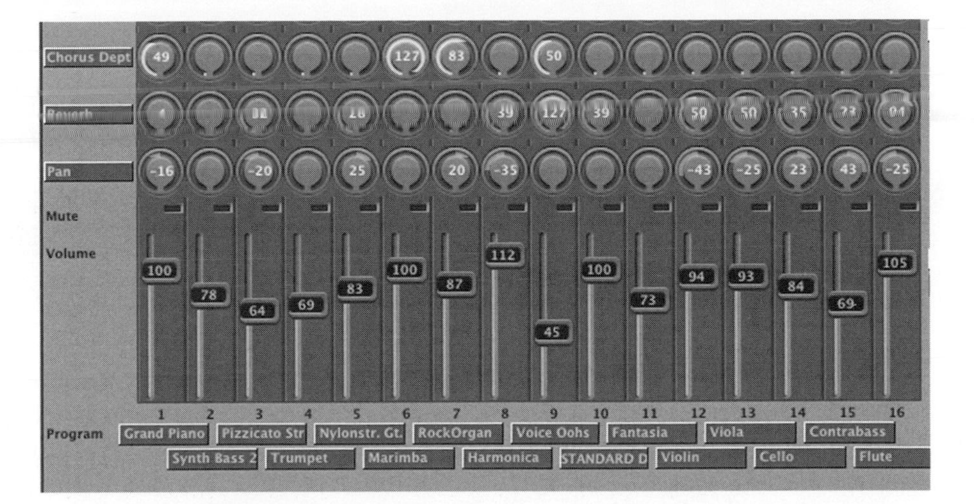

Fig. 5.9.9 – A general MIDI mixer.

- *Volume* is controlled via the fader. By moving the fader higher we increase the volume and by moving it lower we decrease the level.
- *Pan* is normally the dial just above the volume fader. Sometimes it can also be a horizontal bar just above the fader. However, whether it is a dial or bar it works in the same way. Click inside the object and drag a small dial or line to the left or right. By doing this we are sending the signal to the left or right speakers.

You may also have access to other controllers which are often available from this type of window. For example:

- *Reverb* (controller number 91) adds ambience to a sound. This allows you to determine how much reverb is added to an instrument on a particular MIDI channel.
- *Chorus* (controller number 93) 'thickens' the sound. This allows you to determine how much chorus is added to an instrument on a particular MIDI channel.

How effective these parameters will be will depends on how well your sound source interprets controller messages, and if the sound source itself is capable of producing reverb and chorus.

note ▶

You need to be aware when making adjustments to controller information outside an edit window using the MIDI mixer, or in the main Arrange window, that you are not actually inputting information into the sequencer, you are simply sending a value to the sound source.

Transmitting controller data from a keyboard

It is usually possible to send controller data from your keyboard. Most MIDI keyboards have a data wheel or control strip located somewhere on the keyboard designed to send out controller messages. Before you can do this you will need to assign it to a controller number.

note ▶ Avoid using the sprung wheel or dial on a controller keyboard, as this is usually set up to control pitch bend.

Recording controller data in real time

To record controller information as the music plays, you simply need to assign a controller number to your wheel or dial, such as volume or pan and then drop your sequencer into Record. Instead of 'playing' notes, you would move the data wheel or dial. These movements will be recorded and show up in the part/region.

tip ▶ When recording controller data in real time from a keyboard you may want to record it on a separate track. Providing it is set to the same MIDI channel and sound source, it will allow you to easily delete or edit the controller data separately from the note data.

Exercise 5.10 Program change

Before attempting this exercise we suggest you complete Exercises 5.1, 5.2 and 5.5. This exercise will cover the following:

- Transmitting a program change message
- Automatically changing sounds throughout a song
- Separating note and program change data
- Recording program change messages in real time

Most sound-producing devices, such as sound modules and virtual instruments, usually contain lots of different sounds. Each of these individual sounds can usually be selected or recalled manually either by pressing a button, turning a dial or making a selection from a menu.

A program change message is a special type of MIDI message that allows you to select and recall these different sounds using MIDI. Each individual sound will have its own unique number, ranging from 0 to 127, so it's simply a matter of sending out a number on one MIDI device to match a certain sound on another MIDI device. Note: in order to successfully transmit and receive a program change message, the MIDI channels on both devices will need to correspond.

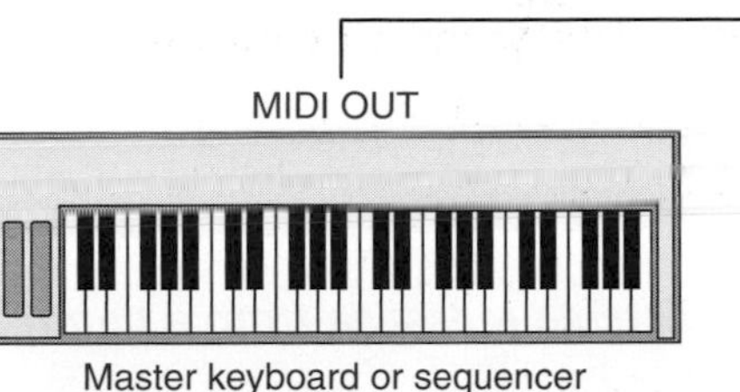
MIDI OUT
MIDI IN
Master keyboard or sequencer sends out a program change number
Sound-producing MIDI device changes sound

Fig. 5.10.1 – Diagram showing how a program change message can be transmitted.

note ▶ A program change message can be sent to any MIDI sound source, such as a sound module or virtual instrument, to change its sound.

The exercise ...

In this exercise we are going to explain how to *insert* program changes into a MIDI sequencer to automatically change the sounds throughout a song.

Why you may need this option

Even with 16 MIDI channels and 16 different sounds available at the same time it is still possible to run out of instruments. However, it is unlikely that all 16 instruments will be playing at the same time, so if a new sound needs to be introduced then it could share the same MIDI channel as another sound and simply switch sounds at different points in the arrangement. For example, halfway through a song you may want the sound playing on MIDI channel 1 to swap from being a bright piano to a trumpet. This would create the illusion that there are more than 16 separate instruments or simply help create some variation.

producer says ▶ Each MIDI channel can only play one instrument at one time, so if you run out of channels then creating program changes can help create the illusion of having more than 16 instruments throughout the same piece of music.

Transmitting a program change message

A program change message is usually transmitted from a controller keyboard or MIDI sequencer, as this provides the most convenient route for a program change message to be sent to a sound-producing device. If you are using a MIDI sequencer with its own built-in sound source or virtual instruments, then the process is just the same.

Transmitting a program change message from a controller keyboard

One of the simplest ways to transmit a program change message is from a controller keyboard. This is normally done by pressing one of the numbered buttons that is used to change the keyboard's internal sounds. If you are using a dedicated controller keyboard with no internal sounds, you may have to select a special program change mode and then type in a number. Sometimes this is done by using specified notes on the actual keyboard itself. As soon as this button is pressed, a program change message gets sent out from the controller keyboard.

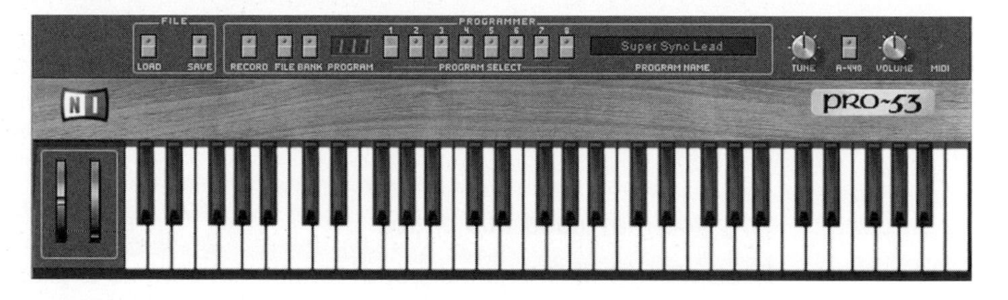

Fig. 5.10.2 – Program change buttons are usually located above the keyboard.

In order for a keyboard to successfully transmit a program change message to another MIDI device, it will need to be connected to it and the MIDI channels on each device will need to correspond. You may also need to check if the program change option on the keyboard is enabled and that it is not filtered out on the sound source (see MIDI filters, p. 133).

note ► A program change message can be transmitted from a controller keyboard to a sound-producing device to select a different sound.

Using a sequencer to transmit a program change message

MIDI sequencers also allow you to output program change messages to any sound sources used by the computer. This provides a quick and easy way of changing sounds from within the sequencer.

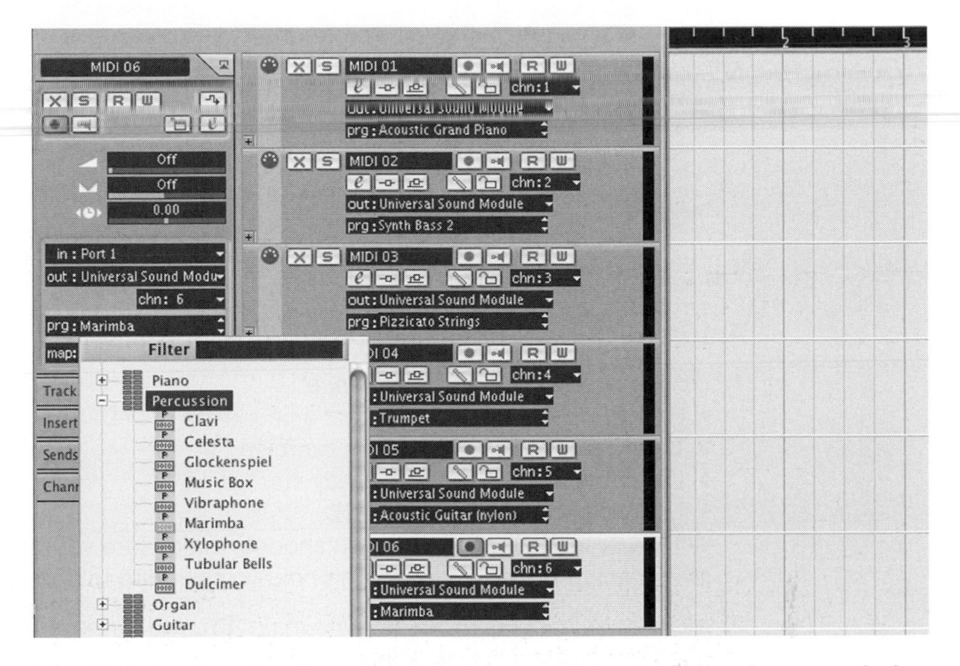

Fig. 5.10.3 – Sending a program change message from the Arrange window in Cubase.

Located on most Arrange windows will be a program change parameter box for each individual sequencer track. This will allow you to send out a program change message to the sound source used by that track.

tip ► Program change can make it easier for you to select the sounds you want to use.

Different numbering systems

Be aware that some MIDI devices start their program change numbering system from 0, while others start from 1. This basically means if you're using two separate devices that don't match up, your numbering system will be one number out. For example, sending out program change number 1 from a sequencer may dial up sound number 2 on an external sound module.

tip ► Another advantage of using program change within a sequencer is that all your sound assignments are stored within the song.

MIDI filters

Most sound-producing devices have a MIDI filter that allows you to choose whether they actually receive program changes or not. Always make sure this is set up correctly if you want the device to respond to program changes.

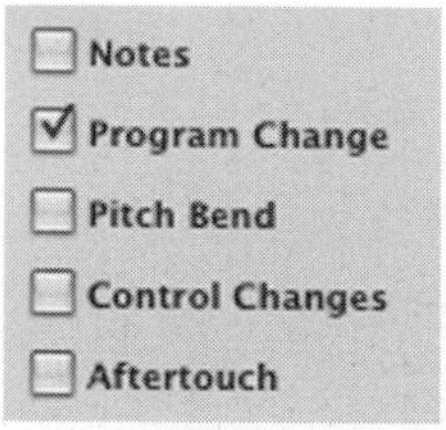

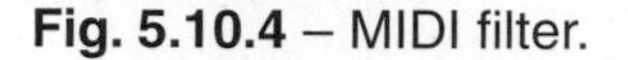

Fig. 5.10.4 – MIDI filter.

Automatically changing sounds throughout a song

1. *Create or load in a MIDI sequence*:
   - Create a new project and record a sequence of MIDI notes
   - Use the data from a previous exercise
   - Load in a project you've already started
   - Load a tutorial file from our CD.

   Whatever method of starting you choose, make sure you have at least one MIDI track with eight bars worth of music assigned to a sound source.
2. *Make sure that you are viewing the main Arrange window.* Most sequencers open up on this page as default.
3. *Locate the song position to bar 1.* This is where we are going to insert the first program change message.

Fig. 5.10.5 – MIDI data selected in the Arrange window.

4. *Select a MIDI part and open the List editor.* We recommend using the List-style editor as it provides a very accurate way of displaying and entering program change information. Inside the List edit window you will often find a display filter that will allow you to choose the type of MIDI data you want to see. Make sure you can view other types of MIDI data other than just notes.

Fig. 5.10.6 – MIDI filter.

5. *Insert a program change.* Look at the tools you have available. The pencil will allow you to draw changes onto the screen. The rubber will allow you

to delete program changes. Use the pencil tool to insert the program change at the selected song position. You may first have to choose the type of data you want to insert from a menu or simply click on the program change icon.

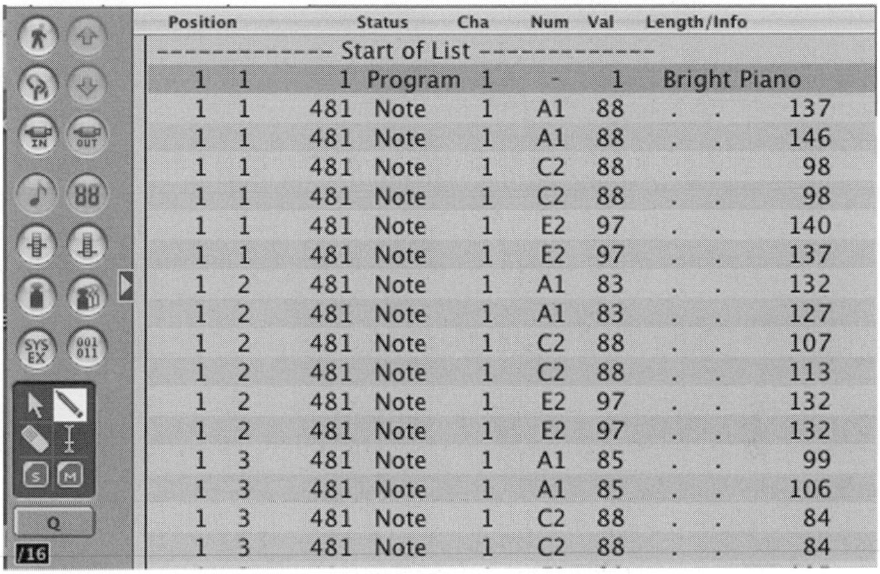

Position	Status	Cha	Num	Val	Length/Info
------------- Start of List -------------					
1 1 1	Program	1	-	1	Bright Piano
1 1 481	Note	1	A1	88	. . 137
1 1 481	Note	1	A1	88	. . 146
1 1 481	Note	1	C2	88	. . 98
1 1 481	Note	1	C2	88	. . 98
1 1 481	Note	1	E2	97	. . 140
1 1 481	Note	1	E2	97	. . 137
1 2 481	Note	1	A1	83	. . 132
1 2 481	Note	1	A1	83	. . 127
1 2 481	Note	1	C2	88	. . 107
1 2 481	Note	1	C2	88	. . 113
1 2 481	Note	1	E2	97	. . 132
1 2 481	Note	1	E2	97	. . 132
1 3 481	Note	1	A1	85	. . 99
1 3 481	Note	1	A1	85	. . 106
1 3 481	Note	1	C2	88	. . 84
1 3 481	Note	1	C2	88	. . 84

Fig. 5.10.7 – Program change message inserted at bar 1.

6. *Choose the program change number.* The program change number you choose will determine the instrument that gets selected. Sometimes you will have to edit the number after it has been inserted into the list.
7. *Locate the song position to bar 5.*
8. *Insert another program change.* Using the pencil tool insert another program change at the selected song position. Choose a different sound from bar 5.

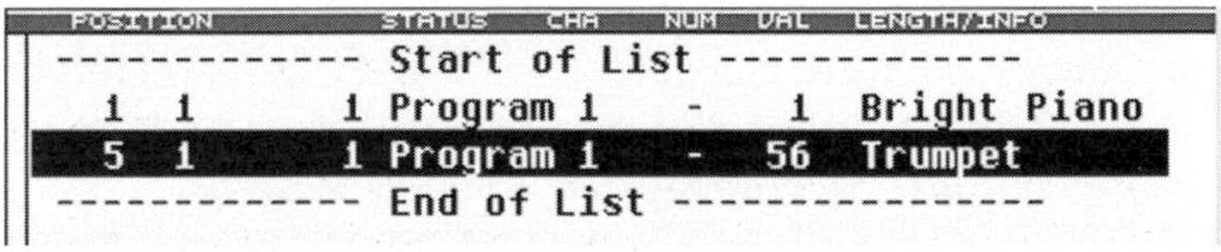

POSITION	STATUS	CHA	NUM	VAL	LENGTH/INFO
------------- Start of List -------------					
1 1 1	Program	1	-	1	Bright Piano
5 1 1	Program	1	-	56	Trumpet
------------- End of List ---------------					

Fig. 5.10.8 – Two program change messages displayed in a list.

9. *Set up a cycle.* Setting up a temporary cycle within the sequencer will make it easier to hear the transition between the two sounds.
10. *Locate back to the beginning of the song and press Play.* You should now hear the MIDI data playing back. As the song position reaches bar 5, the sound should change and when it reaches bar 1 the sound should change again.

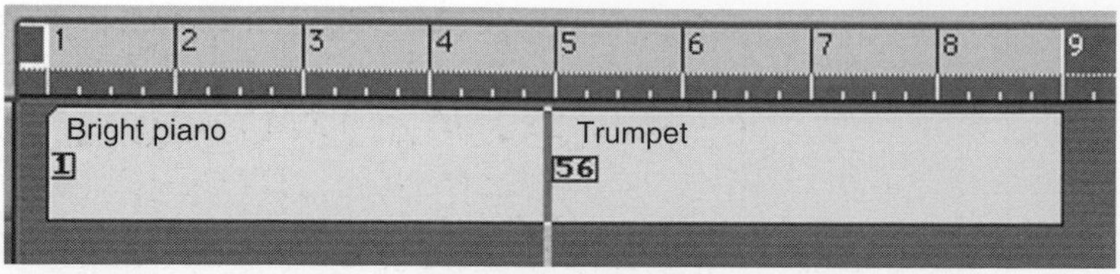

Fig. 5.10.9 – Program change messages displayed on the Arrange window.

Remaining on the last sound selected

When sending out a program change message from a sequencer in this way, the sound source will remain on the last message received, so if you stop the sequencer at bar 6, for example, the sound will remain a trumpet until it passes bar 1 again.

note ▶ Changing the sound does not affect the MIDI data in any way – it simply allows you to play it back using a different sound.

Additional things to try using program change

Separating note and program change data

Once you have got the basic idea of how to insert program changes into a sequencer, try adding program change messages on a separate track from the actual note data. This will make it easier to see and edit. Just make sure that both tracks are on the same MIDI channel and are routed to the same sound source.

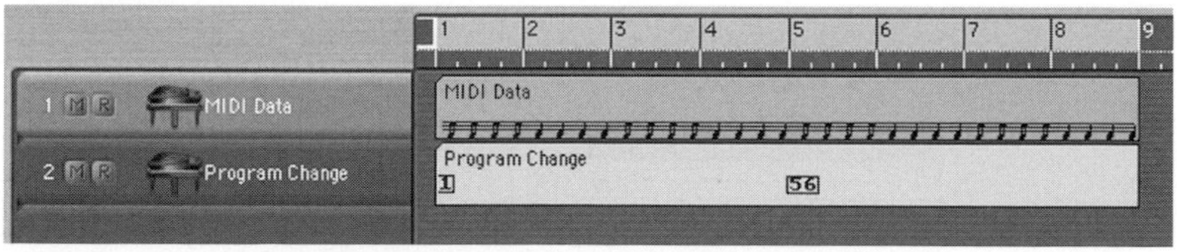

Fig. 5.10.10 – Note and program change data placed on separate tracks.

note ▶ If you have several tracks routed to the same MIDI channel, such as all the drums on MIDI channel 10, a program change will change the sound for all these tracks.

Recording program change messages in real time

Now you have learnt how to insert program change messages using the List editor, why not try recording them into the sequencer in real time. Simply drop your sequencer into Record and press the program change button on your keyboard or MIDI device at the correct time. The program change message will be recorded in the same way as note data.

GM mixer

Program changes can be sent from this window too. If your sequencer has a general MIDI mixer it will be possible to send out program change messages for all 16 MIDI channels. Note that in order to do this you may have to first create and set up the GM mixer.

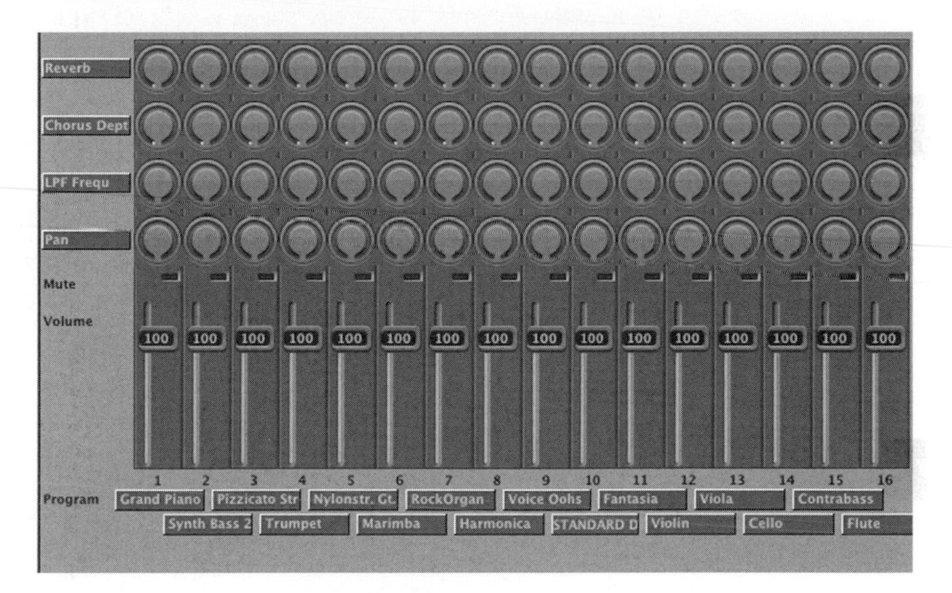

Fig. 5.10.11 – A general MIDI mixer.

Communicate with other MIDI devices

You may find other types of MIDI devices that are capable of transmitting or receiving program change messages. For example, it may be possible to send a program change message to an external effects unit (providing it's got MIDI). This would allow you to automatically change and recall the actual effect that is being used. This would be useful if you needed a sudden burst of reverb at a certain point within a song or needed to switch from a delay to chorus type effect, for example.

Another area where you may find program change useful is when using a MIDI equipped mixing desk that has a snapshot recall facility. Each snapshot or scene allows you to store certain settings for the mixing desk. It is then possible to use program change to recall these scenes and therefore automate these settings.

General MIDI sounds

Some sound sources correspond to a set list called GM sounds. These are basically 128 different sounds that are arranged in a predetermined order. This means the program change numbers will always correspond to the following instruments in this table (see 'General MIDI' in Chapter 3).

0	=	Grand Piano	43	=	Contrabass	86	=	5th Saw Wave
1	=	Bright Piano	44	=	Tremolo Str.	87	=	Bass&Lead
2	=	ElectricGrand	45	=	Pizzicato Str	88	=	Fantasia
3	=	HonkyTonkPno.	46	=	Harp	89	=	Warm Pad
4	=	E. Piano1	47	=	Timpani	90	=	Polysynth
5	=	E. Piano2	48	=	Strings	91	=	Space voice
6	=	Harpsichord	49	=	Slow Strings	92	=	Bowed Glass
7	=	Clavinet	50	=	Syn. Strings1	93	=	Metal Pad
8	=	Celesta	51	=	Syn. Strings2	94	=	Halo Pad
9	=	Glockenspiel	52	=	Choir Aahs	95	=	Sweep Pad
10	=	Music Box	53	=	Voice Oohs	96	=	Ice Rain
11	=	Vibraphone	54	=	SynVox	97	=	Soundtrack
12	=	Marimba	55	=	OrchestraHit	98	=	Crystal
13	=	Xylophone	56	=	Trumpet	99	=	Atmosphere
14	=	Tubular-Bell	57	=	Trombone	100	=	Brightness
15	=	Dulcimer	58	=	Tuba	101	=	Goblin
16	=	DrawOrgan	59	=	MutedTrumpet	102	=	Echo Drops
17	=	PercOrgan	60	=	French Horn	103	=	Star Theme
18	=	RockOrgan	61	=	Brass 1	104	=	Sitar
19	=	Church Organ1	62	=	Synth Brass1	105	=	Banjo
20	=	Reed Organ	63	=	Synth Brass2	106	=	Shamisen
21	=	Accordion Fr	64	=	Soprano Sax	107	=	Koto
22	=	Harmonica	65	=	Alto Sax	108	=	Kalimba
23	=	TangoAcd	66	=	Tenor Sax	109	=	Bag Pipe
24	=	Nylonstr. Gt.	67	=	Baritone Sax	110	=	Fiddle
25	=	Steelstr. Gt.	68	=	Oboe	111	=	Shanai
26	=	Jazz Gt.	69	=	English Horn	112	=	Tinkle Bell
27	=	Clean Gt.	70	=	Bassoon	113	=	Agogo
28	=	Muted Gt.	71	=	Clarinet	114	=	Steel Drums
29	=	Overdrive Gt.	72	=	Piccolo	115	=	Woodblock
30	=	Distortion Gt	73	=	Flute	116	=	Taiko
31	=	Gt.Harmonics	74	=	Recorder	117	=	Melo Tom
32	=	Acoustic Bs.	75	=	Pan Flute	118	=	Synth Drum
33	=	Fingered Bs.	76	=	Blown Bottle	119	=	Reverse Cym.
34	=	Picked Bs.	77	=	Shakuhachi	120	=	Gt FretNoise
35	=	Fretless Bs.	78	=	Whistle	121	=	Breath Noise
36	=	Slap Bass 1	79	=	Ocarina	122	=	Seashore
37	=	Slap Bass 2	80	=	Square Wave	123	=	Bird
38	=	Synth Bass 1	81	=	Saw Wave	124	=	Telephone 1
39	=	Synth Bass 2	82	=	Syn. Calliope	125	=	Helicopter
40	=	Violin	83	=	Chiffer Lead	126	=	Applause
41	=	Viola	84	=	Charang	127	=	Gun Shot
42	=	Cello	85	=	Solo Vox			

Fig. 5.10.12 – General MIDI list of sounds.

Exercise 5.11 Creating a drum beat

Before attempting this exercise we recommend you complete Exercises 5.1, 5.2, 5.5 and 5.6. This exercise will cover the following:

- Learning how to create empty regions/parts
- Opening the edit window
- How to input data inside a part/region
- Adding another drum sound.

The exercise ...

What we want to do in this exercise is to show you how to create a drum beat without using a controller keyboard. This way we can prove to ourselves that it is possible to input MIDI information into a sequencer, albeit slowly, without having great keyboard skills. This process will also help you become more familiar with the different types of edit windows and help you edit any existing material you may have already created.

Most edit windows allow you to make changes to the notes you've played in some way; for example, it's possible to change their pitch or make them different lengths and so on. In this example we are going to concentrate on creating some new MIDI notes and placing them in time to create a drum beat.

- *Load your sequencing program and create a new project.*
- *Make sure that you are viewing the main Arrange window.* Most sequencers open up on this page as default.
- *Create four MIDI tracks.*
- *Make sure that each MIDI track is assigned a sound source.* As we are going to be creating a drum beat, make sure that a drum or percussion sound is selected for each track. In our example we are using a Universal Sound Module and have selected MIDI channel 10 in order to access the drum sounds.

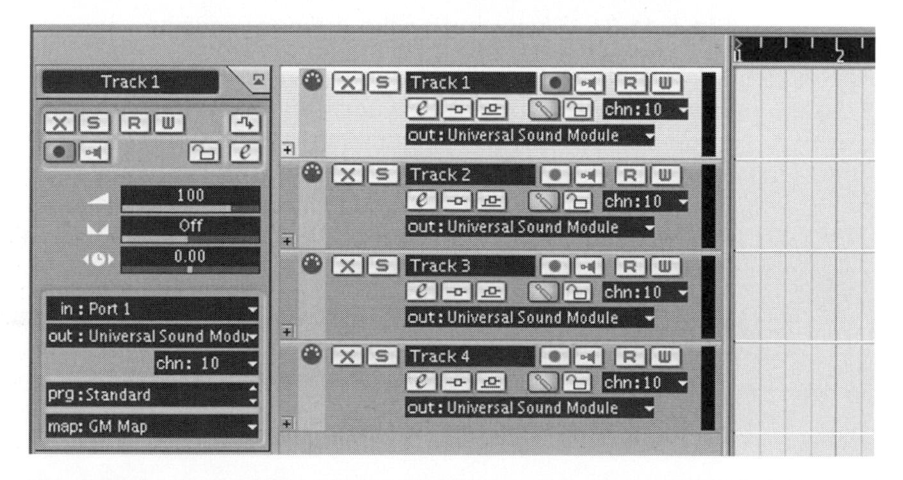

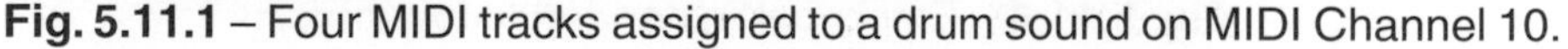
Fig. 5.11.1 – Four MIDI tracks assigned to a drum sound on MIDI Channel 10.

note ► MIDI channel 10 is where the drums are allocated on GM-compatible devices.

- *Look for your tools.* These are normally based around an arrow, pencil, rubber and scissors. Choose the pencil tool.

Fig. 5.11.2 – Toolbox.

Learning how to create empty regions/parts

Make sure the first track is selected. Using the pencil, click on the Arrange window in the space to the right of track 1 between bars 1 and 2. This should result in a block appearing. Blocks are referred to in different ways depending on which sequencer you use. Some call them parts, while others call them regions or objects (throughout this book we will refer to them as parts). A part usually contains MIDI notes that produce music. However, at the moment there are no notes in the part so you won't hear any sound.

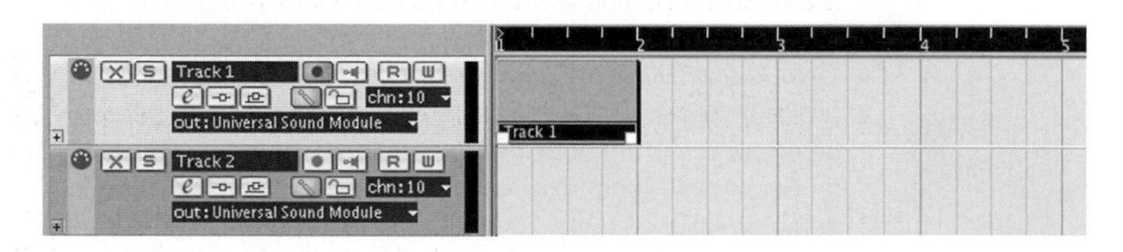

Fig. 5.11.3 – Empty part on the Arrange window.

Extend the part so that it lasts a full four bars. You may do this using either the arrow or the pencil. Move the mouse to the bottom right corner of the part and look for a small arrow or for the display to change. Hold the mouse down and drag the part to the right until it is at the beginning of bar 5. To check how long it is make sure you are looking at the time ruler at the top of the screen.

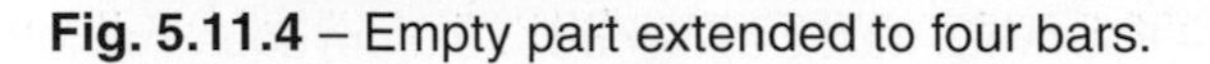

Fig. 5.11.4 – Empty part extended to four bars.

note ► Most sequencers require you to create a part in order to open an edit window.

Opening the edit window

Highlight the empty part by clicking once to select it and then, using the appropriate drop-down menu, open the Piano-roll or Drum editor. The other way of achieving this is to double-click on the part (click twice with the left mouse button quickly). This should open out into your Piano-roll/Matrix-style editor. If this doesn't open the edit window you require and instead opens into your Score editor, for example, explore your sequencer preferences, where you can usually change what happens when you double-click on a part.

note ►

Piano-roll vs Drum editor

The Drum editor and the Piano-roll-style editor are very similar. The main difference is whether the vertical axis is showing the actual drum names or simply a keyboard.

The Piano-roll-style editor

Make sure the current sequencer's song position is set to bar 1, so it matches the same position as the part you are looking at.

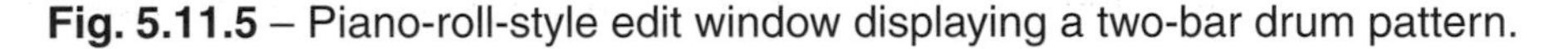

Fig. 5.11.5 – Piano-roll-style edit window displaying a two-bar drum pattern.

The Drum editor

If you open the drum editor, you will see a grid displaying the bars, beats and smaller divisions of the beats. You can change the size of the divisions by experimenting with the edit resolution or Quantize options inside the editor. This means changing the number, which normally looks like 1/4, 1/8 or 1/16. The higher the number, the more subdivisions of the bar are created.

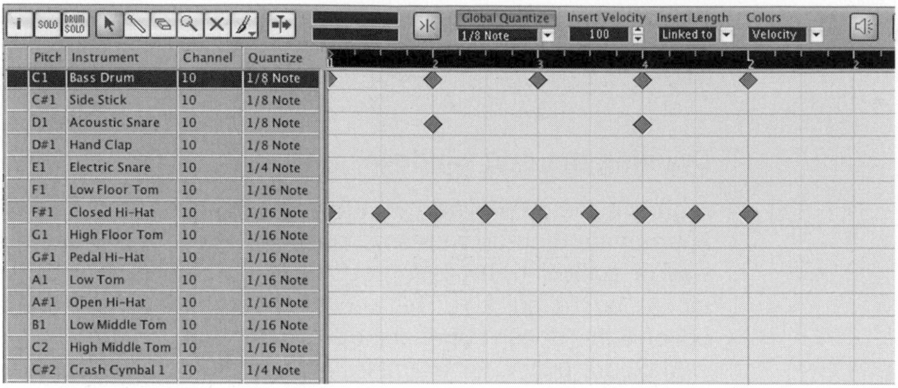

Fig. 5.11.6 – Drum edit window.

Even though we do not want to quantize anything (see Chapter 4), just altering this option affects the lines and grid lines on the screen. For example, by selecting a 1/4 value we see the bar broken up evenly by four lines. By selecting 1/16th value we see it evenly broken up by 16 lines, and so on.

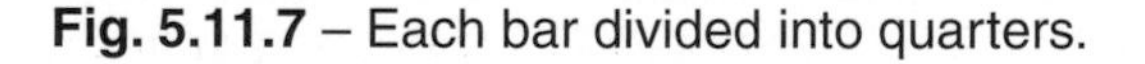
Fig. 5.11.7 – Each bar divided into quarters.

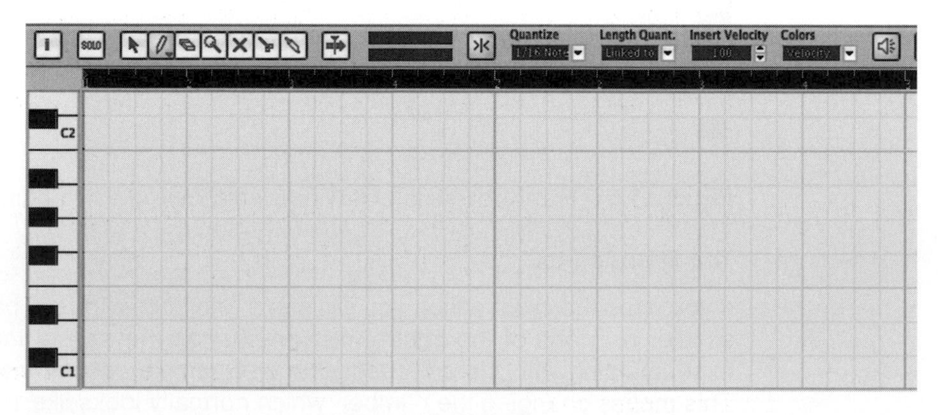

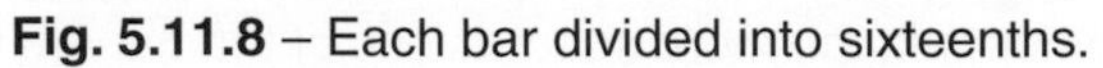
Fig. 5.11.8 – Each bar divided into sixteenths.

Look for a Snap option in the edit window. The Snap value will restrict the movement of the notes by a set value and will make sure that any notes you create or edit will move to the nearest line and snap into position. Try setting this to 8ths to start with, as you can always adjust it later.

How to input data inside the part/region

The edit window itself has its own selection of tools, similar to the Arrange window. Choose the pencil.

Fig. 5.11.9 – Tools.

Try clicking on the screen. You should see a note appear and hear it at the same time. Now switch to the arrow tool. Once you have an arrow, click and hold the note you have just created. Drag it up and down. You should hear all the drum sounds being triggered. This is the equivalent of running your finger up and down a keyboard.

producer says ►

When using some sequencers you may find you can trigger the drum sounds using a MIDI keyboard but are unable to hear anything when entering or moving notes in the edit window. If this is the case you may need to enable the MIDI output button. This is a small button that is located in the edit window that usually has a picture of a loudspeaker or MIDI plug icon on it. This button basically allows you to choose whether the sequencer will output a MIDI signal each time you edit or move a note in the edit window.

tip ►

If you make a mistake or want to change anything you draw in with the pencil, simply delete the notes by using the rubber tool.

If you haven't done so already, try and locate a kick drum sound. This is usually located on C1.

Use the pencil to create a note on each beat of the bar (usually there are four beats in each bar). Try and place the notes equally between bars 1 and 3.

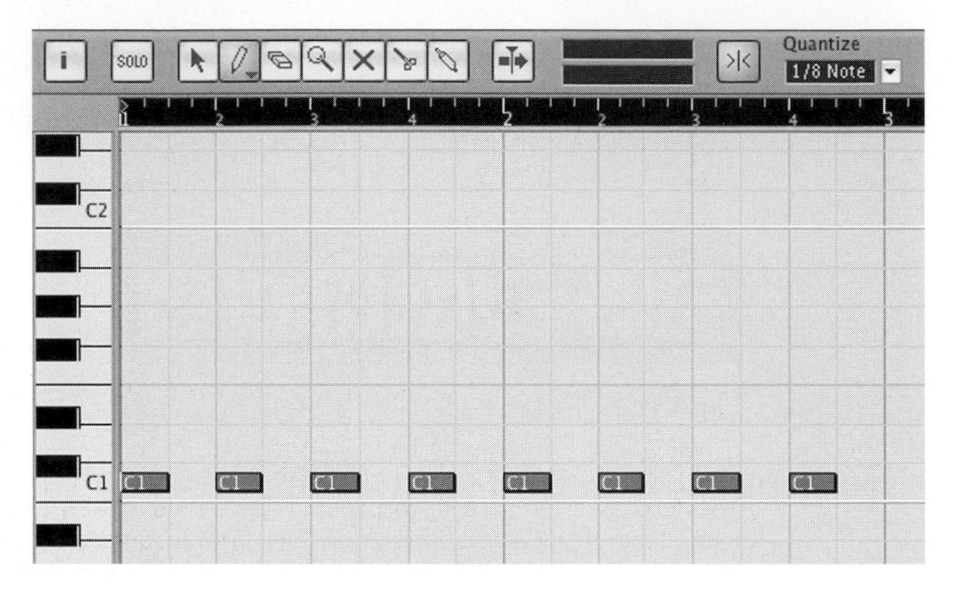

Fig. 5.11.10 – Example two-bar kick drum pattern.

note ► Where you place a note on the grid will determine both the sound and the rhythm.

Press Play on the sequencer to hear back the notes. Make sure the current song position is set to bar 1 or to the position where the notes have been entered. You should now hear a regular kick drum pattern.

Adding another drum sound

Now we have checked everything is working it's time to add another drum sound. Simply recreate the process again but this time create your new part on track 2 and locate a different drum sound, such as a snare drum or hi hat. Also experiment by changing the amount of grid lines to help you create different rhythms.

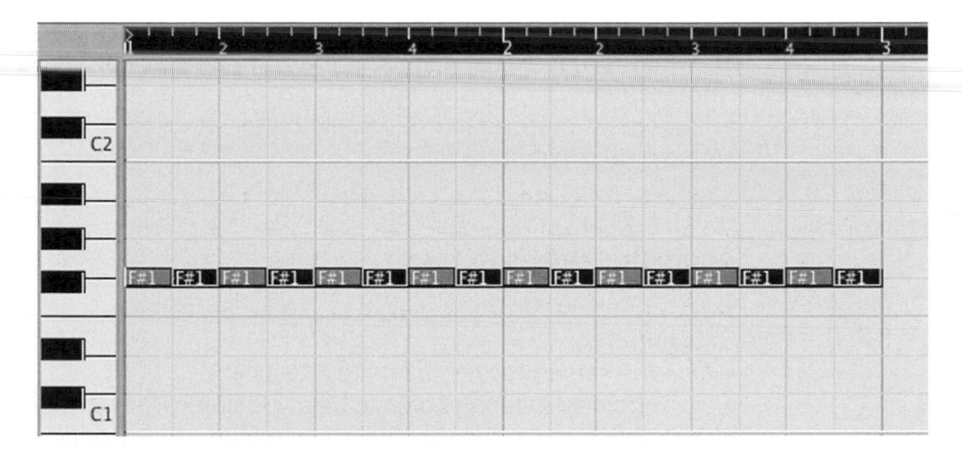

Fig. 5.11.11 – Example Hi hat pattern.

In the example in Fig. 5.11.11 we have changed the resolution to 16ths and entered a fast hi hat pattern on each line.

tip ▶ Remember to assign each new track to the same sound source and MIDI channel, but choose a different note and drum sound.

note ▶ This method of sequencing allows you to create rhythms that would be very difficult to play in real time with the metronome. In many ways you will gain a better understanding of how rhythms work and what they 'look like' by working this way.

Use the other remaining tracks for the other drum sounds. For example, kick on track 1, hi hat on track 2, snare on track 3 and shaker on track 4.

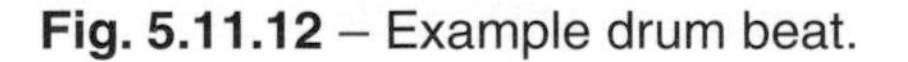

Fig. 5.11.12 – Example drum beat.

In this task we have shown you how to build up a drum track by placing only one drum sound on each track. There is no reason why you couldn't create a whole drum beat on one track inside one part. However, for editing and arrangement purposes, it is usually better practice to separate each drum sound onto different tracks.

Additional things to try

- Try changing the Snap value to create something a bit more interesting rhythmically.
- Use the same process to create a bass line.
- Try randomly entering notes to help you come up with ideas.
- Jump back and forth to see how your rhythms look in different edit windows, such as the Score editor or List editor.
- Now copy the whole section again to create an 8- or 16-bar drum track.
- Set a cycle around the drum sequence so you can practise playing another instrument along with the beat.

Exercise 5.12 Step time recording

This exercise will cover the following:

- Entering notes into a MIDI sequencer using step time
- Changing the edit resolution
- Entering rests.

The exercise ...

Step time recording offers an alternative way of recording MIDI data into a sequencer. Instead of using a metronome and playing in real time, notes can be entered directly into an edit window using a MIDI keyboard.

In order to enter MIDI data this way you will need to enable the step input mode within your sequencer. Each time you enter a note or chord the sequencer will jump forward to the next step. The distance it jumps is determined by the Snap or edit resolution. If this is set to 8ths, for example, you will be able to enter 8 notes or chords in each bar.

This function is often overlooked and can provide the user with an effective way of recording very complex sequences that would be impossible to play in real time with the metronome.

1. *Load your MIDI sequencing program.* Make sure that you are viewing the main sequencing window for recording and arranging MIDI data.
2. *Select a MIDI track.* Create a new MIDI track or use an existing one.

3. *Assign a sound source.* Route the selected track to a sound source such as a piano sound.
4. *Enable the record ready switch.* Make sure you can hear a sound when you play your controller keyboard (see Exercise 5.1).
5. *Move the Song position marker to the beginning of bar 1 (or to the location where you want to enter the notes).* This can usually be achieved by adjusting the song position value into the transport window or by dragging the Song position marker.
6. *Create an empty object.* On the main Arrange window use the pencil tool to create an empty MIDI object between bars 1 and 2. Usually, a MIDI object has to exist in order to open an edit window.

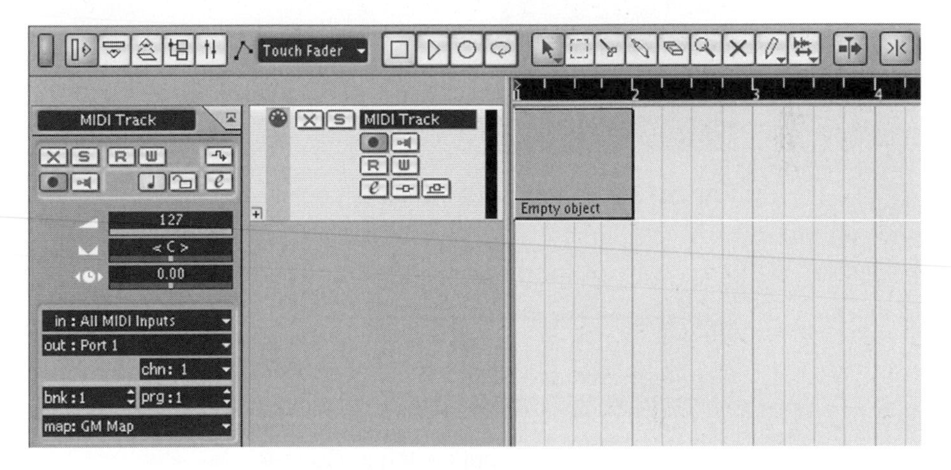

Fig. 5.12.1 – Empty object on the Arrange window.

7. *Open the Piano-roll-style edit window.* This is usually referred to as the Matrix or Key editor. Make sure you are viewing bars 1 and 2.

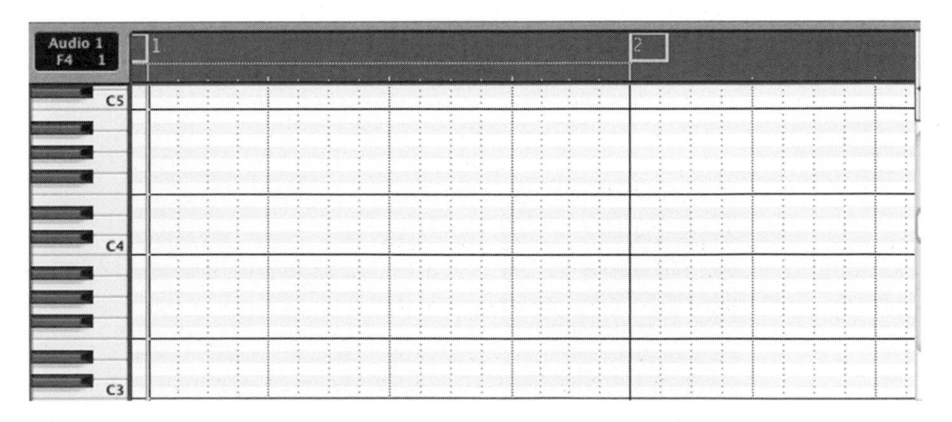

Fig. 5.12.2 – Piano-roll-style edit window.

Entering notes into a MIDI sequencer using step time

1. *Within the edit window locate and enable the step recording switch.* This is usually a button that looks like a MIDI plug or a foot.

Fig. 5.12.3 – Step input switch in Logic.

Fig. 5.12.4 – Step input switch in Cubase.

2. *Enter the amount of divisions or steps you want to use per bar.* This value will determine the length of each note and the distance between each step. It can be adjusted at any stage, making it possible to switch between different note lengths in the same bar. Try selecting 8ths to start with.

   This value is usually located on the top or left-hand side of the window.

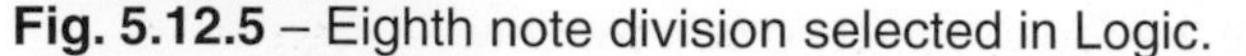

Fig. 5.12.5 – Eighth note division selected in Logic.

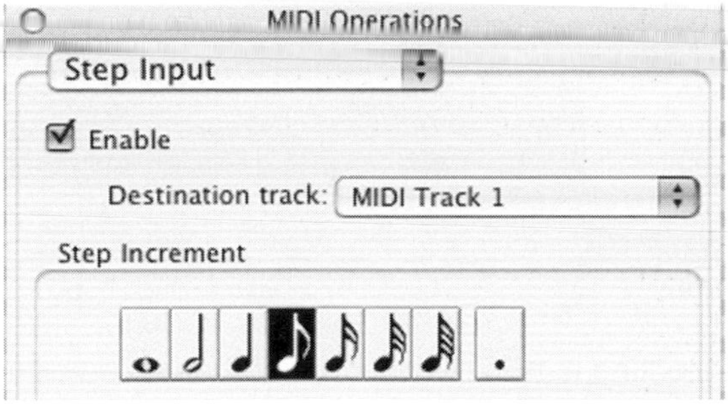

Fig. 5.12.6 – Step input window in Pro Tools.

In Pro Tools the step write enable button and the division selections are located together in the MIDI operations window.

3. *Press a single note or a chord.* As you press a note on the keyboard you should see the notes appear on the screen and the Song position marker will move forward to the next step.

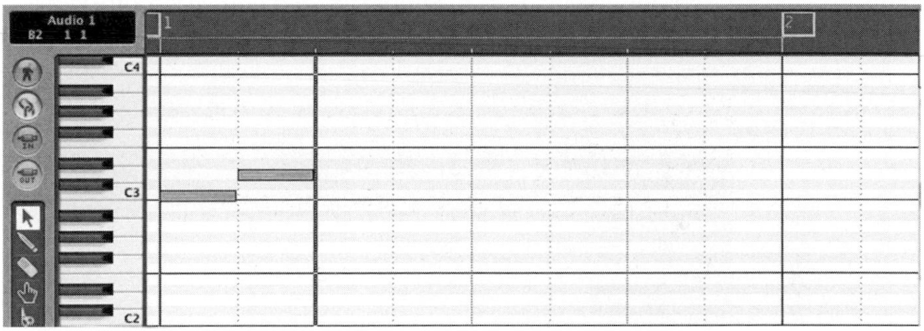

Fig. 5.12.7 – Two 8th notes being entered in step time.

4. *Now press another note.* You will see the song position marker advance to the next step. Each time you press a note, the song position marker moves to the right.

note ▶ The distance the song position marker jumps is determined by the Snap or edit resolution. For example, if this is set to 8ths then it will jump 8 steps in one bar.

Changing the edit resolution

If you have set a resolution of 8ths then you will need to step in 8 notes in order to reach bar 2, whereas if you select a division of 1/16 then you will have to step input 16 notes in order to reach bar 2. Before you reach the end of bar 1, try changing the edit resolution to a different value.

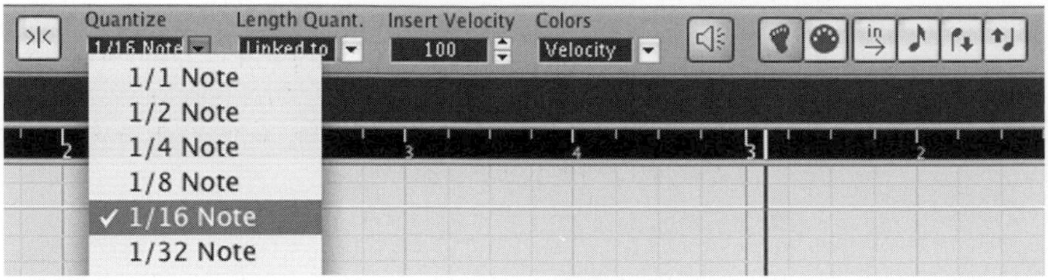

Fig. 5.12.8 – Selecting a 16th resolution in Cubase.

producer says ▶

> **Plenty of time**
>
> As you are not using the metronome or referencing to the sequencer's tempo, the time distance between each note is irrelevant. This means you could enter two notes, leave the room for half an hour and when you came back the song position would be at the same point (providing someone else hadn't interfered with your computer).

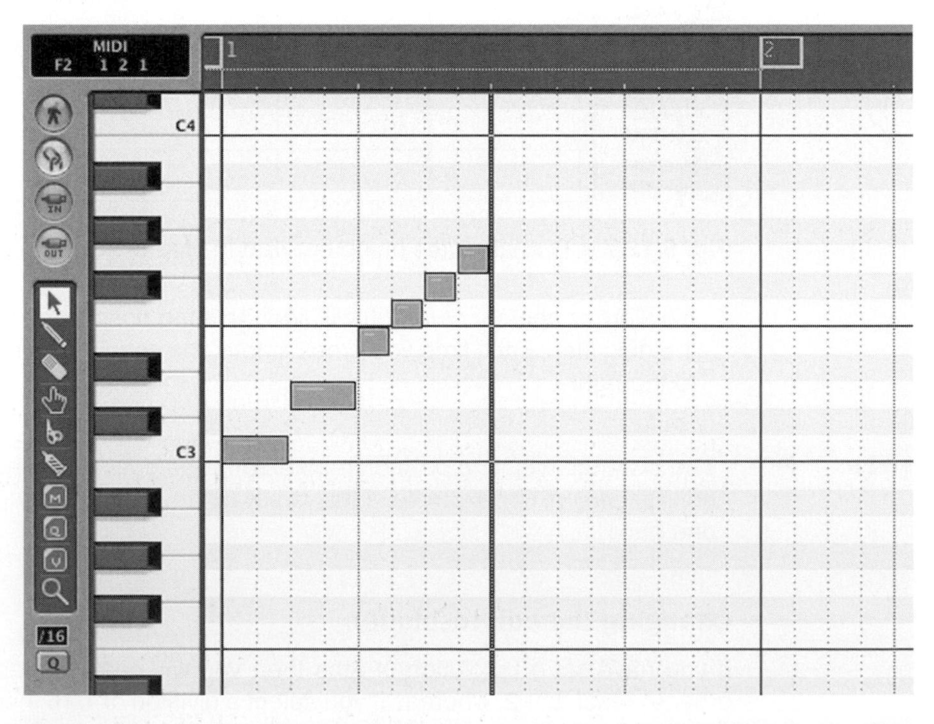

Fig. 5.12.9 – In this example the first two notes have been entered using 8ths and the last four notes using 16ths.

Entering rests

You can also advance a step without inputting any notes at all. This will allow you to enter spaces or rests between notes. The value of the rest will be determined by the resolution set. You can usually enter a rest by pressing a dedicated key or using a foot pedal.

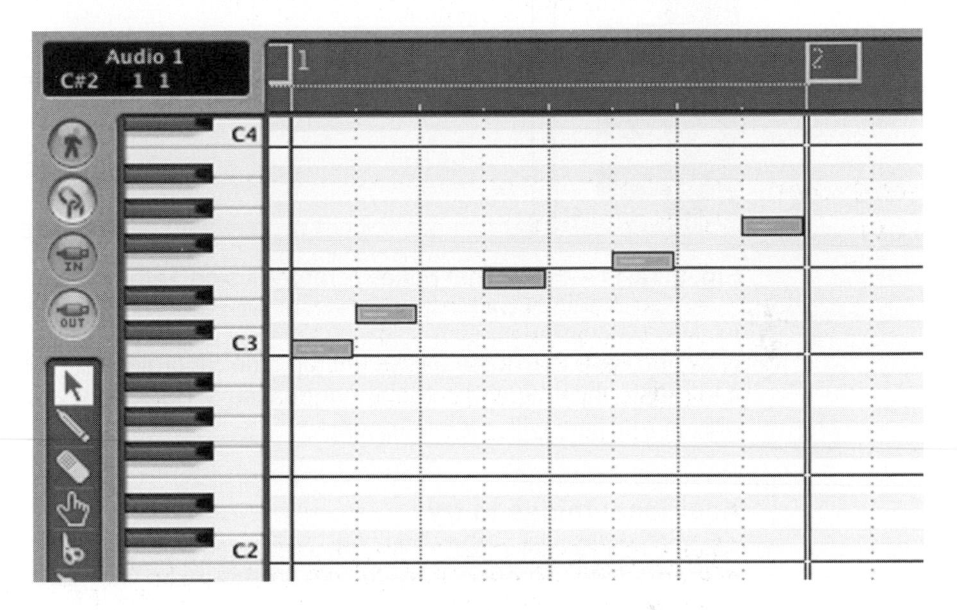

Fig. 5.12.10 – 1/8th note sequence with rests.

note ► The value of the rest will be determined by the resolution set.

Entering complex sequences

Once you are able to adjust the length of each note and enter rests, you should start to see how powerful step recording can be. It can, for example, be used to create very fast rhythms or sequences that would be difficult to play in real time.

Creating a fast hi hat pattern

You can also use step time recording with drum sounds. In the following example we assigned the MIDI track to a GM drum sound. We then selected a hi hat sound on note F#1 and changed the resolution to 16ths. This allowed us to enter 16 notes into each bar.

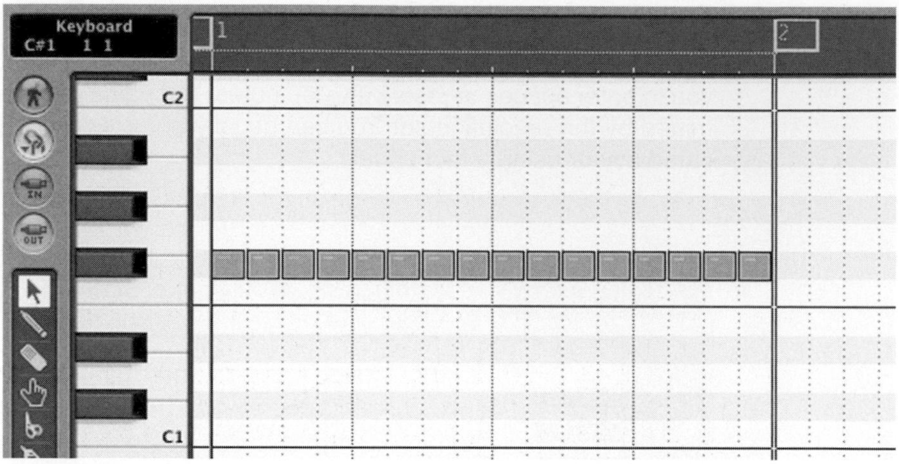

Fig. 5.12.11 – Fast hi hat pattern created using 16th notes.

Now you have learnt the basics of step time recording you will be able to use it to create rhythms and complex sequences of notes that would be difficult to play in real time with the metronome.

Additional things to try

- Try randomly entering notes to help you come up with ideas for a bass line, for example.
- Experiment inserting different combinations of notes with rests to create rhythms. You may be surprised by the result.
- Try inputting chords.
- Create an arpeggio by carefully entering individual notes from a chord.

Exercise 5.13 Changing the time signature

Before attempting this exercise we recommend you complete Exercises 5.1 and 5.2. This exercise will cover the following:

- Creating a fixed time signature
- Changing the time signature
- Inserting time signature changes throughout a song
- Deleting and editing time signature changes.

Music is structured in bars whether it is written on paper or created on a computer. The time signature determines where the bar lines fall within your sequencer. For example, the default setting 4/4 means there are four crochet beats contained in each bar. This means the fifth beat would have to be in the next bar, as only four beats are allowed in a bar.

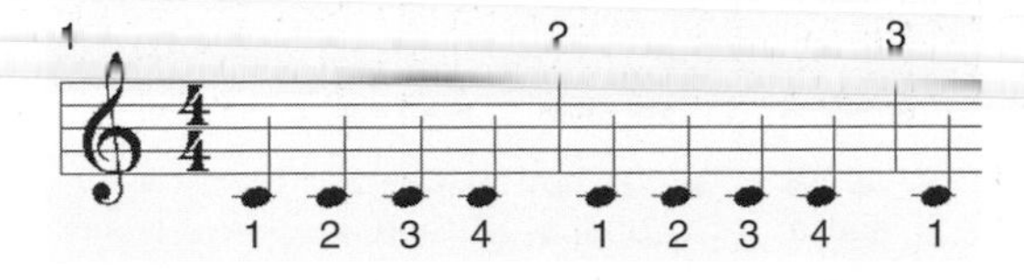

Fig. 5.13.1 – Four crochet beats in each bar.

So why is it important to learn how to change the time signature? Well most of the popular music we listen to is based around a 4/4 rhythm, also known as common time. However, the waltz, for example, has a fundamentally different rhythmic pulse. This is achieved by having three beats in a bar (3/4) instead of four. The first beat is often made louder so it's easier to identify the beginning of each bar. To help you understand this try clapping and counting to three; remember to accent the first beat. Make sure the distance between each clap is equal, as this will provide a regular pulse.

Example of a three-beat bar:

1 2 3 1 2 3
accent accent

Now try clapping and counting to four, and remember to accent the first beat.

Example of a four-beat bar:

1 2 3 4 1 2 3 4
accent accent

Notice how changing the time signature alters the rhythmic pulse and gives the rhythm a totally different feel.

What each value means

A time signature consists of two numbers and is normally set to 4/4 as default on most sequencers. The bottom number determines the note value the timing of a bar is based on. For example, 4/4 means you have four regular crochet beats in each bar. Therefore, crochets or quarters are the reference, so adjusting the top number simply determines how many crochets or quarters are allowed in each bar. 6/8 is slightly more complicated as it indicates six quavers per bar. Therefore, quavers or eighths are the reference, so adjusting the top number determines how many eighths are allowed in each bar. You may find it easier to count along to this style of rhythm by counting 1 and a, 2 and a, etc.

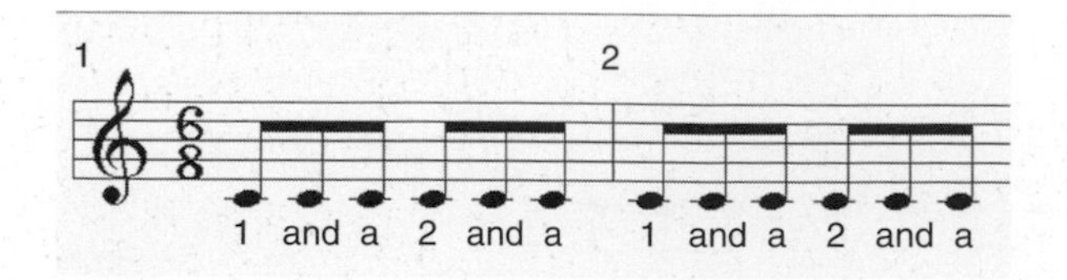

Fig. 5.13.2 – A 6/8 rhythm.

The exercise ...

In this exercise we are going to explain how to adjust the time signature and set up some practical examples for you to listen to. You will therefore need to be able to hear your sequencer's metronome click. You could also consider loading or creating a MIDI sequence as well if you prefer to hear the metronome click in context with some MIDI data.

Creating a fixed time signature

Locate the time signature display. This may be displayed on the transport bar or across the top of the main Arrange window.

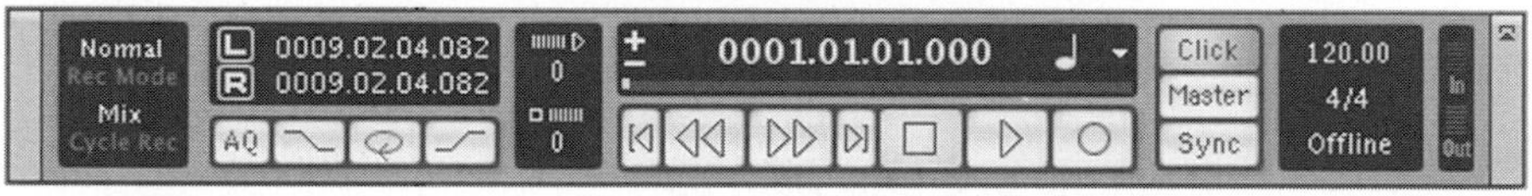

Fig. 5.13.3 – Transport bar displaying a 4/4 time signature.

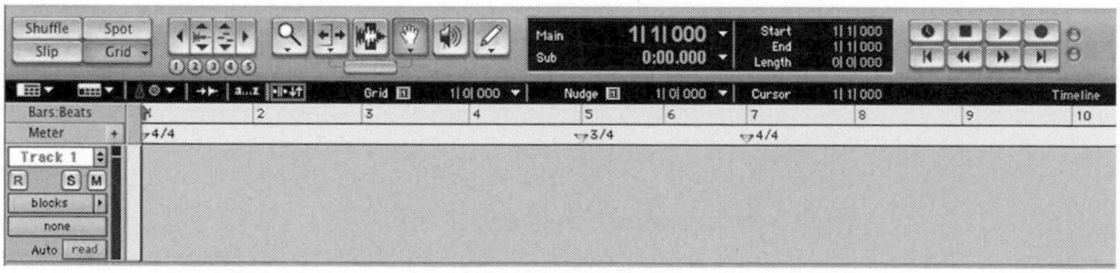

Fig. 5.13.4 – Time signature information located across the top of the main Arrange window.

Global and master tracks

Some sequencers display time signature information using a special global or master track. Any changes made in this type of edit window will affect the whole song. Note: you may have to open this window from a menu.

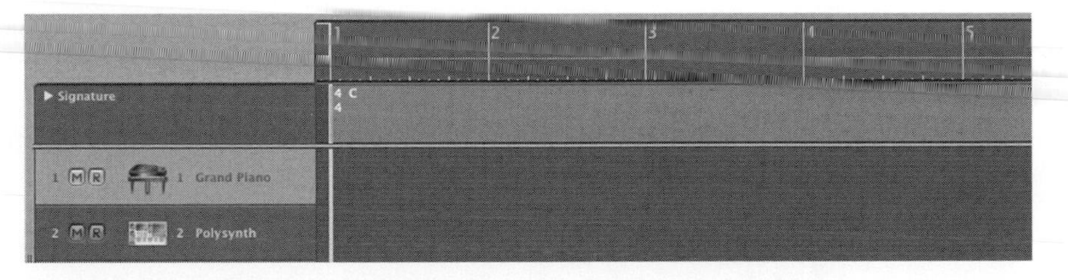

Fig. 5.13.5 – Logic mastertrack.

Make sure the time signature for the song is set to 4/4. This is normally the default setting.

Switch on the metronome/click. This is usually located in the transport bar. Make sure the metronome is actually set up to produce a sound. See 'Setting up the metronome' box.

Fig. 5.13.6 – Metronome enabled in transport bar.

Setting up the metronome

In order to produce a sound from a metronome, you may have to first specify how you want to do this. There is usually an option for the metronome to produce a speaker click. This allows the sound to come out of the computer's own internal speaker. Alternatively, you may be able to produce a MIDI click. A MIDI click will output MIDI data to a specified sound source, such as a sound module or sound source inside the computer. This normally needs setting up and is similar to routing a MIDI track to a sound source.

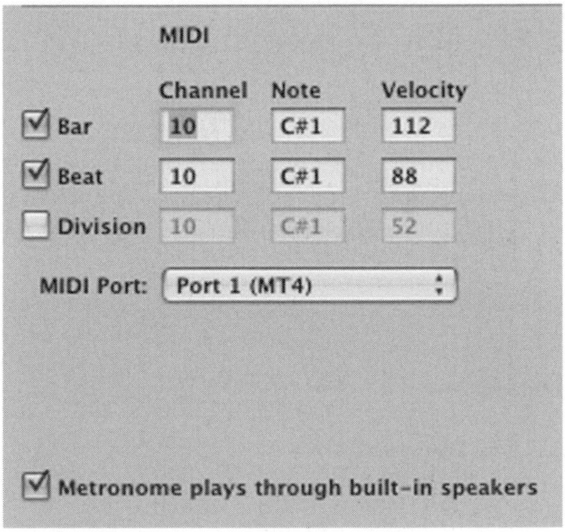

Fig. 5.13.7 – Metronome settings.

Press Play and listen to the metronome. The metronome should provide a regular pulse at the selected tempo (the default tempo is 120 beats per minute). Notice how it makes the first click of every four slightly louder. This is emphasizing the first beat of every bar. Try watching the song position marker, as this will help you to identify the beginning of each bar.

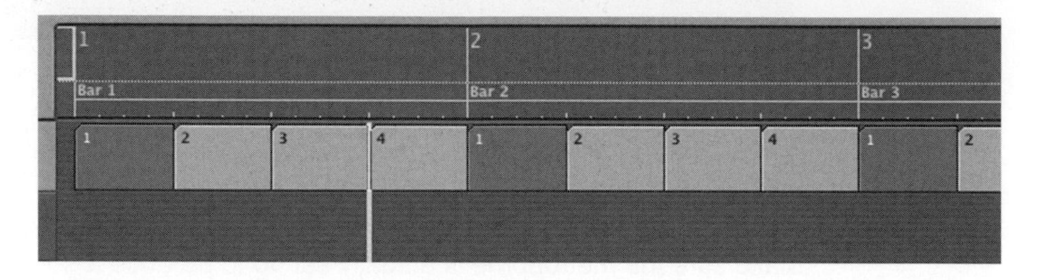

Fig. 5.13.8 – Bar divided into four beats.

Changing the time signature

1. *Press Stop and locate back to the beginning of the song.* Make sure you have located exactly back to the beginning of the song.
2. *Change the time signature to 3/4.* Use the transport bar or menu.

Fig. 5.13.9 – Time signature set to 3/4.

tip ►

> **Cubase users**
>
> If you are using Cubase, make sure that the master track switch is *off* in the transport bar (see notes on using the master track).

3. *Press Play and listen to the metronome.* Notice how it makes the first click of every three slightly louder. This is still emphasizing the first beat of every bar, but now we only have three clicks per bar. Watching the song position marker will also help you to identify the beginning of each bar.

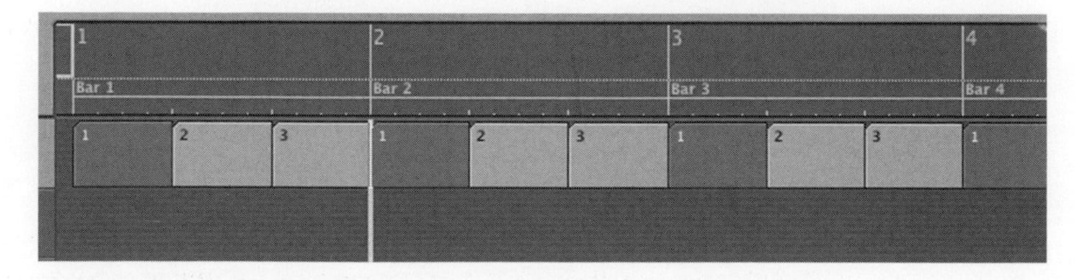

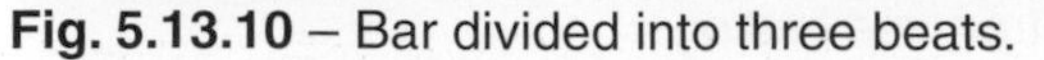

Fig. 5.13.10 – Bar divided into three beats.

tip ►

Always change at bar 1 for a constant time signature

To provide the same time signature for the whole song, always remember to locate the Song position marker back to the beginning of bar 1 before you make any changes to the time signature. On some sequencers, changing the time signature at any other location than at bar 1 will result in a time signature change being inserted at that point into the song.

Fig. 5.13.11

Experiment with different time signatures

In our example we have only looked at the basics of how to change and create a fixed time signature for the whole song. Now it's over to you to experiment using different time signatures. Why not try a 5/4 or 6/8 or 12/8?

note ►

Changing the time signature will visually affect the distance between each bar on the main Arrange window.

Varying the time signature

Sometimes you may wish to vary the time signature throughout a song instead of using just one static value. This may be because you want to introduce a different rhythmic feel at a certain point in a song or experiment with a new rhythm that may not 'fit' with the current time signature. Alternatively, you may just need the occasional half bar.

Opening the Time Signature editor window

The best way to manage any time signature changes is to open up the Time Signature edit window or view a global track. This will give you an overview of all the time signature information for the current song. You may find that time signature information is displayed with tempo information. This will vary depending on the type of software you are using.

Displaying time signature changes graphically

Time signature information can usually be viewed graphically or in a numerical list. The graphic editor normally adopts a timeline technique with height being BPM and width being time (bars and beats, for example). By using a pencil tool one can create time signature changes at any location in the song by drawing them in. This provides a visual 'feel' of how the music moves.

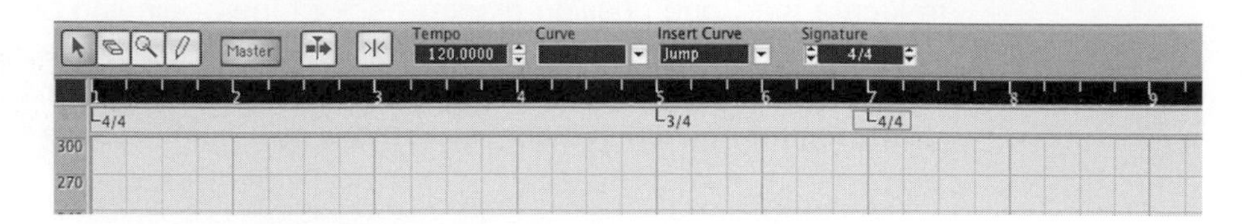

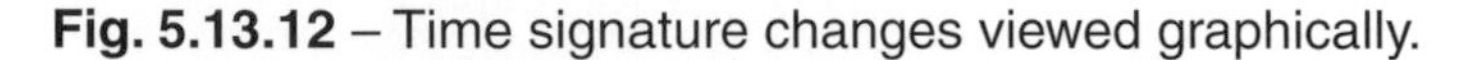

Fig. 5.13.12 – Time signature changes viewed graphically.

The numerical list

The numerical list is a list of time signature changes throughout the song. This is a very precise way of viewing and editing. In this window you have precise control over every time signature change and its position within the song. You can also edit the position within the song of each time signature change.

Position			Type	
			------------- Start of List -------------	
			Signature	4/ 4
			Key	C major
5	1	1	Signature	3/ 4
7	1	1	Signature	4/ 4
			------------- End of List -------------	

Fig. 5.13.13 – Time signature changes displayed in a list.

Deleting and editing time signature changes

It is also possible to delete time signature changes that you may have made. Open up the Time Signature editor and highlight the signature you want to delete. Then use the rubber tool or hit the Backspace key on your computer keyboard.

Inserting time signature changes throughout a song

1. *Locate back to the beginning of the song.* Bar 1, beat 1.

Fig. 5.13.14 – Sequencer located at bar 1.

2. *Open the Time Signature editor window.*

tip ► When inserting a time signature change, always ensure that the song position marker is positioned exactly at the beginning of the bar.

3. *Set the initial time signature to 4/4.* When doing this, ensure that the song line is at the start of bar 1. The initial time signature can usually be set via the transport bar.

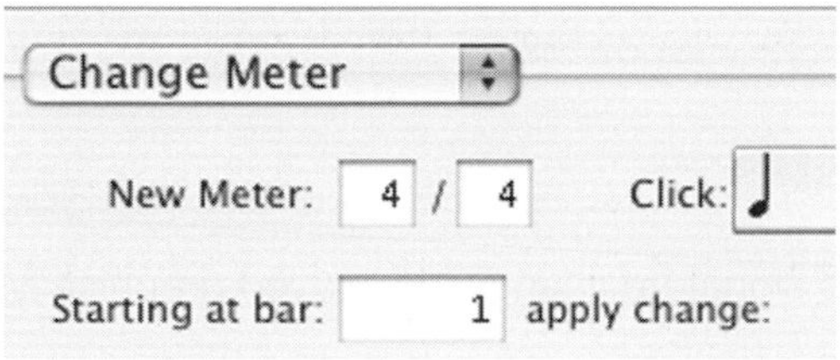

Change Meter
New Meter: 4 / 4 Click:
Starting at bar: 1 apply change:

Fig. 5.13.15 – Time signature set to 4/4 at bar 1.

4. *Move the Song position marker to bar 5.* Make sure it is exactly at the beginning of bar 5.

0005.01.01.000

Fig. 5.13.16 – Sequencer located at bar 5.

5. *Insert a time signature setting of 3/4.* You may have the option of adjusting the time signature via the transport bar or by creating a new entry in the Time Signature edit window. This will vary depending on the type of sequencer being used.

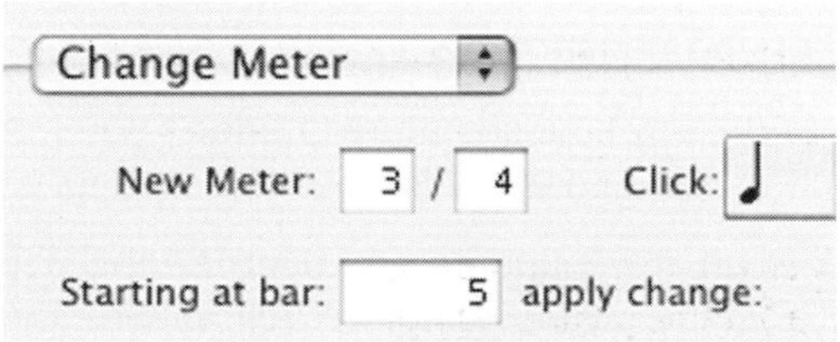

Change Meter
New Meter: 3 / 4 Click:
Starting at bar: 5 apply change:

Fig. 5.13.17 – Time signature set to 3/4 at bar 5.

6. *Move the Song position marker to bar 7.* Make sure it is exactly at the beginning of bar 7.

Fig. 5.13.18 – Sequencer located at bar 7.

7. *Insert or create a time signature of 4/4.* Create a new entry in the Time Signature edit window. This will change the signature from 3/4 to 4/4.

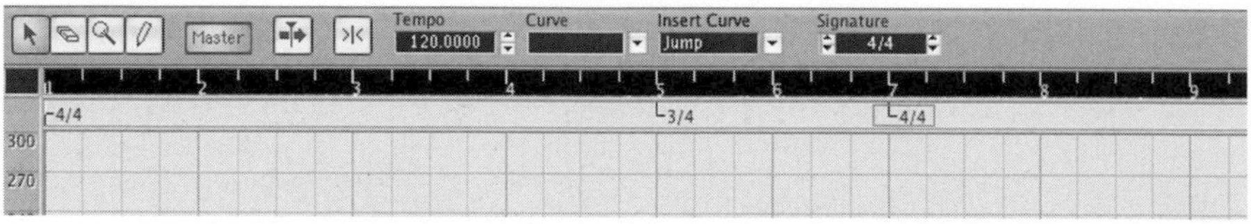

Fig. 5.13.19 – Time signature changes displayed graphically.

8. *Locate back to the beginning of the song.* Ensure that the metronome click is switched on.
9. *Press Play and listen to the metronome.* Listen to see if you can hear your time signature changes. Whilst listening, watch the play line and notice how it moves between the different bars on the Arrange window. You should notice that the bars between 5 and 7 are shorter and have less beats.

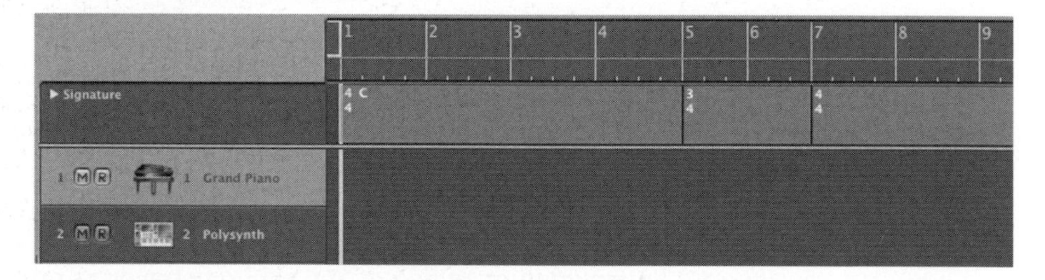

Fig. 5.13.20 – Time signature changes displayed in the Arrange window.

note ▶

Cubase and the master track

If you are using Cubase you will need to be aware of the *master track* button that is located on the transport bar.

If the master track is off, any time signature changes you make in the transport bar will be recognized and apply for the whole song.

If the master track is on, it will update and override the value displayed in the transport bar and the whole song will follow whatever settings are included in the master track.

The master track is a separate edit window that needs to be opened, as it's normally invisible. Only use the master track if you plan to create time signature changes throughout the song.

Additional things to try

Now you have got the basic idea of how to change the time signature, why not try experimenting and create some music using different time signatures you have never tried before. This may help inspire you to create different rhythms and melodies and can be useful when you hit a dead end with an idea and want to 'shake up' the music and explore new ideas.

Exercise 5.14 Changing the tempo

Before attempting this exercise we recommend you complete Exercises 5.1 and 5.2. This exercise will cover the following:

- Creating tempo changes
- Viewing tempo information
- Creating tempo changes throughout a song.

All music has a tempo whether it is written on paper or created on a computer. Many records or pieces of music use the same tempo throughout and therefore have a fixed tempo. However, in some pieces of music the tempo varies slightly or changes in a dramatic way. This can totally change the feel or mood of a piece of music and provide an interesting variation.

Changing the tempo

Experiment by adjusting the tempo to see how it changes the feel of your music. Normally, you will be able to set the tempo in the transport bar or via a menu in the main Arrange window. There will be a number that is normally set to 120 as default that can be increased or decreased to speed up or slow down the tempo. In most sequencers this is also possible during playback. Here are some examples to get you started. Dance music tends to have a fixed tempo from anywhere between 120 and 134 BPM, whereas a slow ballad would be around 70–90 BPM.

Fig. 5.14.1 – Transport bar with tempo set to 120 BPM.

By increasing the number we speed up the playback of the sequencer and by decreasing it we slow it down. Note: slowing down the tempo on a sequencer means it will take longer to get from bar 1 to bar 2.

note ▶

If you are using Cubase

To make changes to the tempo in the transport bar, make sure that the master track switch is *off* in the transport bar (see 'Cubase and the master track', p. 165).

When to commit to tempo changes

You can change tempo at any stage of the sequencing process – before recording, at the start of the creative process or while listening back. If you set a certain tempo to record with, try playing back at a different tempo to see how it changes the feel of the music.

tip ▶

How adjusting the tempo can help you

You may find it easier to record a complex sequence of notes at a slower tempo. Simply slow down the sequencer's tempo to record and then speed it back up again for playback.

The exercise ...

In this exercise we are going to need a short section of music of around eight bars long in order to experiment with tempo. If you don't have anything available you can simply use the sequencer's metronome.

Alternatively,

- Create something new
- Use a MIDI sequence from a previous exercise
- Play something you know already into the sequencer
- Load a MIDI file
- Load a MIDI file from the book's CD.

Creating tempo changes

This exercise shows you how to vary the tempo throughout a song by inserting tempo changes.

note ▶

Recording to a moving tempo is very difficult and is something you would need to practise. For the purpose of this exercise we will record at a constant tempo and then edit the tempo later.

1. *Set the main locator so it starts at bar 1 and locate the song position back to the beginning of bar 1.* Either drag the locator to the start or click Stop twice.

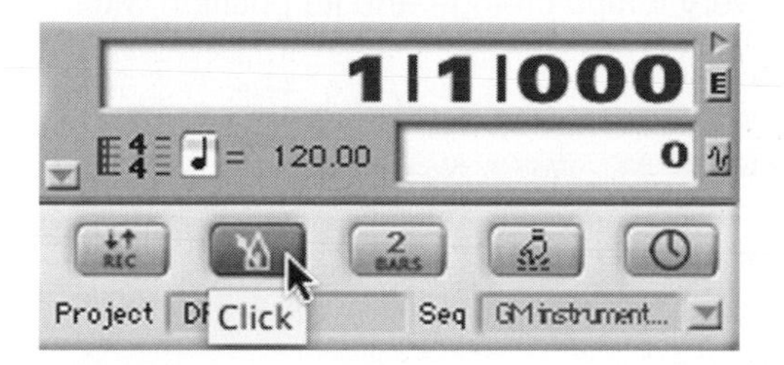

Fig. 5.14.2 – Switching the metronome on.

2. *Switch on the metronome/click.* This is usually located in the transport bar. Make sure the metronome is actually set up to produce a sound (see 'Setting up the metronome', p. 74).
3. *Open the Tempo edit window.* Open your Tempo editor or master track. Depending on the type of sequencer you are using it will usually be possible to view tempo information graphically or as a numerical list.

Viewing tempo information

Graphically

The graphic editor normally adopts a timeline technique with height being tempo and width being time (bars and beats, for example). By using a pencil tool it is possible to create tempo changes at any location in the song by drawing in slopes and diagonals to create gradual changes in speed. This provides a visual 'feel' of how the music moves.

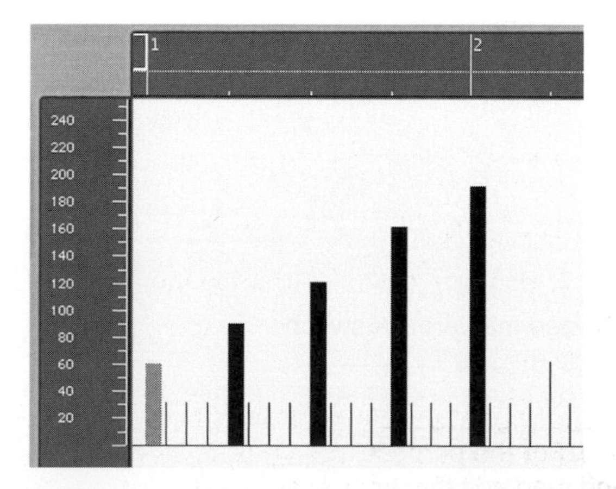

Fig. 5.14.3 – An increase in tempo displayed graphically.

Tempo list

The numerical list provides a list of the tempo changes throughout the song. This is a very precise way of viewing and editing tempo where you have precise control over every tempo change and its position within the song.

------------- Start of List -------------

BAR	POSITION			TEMPO
1	1	1	Tempo	60.0000
1	2	1	Tempo	90.0000
1	3	1	Tempo	120.0000
1	4	1	Tempo	160.0000
2	1	1	Tempo	190.0000

------------- End of List ---------------

Fig. 5.14.4 – An increase in tempo displayed in a list.

Global and master tracks

Some sequencers display tempo information using a special global or master track. Any changes made in this type of edit window will affect the whole song. Note that this window may be hidden and you may have to open it from a menu.

Fig. 5.14.5 – Cubase mastertrack switch.

Cubase and the master track

If you are using Cubase you will need to be aware of the *master track* button that is located on the transport bar.

If the master track is off, any tempo changes you make in the transport bar will be recognized and apply for the whole song.

If the master track is on, the whole arrangement will comply with whatever tempo settings are included within the master track, therefore overriding the value displayed in the transport bar.

The master track or tempo track is a separate edit window that needs to be opened, as it's normally invisible. Make sure that the master track switch is on in order to create tempo changes.

Creating tempo changes throughout a song

1. *Set the tempo to 120.* Set the initial tempo at bar 1 to 120 BPM. This will be the first entry into the Tempo editor and will set the initial speed of the song.

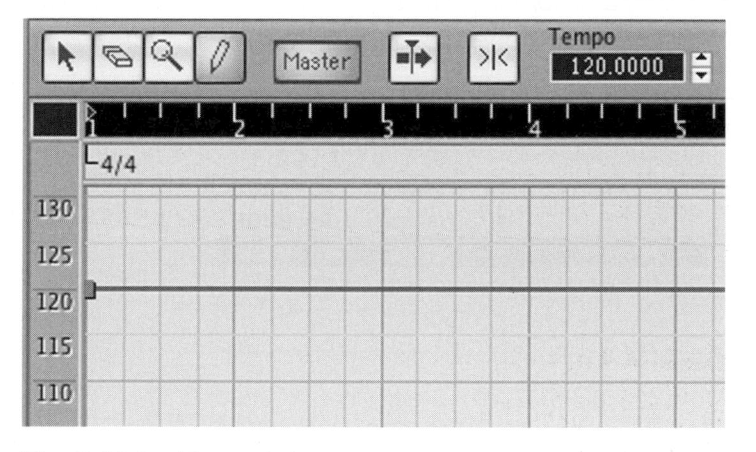

Fig. 5.14.6 – Tempo information displayed graphically.

2. *Move the Song position marker to bar 5.* Make sure it is exactly at the beginning of bar 5.

tip ►

Tempo changes are usually entered at the current song position, so always ensure that the song position marker is positioned exactly on the beginning of the bar.

3. *Insert a tempo change of 140 BPM to increase the tempo at bar 5.*

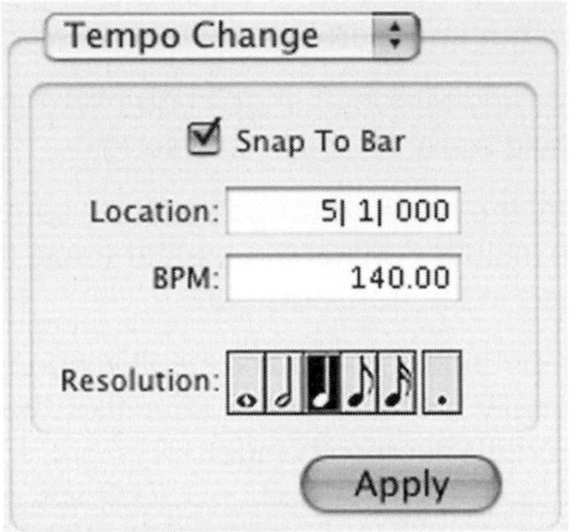

Tempo Change
Snap To Bar
Location: 5| 1| 000
BPM: 140.00
Resolution:
Apply

Fig. 5.14.7 – Inserting a tempo change of 140 BPM at bar 5.

4. *Move the Song position marker to bar 7.* Make sure it is exactly at the beginning of bar 7.
5. *Insert another tempo change to reduce the tempo to around 80 BPM.* You should now have created a tempo map.

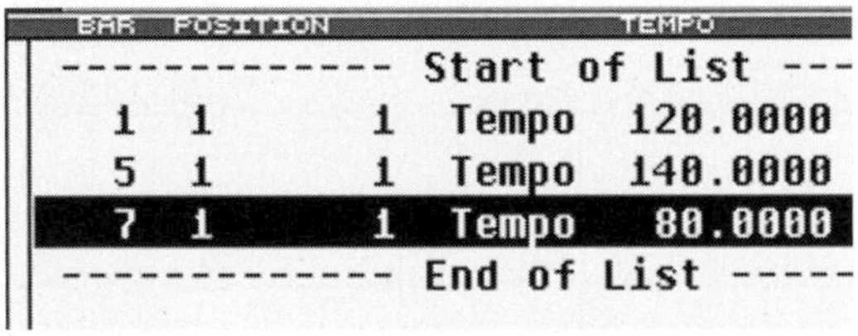

BAR	POSITION			TEMPO
-------------	Start of List ---			
1	1	1	Tempo	120.0000
5	1	1	Tempo	140.0000
7	1	1	Tempo	80.0000
-------------	End of List -----			

Fig. 5.14.8 – Tempo map displayed in a list.

tip ► If you have only used the graphic Tempo editor, take the time to close this and open up the Tempo List editor so you can see how the same information is displayed in a different format.

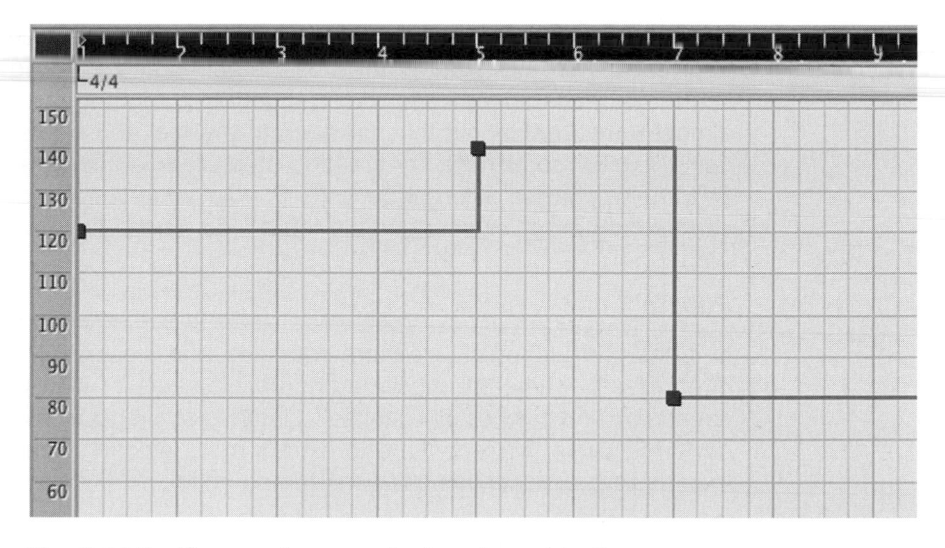

Fig. 5.14.9 – Tempo changes displayed graphically.

6. *Locate back to the beginning of the song.*
7. *Press Play and listen to the tempo changes.* Listen back to your sequence and see how the tempo automatically changes. Whilst listening, watch the play line and notice how it speeds up and slows down as it moves between the different bars on the Arrange window.

Additional things to try

Now you have got the basic idea, here are some other ideas to try.

Tempo crescendos

Using the pencil or cross-hair tool, create a line of tempo events across several bars. Normally, this will be easier to draw in and edit graphically.

- Create a gradual increase in tempo over eight bars
- Try a deceleration, which will result in the tempo gradually slowing down
- Create a gradual slowdown towards the end of a song.

tip ►

Reducing the tempo to record a complex sequence

You may find it easier to record a complex sequence of notes at a slower tempo. Simply slow down the sequencer's tempo to record with and then speed it back up again for playback.

Making your music more human by adding subtle tempo changes

Most popular music has one fixed tempo. In some pieces of music, especially music that is performed live, the tempo may naturally vary slightly. For example, it may gradually speed up to emphasize certain sections such as the chorus to give it a slight kick start, or towards the end to add a feeling of excitement. If the musicians speed up or down together a slight shift in tempo can be almost unnoticeable, but this still adds to the feel and excitement. So when using a MIDI sequencer, try creating small tempo changes of a couple of BPM to emphasize different sections of your song. As a general rule, if you can hear them you've changed the tempo too much.

Rubato

In forms of music which are more 'romantic' and 'classical' in nature, there is often rubato, where the music temporarily 'pauses' or slows down and then speeds up and catches up with itself. Listen to some examples to see how the performers use this technique to extract more emotion from the music and then try and simulate this using the tempo editor.

Exercise 5.15 Displaying MIDI on a score

This exercise covers the following:

- Viewing MIDI notes in the graphic editor
- Viewing MIDI notes in the Score editor
- Changing pitch in the Score editor
- Interpreting velocity in the Score editor
- Printing a score.

Before attempting this exercise we recommend you complete Exercises 5.1, 5.2 and 5.5–5.7.

Using the Score editor

The Score editor is a very useful way of viewing your MIDI data in a score format. Let us explore the best way to view and edit our music in the Score editor so that we can get the best print results.

Firstly, the Score editor does not truly show what MIDI data has been played. It generally shows an approximation of what has been played. After all, score writing was designed so that people could pass ideas between each other, so that a composer could hand musicians instructions on what they should play. It was never designed to accurately capture what had been played.

The Piano-roll-style editor, on the other hand, was never designed as a way of communicating what should be played to other musicians, but instead as a way of accurately showing what someone had played via a MIDI input device. There is an exact correlation between what is heard and seen when using this editor. However, with the Score editor, what you see may not be exactly what was played.

Bearing this in mind, one has to ask why you might want to use the Score editor? Well, you may write your music via the Key and Matrix editors, but you may choose to refer back to the Score editor to see how practical it may be for a live musician to play it. Additionally, some people have spent years writing traditional western scores and would like to continue. Hence, they may choose to do all the drafts in the Score and then use the Matrix editor as a way of adding nuance and interest to the music.

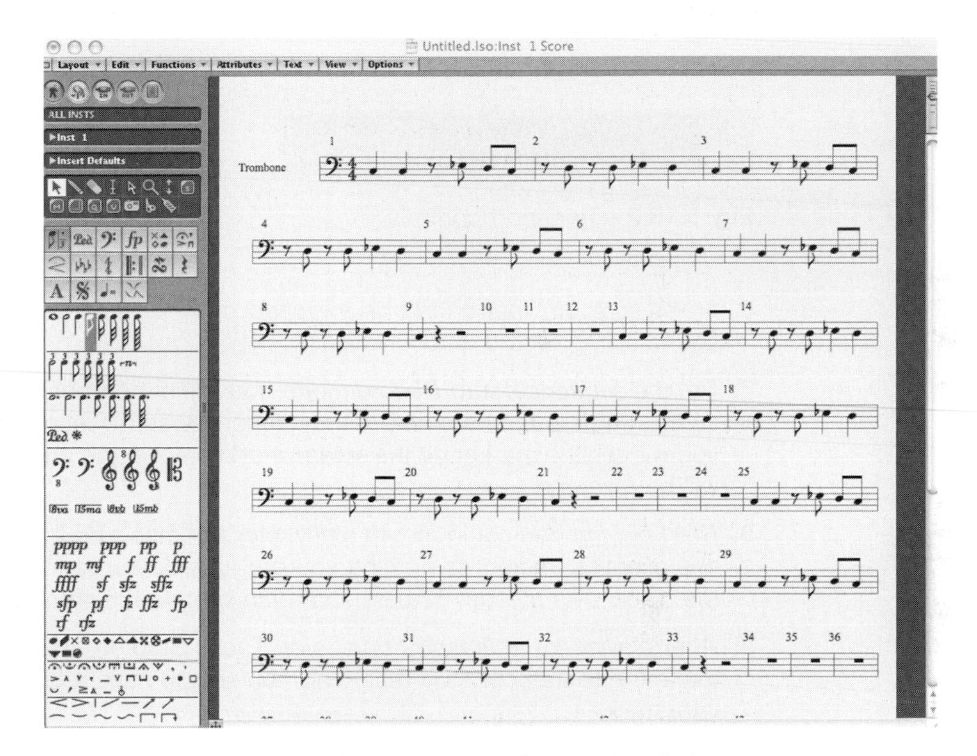

Fig. 5.15.1 – MIDI data displayed in the Score edit window.

The exercise …

1. *Load in a previously created song and go to the next section* ('Viewing MIDI notes in the graphic editor'). Alternatively, create a new project.

2. *Make sure that you are viewing the main Arrange window.* Hopefully it will open a default, which has a selection of audio, MIDI and virtual instrument tracks. If not, delete the tracks shown and create a selection.

3. *Make sure that each MIDI track is assigned to a sound source, such as a Universal Sound Module or a soundcard.* Select a track and assign it to MIDI channel 1. Choose a patch/sound. Try playing your keyboard to check that MIDI signals are being received and sent to the sound source. You

should hear a sound. If you cannot, refer to your sequencer and sound source manuals.

4. *Make sure the main Song postion marker is at the beginning of the piece.* Either drag the locator to the start or click stop twice. This normally makes the locator jump to the beginning.
5. *Now set the speed of the music.* Do this by setting the speed in the transport window. The speed chosen should be appropriate to the type and style of music you wish to record.

Fig. 5.15.2 – Transport controls.

6. *Make sure the metronome is switched on.* This will produce a clicking sound during record and play back.
7. *Put the sequencer into Record mode.* You will get one or two bars of pre-count, which is the time before the sequencer starts recording. After this, you will see the marker on the screen start to move and any notes you play will be recorded.
8. *Start playing.* You could record a new piece of music or, for the purpose of this exercise, record something you are already familiar with. Make sure the piece lasts at least 32 bars and try to keep the rhythm simple.
9. *After playing for at least 32 bars, stop.* Press Stop on the transport bar to make the sequencer stop recording. You should have a long MIDI part on your track.

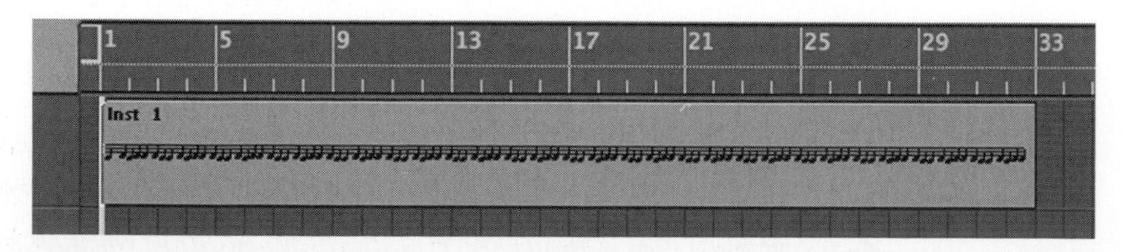

Fig. 5.15.3 – 32 bars of MIDI information.

Viewing MIDI notes in the graphic editor

Have a look at the MIDI part in more detail by opening the Piano-roll-style edit window. See how some of the notes are in time and others are not. You can do this by looking at the notes in reference to the grid lines (grid lines are usually set to 1/16ths). If you have played a simple rhythm in time with the sequencers metronome the notes fall close to the lines. However, in reality most people are rarely 100% accurate so you will normally find the notes will fall slightly before or after the grid lines. Now close the Matrix editor.

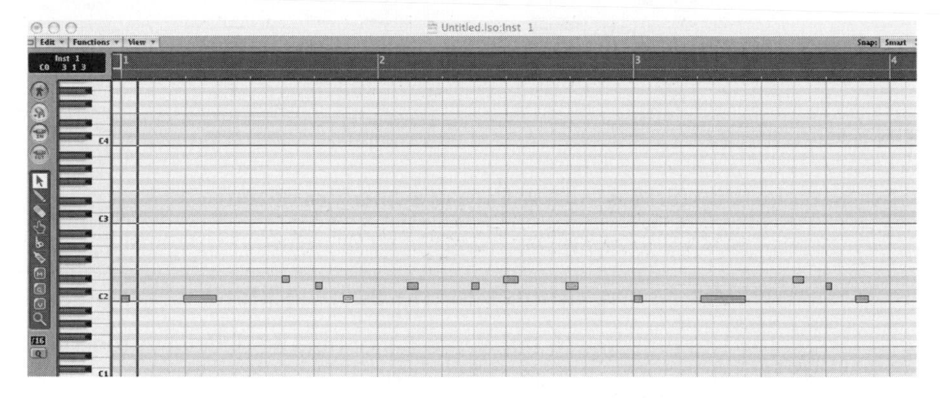

Fig. 5.15.4 – MIDI notes displayed in the Piano-roll-style edit window.

Viewing MIDI notes in the Score editor

Now select the same part/region and open the Score editor. This will display one of two things. It will either show a very complicated piece of music with short notes and rests which is difficult to read. Or it will display a perfect score with a simple rhythm which has no rests smaller than a quaver rest. An example of a difficult to read score that is unquantized is shown in Fig. 5.15.5.

Fig. 5.15.5 – Difficult to read score (unquantized).

Fig. 5.15.6 – Easy to read score (quantized).

Displaying timings in the Score editor

The reason why the Score editor may be difficult to read is due to the choice of settings. It is possible to have the score almost recreate what you have really played. However, traditional manuscript was never designed for this, as it could become too complex for a musician to play. For example, jazz scores are never written how they are played, but are instead a template or starting point for the musician. Even for classically trained proficient users of traditional western manuscript, reading the Score editor in this form would be difficult.

On the other hand, the Score editor may be very easy to read due to the choice of settings made. It is possible to adjust the score visually without changing what has been played. When you listen back to the track it will sound the same as when it was recorded. However, what you see in the editor will instead be a simplified version. This way it is easy to print the score and hand it to another musician so that they can then play something similar to the original.

producer says ▶

Quantizing the score

Of course, if what you played was simple and regular to begin with *and* you have quantized the notes on the Arrange window (to 1/8 or 1/16), then the score will accurately show what is being heard without the need for simplifying the music.

Changing pitch in the Score editor

Use the arrow tool to select the first note in the part/region. Click and hold on the note. When it changes colour it means it is selected. Drag the note up or down to change its pitch. Try not to drag it side to side, as this will affect its timing. Now close the Score editor.

Fig. 5.15.7 – Tools.

Now open the graphic editor. Notice how the first note has moved. This demonstrates that editing the pitch in one edit window affects the data in another edit window.

Interpreting velocity in the Score editor

Remember that other MIDI information such as velocity will not be added to the score so if you need to add additional information, such as accents and dynamics, you will need to insert this information manually in the score. This can be achieved by adding symbols directly to the Score edit window.

- A high MIDI velocity is the equivalent to a score accent
- A MIDI volume curve is the equivalent of the crescendo and decrescendo symbols
- A MIDI volume change (controller number 7) is equivalent to dynamic symbols such as f, pp, mp, etc.

Score display options

Depending on the type of sequencer you are using you will find various functions that are unique to score edit window.

Below is a list of some of the most commonly used functions that you should experiment with when trying to simplify a complex looking score.

- Quantize – tidies the timings so that they conform to a set resolution
- Smallest rest or value on screen – allows you to determine the smallest rest value displayed throughout the score
- Overlapping notes – this determines whether tied notes (held notes) are allowed to cross over into another bar.

Printing a score

Before you print your score, check that each track is displayed in a format which makes it easy to read and therefore perform. Make sure that each instrument is being displayed at the right pitch for an instrumentalist to play and that the rhythms are clear and easy to read. If this is all in order you can progress to printing.

Displaying multiple parts

You may need to print or display multiple parts. To do this, highlight the parts or tracks you want to display in the Arrange window (Multiple parts can be selected by holding down the shift key). Now open the Score edit window and you should see all the selected instruments.

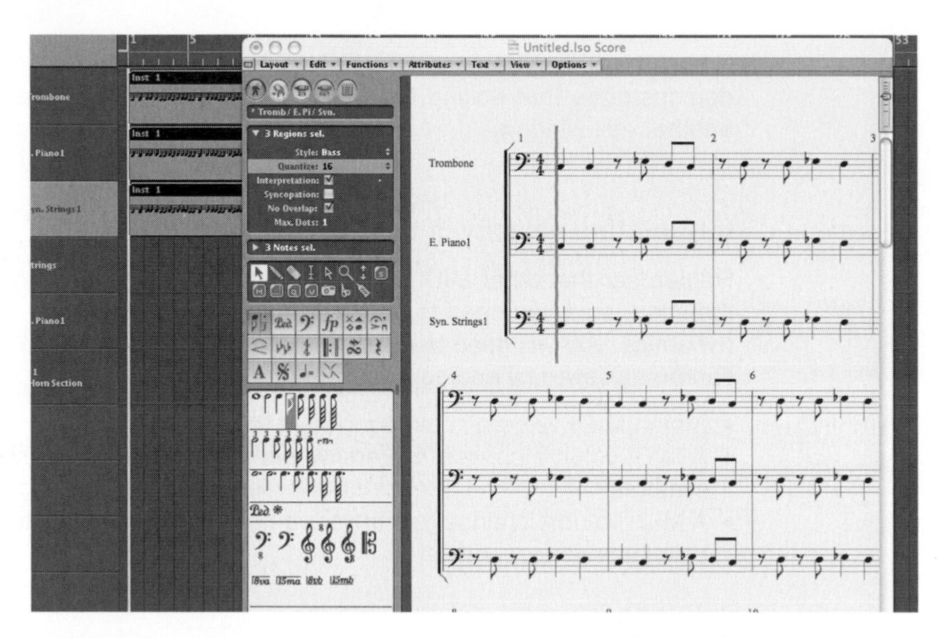

Fig. 5.15.8 – Score edit window displaying three different parts.

producer says ►

> **Printing a score**
>
> If you are planning to print out your music, you may have to make several adjustments to the score. Adjustments may include quantizing and transposing (changing an instrument's pitch in order for it to be displayed correctly in the score for musicians to read).
>
> You could consider making these adjustments in a separate song file by choosing Save As from the File menu and then entering a different file name such as 'Score printout'. This will allow you to make any adjustments to the score without affecting the original performance data.

6

MIDI Sequencing Projects

This chapter focuses on composing a piece of music using a MIDI sequencer. To help you get started we have included several step-by-step examples in different musical styles.

Project 6.1 Ground bass

Before attempting this project we recommend you complete Exercises 5.1–5.7, 5.9, 5.11 and 5.12, *and refer to Chapter 4 to brush upon anything you're not sure of.*

Introduction

A ground bass is a repeating bass line that provides a rhythm to build other instruments onto. Traditionally this technique was used in various aspects of baroque music. However many contemporary forms of music such as garage, hip-hop, R&B and funk are also based around a repeating bass line and use a similar technique.

The notes and rhythms used to create the bass line are often relatively simple and form short phrases. For example, a bass line may simply consist of just four notes played on each beat of the bar.

The way the other instruments are structured on top of the bass line is also important as this provides a way to introduce some variation in the composition. Chord sequences are commonly used to define different sections, or varying melodies over the same chord sequence. Other factors such as percussion and counter-melodies can also be used to create variety.

The project …

- **Stage 1.** Creating a bass line
- **Stage 2.** Creating a drum track
- **Stage 3.** Creating a pad
- **Stage 4.** Creating a melody
- **Stage 5.** Arranging the song
- **Stage 6.** Mixing and adding effects.

Getting started

Load your MIDI sequencing software. Create a new project or choose Save As from the file menu. Navigate to a suitable location on the computer's hard drive where you want to save your new song or project. Give the song or project a name, e.g. 'Ground Bass 1' then press Save.

Fig. 6.1.1 – Selecting Save As from the file menu.

tip ▶ Always remember to name the file you are working on and to save it in a location on the hard drive where it can be easily located.

Stage 1. Creating a bass line

Using your sequencer *create a new MIDI track and assign it to a bass sound.* The options you will have to produce a sound will vary. Sounds may be produced internally within the sequencer or externally using a stand-alone sound module.

tip ▶ Avoid using MIDI channel 10, as this is where the drum parts are usually located.

Try looking for a sound that is deep and that has a fast attack and decay time, i.e. it starts immediately and finishes once you let go of the key. Experiment selecting different sounds and you will find that some sounds work better than others as bass sounds. When using a *GM sound device* you should find the bass sounds are located between 041 to 049.

tip ▶

You don't have to select a sound called bass. Try using the lower end of the keyboard to play any sound. You could, for example, use the bass notes of an organ sound, so be creative.

A ground bass is usually structured around a repeating bass line with a simple rhythm. You therefore don't have to use lots of notes to make the bass line sound interesting, as the rhythm is equally as important as the notes. Playing the same note over and over is fine if the timing is interesting and you are creating a groove.

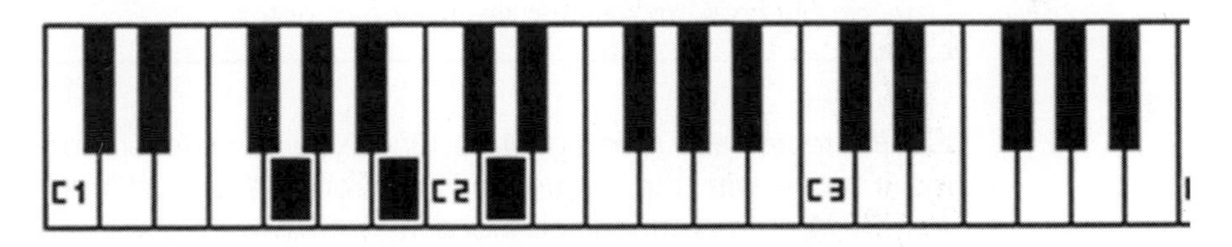

Fig. 6.1.2 – Notes used in a G major chord.

Try selecting 3 or 4 notes from a chord or scale. Now try playing each note while counting to four so you place a different note at the beginning of each bar. To speed up the sequence, try only counting to two while holding each note, so you place a different note every two beats. Now try playing the notes in a different order (e.g. A, B, D, G or D, B, A, G).

producer says ▶

When creating a ground bass many people simply fall upon the well-tried Pachabel's Cannon sequence. This is CGAEFCFG. Examiners will frown upon this, as it is not exactly original. It has been used over and over again in many types of music, including in popular hip-hop and pop music. Even though you might create a great piece of music using this sequence, try creating an original ground bass.

Rhythms are just as important as notes in a bass line, so try and be creative. Experiment playing different rhythms using the same notes. Listen to the example on the CD.

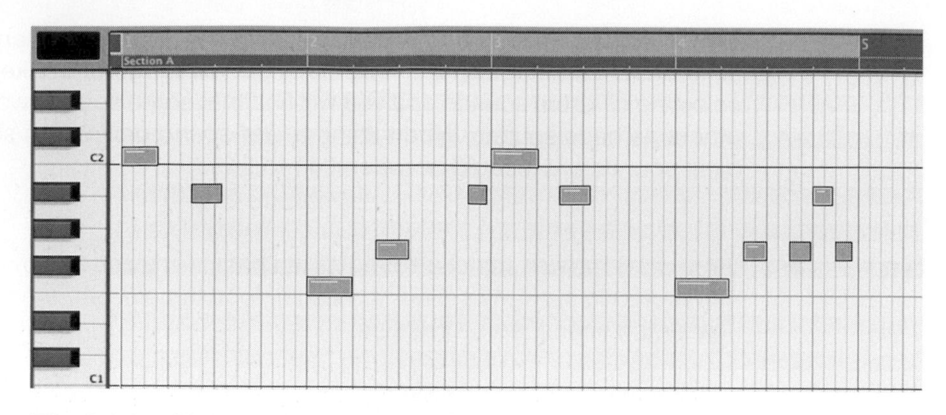

Fig. 6.1.3 – Example four-bar bass line.

producer says ▶

> Try and find your own combination of notes that sound interesting together. After all, it's better to create something original sounding than to simply copy an existing sequence of notes.

Set the tempo and time signature. Once you have developed an idea, try playing it along with the sequencer's metronome. Adjust the tempo to around 110 BPM or to a speed you feel comfortable with. Switch the metronome on and press Play on the sequencer. You should now hear a regular pulse or beat at the chosen tempo to play along with.

Fig. 6.1.4 – Tempo and time signature information.

Practise playing your bass ideas along with the metronome to see how they work at the selected tempo. Make sure you are playing in time with the metronome beat.

Once you're clear in your mind what you're going to do, *stop the sequencer and return to the beginning of the song and drop into Record.* Wait for the count-in before you start playing.

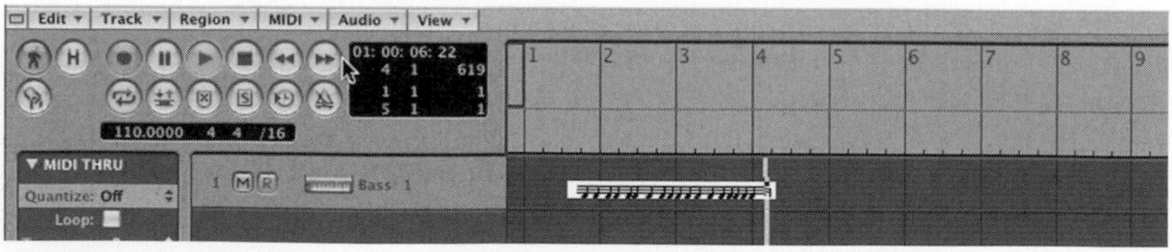

Fig. 6.1.5 – Sequencer recording MIDI information.

Play until you have successfully recorded your bass ideas. Try not to stop recording if you make any mistakes. Just keep playing, as you can edit small mistakes out later.

Stop the sequencer and return to the beginning of the song. If there are lots of mistakes when you listen back, delete what you don't want to keep and then arrange what's left into a continuous part or sequence. Ideally you should aim to create a section that works as a one-, two- or four-bar loop.

tip ▶ It is usually a good idea to record any ideas you may have into a sequencer even if they are not exactly what you want, as once they have been recorded they can be edited to perfection.

Editing MIDI

The scissors can be used to separate the sections you want to keep from the sections you want to remove. You can then use the rubber to remove any unwanted sections. You will then need to arrange what's left into a continuous part or sequence and then move this sequence so it plays back from bar 1. You should now be able to repeat this section several more times.

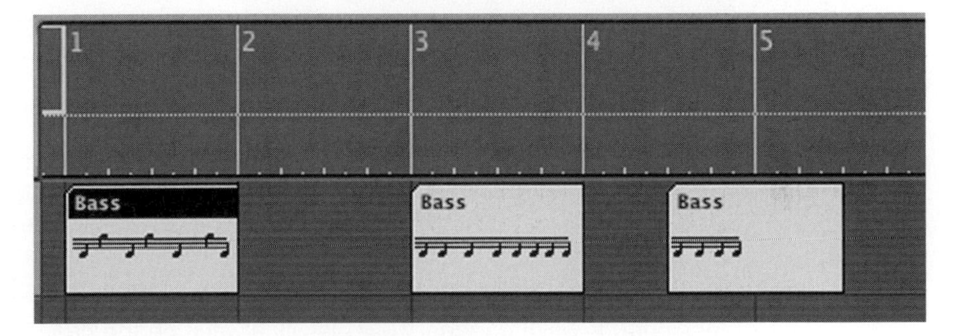

Fig. 6.1.6 – Removing unwanted sections from a bass sequence.

tip ▶ Repeating one small section may inspire you to add and build up a more complex sequence.

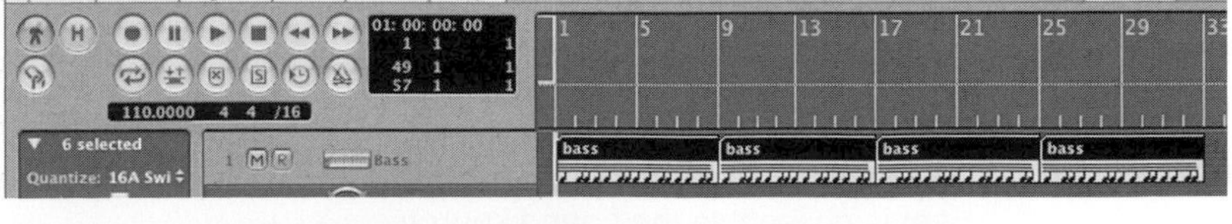

Fig. 6.1.7 – Edited eight-bar sequence repeated four times.

Now name the newly created MIDI track and press Save to update your file to disk.

Stage 2. Creating a drum track

Create four additional MIDI tracks and assign them to a percussion or drum sound. Drum sounds are usually assigned to MIDI channel 10, so ensure each track is correctly routed. This gives you much better control when editing and quantizing individual sounds.

Fig. 6.1.8 – Additional MIDI tracks assigned to drum sounds.

note ►

When using a GM-compatible sound module, the drums will always be assigned to MIDI channel 10. This will always be the case if you have the unit set to GM or XG mode.

Select one of the MIDI drum tracks and enable the record ready switch. If you now play your MIDI controller, you should find a different drum sound on each key of the keyboard. When using a GM drum kit you should find the bass drum on C1 and the snare drum on C2.

note ►

If you're using a non-GM sound source, make sure a drum or percussion set is chosen as the sound source.

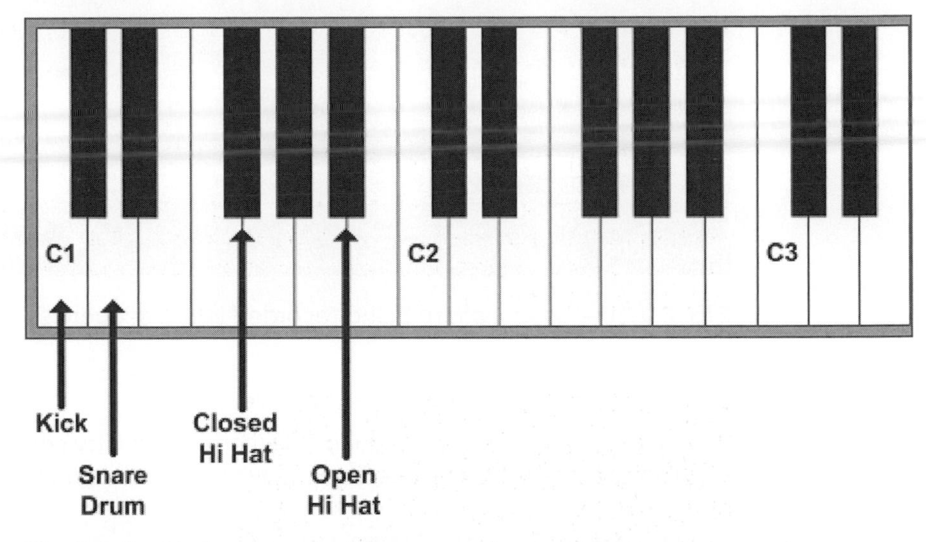

Fig. 6.1.9 – Typical layout of drum sounds on a keyboard.

producer says ▶

If you are finding it difficult to come up with any ideas, you could always try listening to CDs of other people's music. Sometimes this can give you inspiration to create your own music in a similar style.

Try playing different keys on the keyboard until you find a sound you like. For example, try starting with a hi hat or bass drum sound. You now need to develop a rhythm pattern that will work alongside the bass line, so press Play on the sequencer and play the sound you have chosen, along with the bass line, and try and develop a rhythm idea. In our example we started with a hi hat pattern then added a kick drum and finally overdubbed a snare drum.

tip ▶

It is also possible to create a drum track by using the pencil tool to enter notes directly into the Matrix or List edit windows. The Drum edit window can also be used.

Once you're clear in your mind what you're going to do, *stop the sequencer and return to the beginning of the song again*. Make sure the loop function is not activated and then drop into Record. Wait for the pre-count before you start playing.

note ▶

When using a MIDI keyboard to enter information into a sequencer it is important to remember that what you play is exactly what is going to heard back, so always ensure that your performance is the best you can achieve.

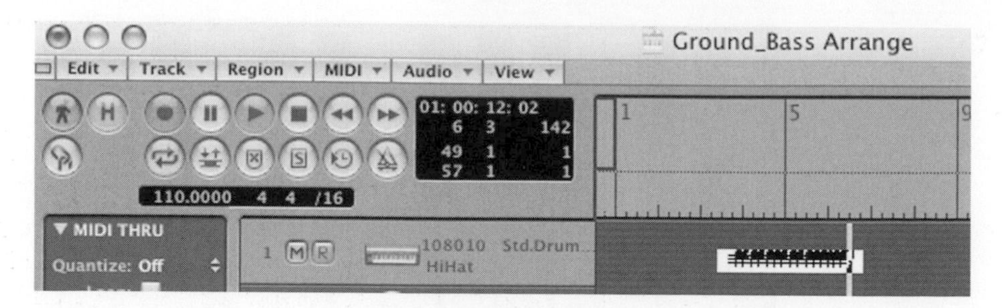

Fig. 6.1.10 – Drum pattern being recorded into a sequencer.

Play until you have recorded a good selection of ideas. Try not to stop recording if you make any mistakes. Just keep playing, as you can edit small mistakes out later.

Fig. 6.1.11 – Example hi hat pattern.

After 16 bars or when you feel like you have recorded something worth keeping, *press Stop and locate to the beginning of your composition.*

Press Play to listen back to what you have recorded. Is it in time? Are you happy with the rhythm? If not, delete the part and re-record it again. It is best to listen back while the metronome is switched off, as this will help you to decide if the drums and bass line fit together.

You could consider *quantizing* what you have just played, as this may help improve the timing of the notes you have played. You may also want to view the drum track in an edit window so you can see the timing of each note.

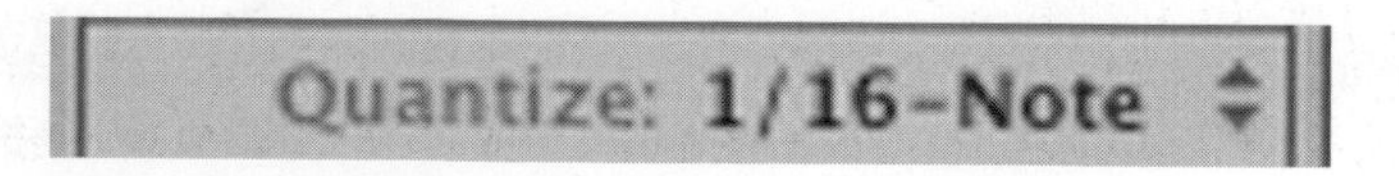

Fig. 6.1.12 – Quantize value set to 16ths.

producer says ▶

> As you build up several drum tracks, the overall timing may start to drift and sound messy as each individual track effectively becomes slightly out of time with the next. We therefore recommend that you try quantizing the drum and bass tracks even if you are a very good keyboard player as this will automatically move the notes you have played exactly to the nearest selected gridline. This can be very effective in making your music sound more in time and can save hours of endless fiddling and tweaking. However, quantize does not always automatically put your music in time and selecting correct Quantize value is essential (see Exercise 5.6).

Once you have got a recording you are satisfied with or a section worth keeping, separate the sections you want to keep by cutting/trimming the data with the scissors.

Fig. 6.1.13 – Editing a hi hat pattern.

Delete what you don't want to keep from the sequencer then arrange what you want to keep into a continuous part or sequence and glue together. Then move this sequence so it plays back from bar 1.

Select the next MIDI track and find another drum sound you want to use, e.g. a kick or snare drum sound. Now build up several drum tracks to create a more complex rhythm. The example in Fig. 6.1.14 shows three MIDI tracks, each containing a different drum sound.

note ▶

> Remember, each drum sound can be placed on a different sequencer track, but all drum sounds usually share the same MIDI channel.

Fig. 6.1.14 – Example bass and drum arrangement.

Once you have recorded and edited several drum tracks, build up an arrangement by copying and pasting. Your composition should now last approximately 1–2 minutes. You can check this by moving the Song position marker to the end of the composition and looking at the time code display. For example, at 110 BPM, 45 bars of music lasts for 1 minute 36 seconds.

Fig. 6.1.15 – Transport bar showing a song's duration at 45 bars.

You have now defined the style of the piece and crafted the foundation of your ground bass. You should now have a drum beat and bass line that plays to the end of the composition. Make sure that you have named any newly created MIDI tracks and then press Save to update your file.

tip ▶ It is important to name each track so they can be easily identified.

Stage 3. Adding a pad

Create a new MIDI track and assign it to a piano, guitar or organ sound. Ideally you should try and use a sound that will fill out the mid range frequency of your composition, i.e. above the bass line but below the melody.

This part usually consists of several notes that are played at the same time to form a chord. As you have already used recorded the bass line, you need to find chords that will work well with this and avoid chords that clash and sound discordant. A good starting point is to try and identify the root notes used in the bass line and then try using chords based around these notes (a root note is the main or most commonly used note). For example, if the root note of the bass line is the note A, try playing a chord based around this note, e.g. A minor (A, C, E) or A major (A, C#, E).

note ▶ The type of chords you choose will often determine the feel or mood of a composition. For example, major chords will usually create a happier feel than using minor chords.

Once you have found a chord that works with the bass line, try finding another chord and then move between them to create a chord sequence. For example, a chord sequence could be created using a combination of major and minor chords (A minor, G major, D minor, C major). This is only one example, so as always experiment. You may find unusual combinations of chords work equally as well over the bass line. Practise playing your ideas along with the drums and bass line to see how they work together. Usually you will find chords change more slowly than the bass line, at the beginning of each bar or every two beats, so holding down a chord over a faster changing bass line is common.

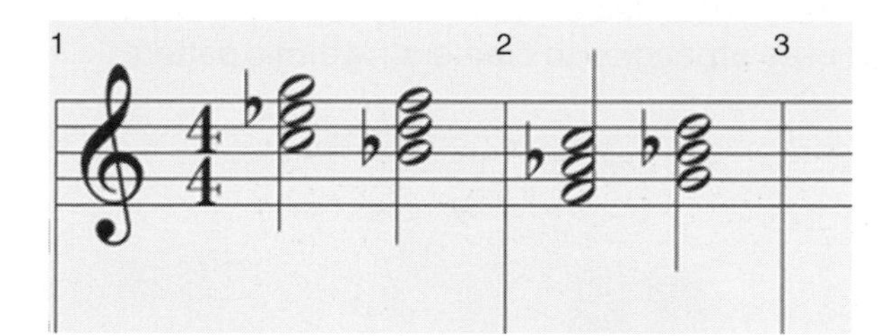

Fig. 6.1.16 – Example chord sequence.

note ▶ Moving between different chords will create a chord sequence. Most popular chord sequences are usually two or four bars long.

Once you're clear in your mind what you're going to do, stop the sequencer and return to the beginning of the song, then *drop into Record*. Play until you have successfully recorded your ideas.

Stop the sequencer and return to the beginning of the song. Now listen back to what you have just recorded and arrange what you want to keep into a continuous part or sequence.

Creating variation and defining different sections

Once you have a chord sequence you are happy with, you need to consider how you can create some variation throughout the arrangement and define different sections. It's possible to use the same drum and bass parts over and over again, but change how the chords are actually played every four or eight bars. This will create the illusion that each section is changing, even if the rhythm section remains the same.

For example, break the song down into four- or eight-bar sections and name each section so that it is easy to identify and locate it. As you have already developed your first chord sequence (the A section), you need to concentrate on the next section (the B section). Here are some examples you can use to create a new section:

- *Use the same chords but vary the style of playing*. If you have already used sustained chords, try using broken chords or arpeggios to create a rhythmic pattern.
- *Use the same chords but change their inversion*. Try changing the inversion of each chord.
- *Create different chords with the same bass line*. Try a different chord sequence over the same bass line.

In the example in Fig. 6.1.17, we have created three different sections:

- Section A uses broken chords to create a rhythm
- Section B uses sustained chords
- Section C uses arpeggios to create a rhythmic pattern.

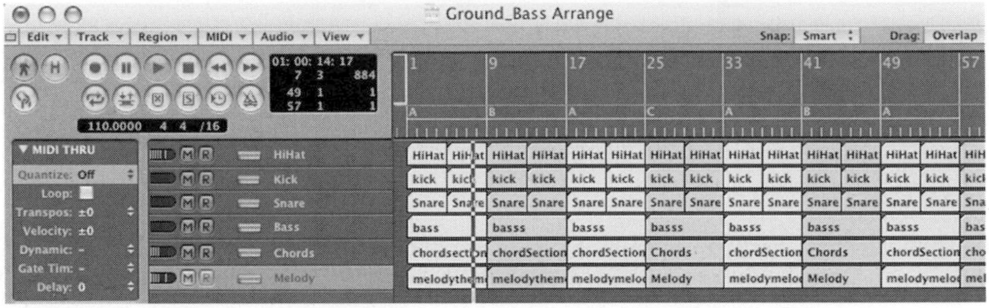

Fig. 6.1.17 – Arrangement based around three different sections.

note ► This is usually a very creative stage, as the sounds and style of the chords you play will often completely change the feel of the track.

Markers

Using markers allows you to easily identify each section on screen. They also allow you to locate to different sections of the song more easily.

Fig. 6.1.18 – Markers defining different sections of an arrangement.

Stage 4. Creating a melody

Create a new MIDI track and select a lead or solo sound such as synth brass or square wave. You need to use a sound that will cut through when played over the chords and bass line.

The melody needs to work with the chord structure and change with the different sections of the song, so to get yourself started you could try using some of the notes being used in the bass line or chord sequence and simply play them one or two octaves higher on the keyboard.

Fig. 6.1.19 – Example melody.

producer says ▶

Melodies can be very simple. Some of the best-known melodies only contain a few notes and are rhythmically simple and can easily be whistled, or sung back and remembered.

Practise playing your ideas along with the drums, bass and chord sequence to see how they work together. Once you have a melody that works, record it into the sequencer. Once you've recorded your melody, edit and quantize it, then build it into an arrangement by copying and pasting.

A melody is the top line and often the main focus of a piece, so it needs to be interesting. Think about how you can make it unique – introduce some originality into your music.

note ▶

The melody will need to vary over the different sections of the song, so you may need to create several different melodies or variations.

Stage 5. Arranging the song

The arrangement needs to sustain interest over time. Listening to the same eight- or 16-bar loop will soon become boring if the musical content is basically the same throughout. You therefore need to use your arranging skills to create some variation. You could, for example, try muting out certain instruments such as the bass or pad for four to eight bars at a time or only allow certain combinations of instruments to play together during certain sections.

tip ▶

Try muting out certain instruments to create variation.

If you have already defined two separate sections in your arrangement you can extend this by simply repeating both sections again. The arrangement would then look something like this:

Example arrangement using two different sections: A → B → A → B

If you have defined three separate sections in your arrangement, it's possible to build up a more complex arrangement. For example, instead of playing back each section in order, you could repeat section A so that it plays again after section B and then add section C. You could then repeat the first three sections again:

Example arrangement using three different sections :
A → B → A → C → A → B → A

Creating an ending

Decide how you want to end your composition. You could, for example, gradually drop each instrument out one by one so you are only left with one instrument playing at the end. Alternatively, you could create a 'fade out'. This is where you loop a section near the end of the composition indefinitely and then after a while simply reduce the master volume on the mixer so that the whole piece of music gradually decreases in volume until nothing is heard.

Stage 6. Mixing and adding effects

At this stage you may want to check if you can hear each instrument clearly and adjust any levels accordingly. You will have created a rough mix while you have been recording, so this stage is really to improve on that.

tip ▶

Be aware that adjusting the volume of the drum track may affect all the drum sounds. This is because all the drums are usually on the same MIDI channel. One solution is to adjust the velocity being sent to each individual drum sound.

You may want to add some reverb to the drums or melody sound to create some ambience and distance the sound. A chorus effect could be used to create a rich texture to thicken the chords or bass sound.

producer says ▶

The last piece of advice on this project is something which we will reiterate throughout the book – that of ideas. No matter how much you are taught, no one can write the music for you!

Well done! You have completed the ground bass project. Make sure you have saved your work and logged any settings you may need.

Project 6.2 Blues

Before attempting this project we recommend you complete Exercises 5.1–5.7, 5.9, 5.11 and 5.12. *You may also need to refer to Chapter 4 to brush up on anything you're not sure of.*

Introduction

Blues is a form of African-American music from the USA. However, as with many popular forms of music, different cultures have adopted it and now perform it. The lyrical content of this music traditionally concerned bad events or situations which people were facing in life. This was often emphasized by using 'blue notes', or flattened notes. These are notes that clash or are outside a more traditional major scale, which often emphasize the feeling of depression in the music. As African-American people's lives changed, so did the music. Hence blues from the Deep South of the USA has a very different sound to that of the North in urban areas such as Chicago.

note ▶ 'Blue notes' are flattened notes, which can be used to emphasize the feeling of depression in the music. Usually, these are notes that clash or are outside a more traditional major scale.

Another distinct characteristic often used in blues music is the 12-bar sequence. This is a sequence of three chords that are played back in a predetermined order over 12 bars. It works by moving between each chord at a very precise moment in time and only using chords that are a 5th and 7th above the one you started with. For example if you started in C you would move up to F and then G (see the example blues structure below). Once you become familiar with the structure, a 12-bar sequence, you will see that it can be played in any key providing the interval between each chord remains the same.

To understand how the different elements of a blues track work together, it may help if you can try to imagine how a live band with drums, bass guitar, piano and saxophone would play together. The rhythm section would play the same 12-bar chord sequence over and over again while a solo instrument would provide a melody. When performing the melody it is common to slide between notes and incorporate 'blue notes'. When using a MIDI keyboard it is possible to achieve a similar effect by using the 'pitch bend' wheel or slider or by adding/editing this information in your sequencer.

Chord	Number of bars
C	4
F	2
C	2
G	1
F	1
C	1
G	1

Fig. 6.2.1 – Structure of a 12-bar seqence in C.

The project ...

- **Stage 1.** Creating a drum track
- **Stage 2.** Creating a bass line
- **Stage 3.** Adding chords
- **Stage 4.** Creating a melody
- **Stage 5.** Arranging the song
- **Stage 6.** Mixing and adding effects.

Getting started

Load your MIDI sequencing software. Create a new project or choose Save As from the file menu. Navigate to a suitable location on the computer's hard drive where you want to save your new song or project. Give the song or project a name, e.g. 12-Bar Blues, then press Save.

tip ▶ Always remember to name the file you are working on and save it in a location on the hard drive where it can be easily located.

Stage 1. Creating a drum track

1. *Create a minimum of six MIDI tracks.*
2. *Assign the first three MIDI tracks to play drum or rhythm sounds*, and name them Kick, Snare and Hi Hat.

tip ▶ It is usually a good idea to start sequencing with drum or rhythm sounds, as this will provide a solid timing base to build other instruments on.

3. *Set the tempo to around 120 BPM* (or to a speed you feel comfortable with) and the time signature to 4/4 before recording.

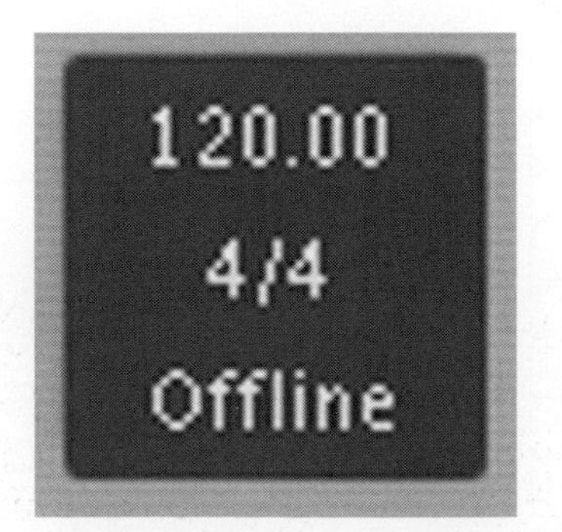

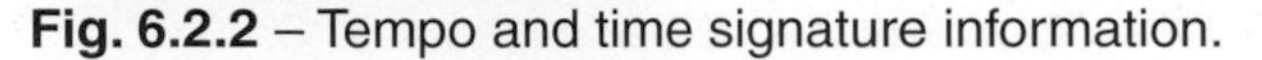

Fig. 6.2.2 – Tempo and time signature information.

4. *Select the first MIDI track and enable the record ready switch.* The record ready switch enables MIDI from the keyboard into the sequencer.

Fig. 6.2.3 – Diagram showing the kick drum track selected with the record ready switch enabled.

Now try playing different keys on the keyboard until you find a kick drum or snare drum sound you want to use (you should usually find a different drum sound on each key of the keyboard).

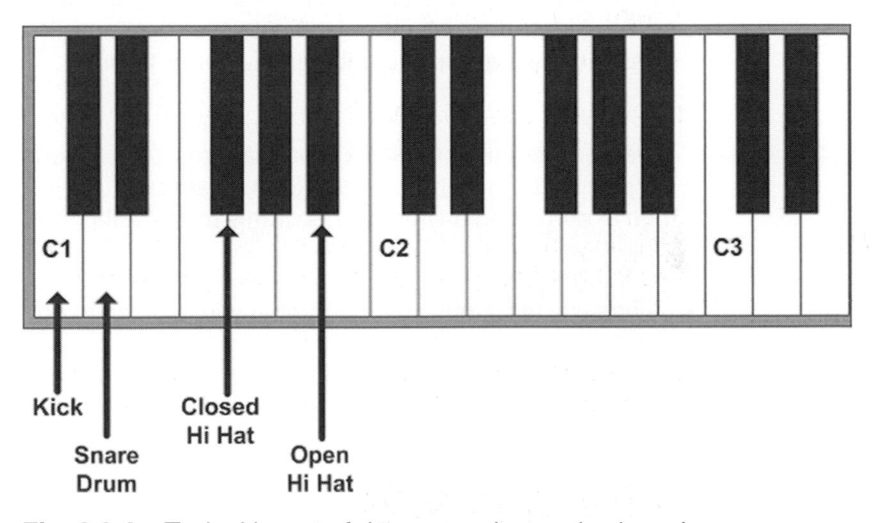

Fig. 6.2.4 – Typical layout of drum sounds on a keyboard.

note ► When using a GM sound source, the drum sounds will always conform to a standard layout across the keyboard. For example, the kick drum will always be assigned to C1, the snare drum to D1 and the closed hi hat to F#1. However, you may find that this layout may vary when using sounds from a drum or percussion set on a sound source that does not conform to the GM standard.

5. *Once you have located the sounds you want to use, practise playing a rhythm in time along with the click/metronome.* The metronome will provide a regular pulse or beat at the chosen tempo to play along with – see notes on setting up a metronome, p. 74.

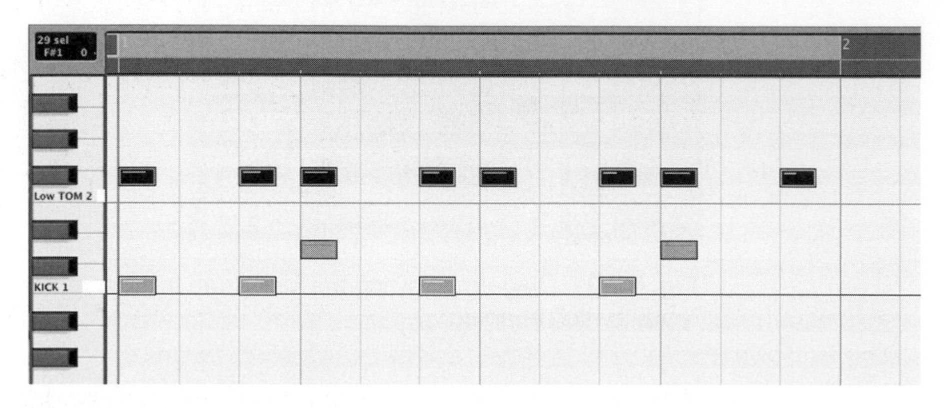

Fig. 6.2.5 – Drum pattern displayed in the Piano-roll-style edit window.

In our example we started with a kick drum playing on beats 1 and 3. We then added a snare drum on beats 2 and 4, and finally added a syncopated hi hat pattern. Note the grid is set to 12ths.

tip ▶ When building up a complex drum pattern, it's often easier to start by placing a kick drum on beats 1 and 3 and a snare drum on beats 2 and 4, and then overdubbing any additional drums to make the rhythm more complex.

6. Once you're clear in your mind what you're going to do, *stop the sequencer and return to the beginning of the song.* You are now ready to record. Make sure the metronome is switched on and press Record on the sequencer.

note ▶ When recording information into a MIDI sequencer always ensure that the click or metronome is switched on.

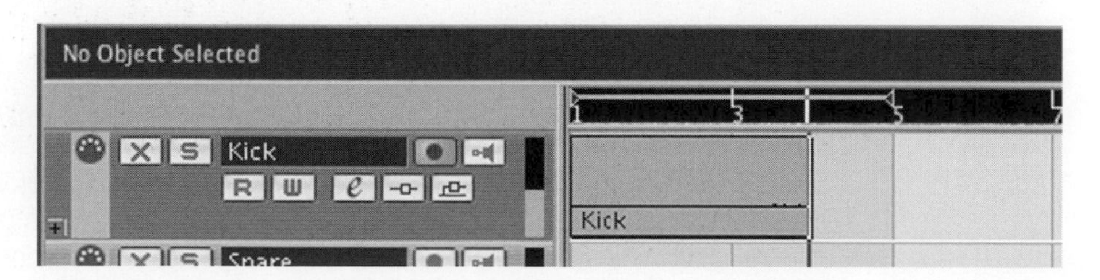

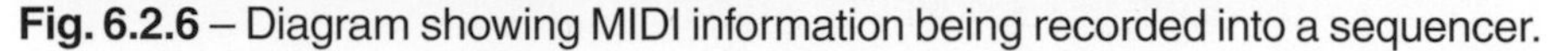

Fig. 6.2.6 – Diagram showing MIDI information being recorded into a sequencer.

7. *Play the keyboard until you have recorded a good selection of ideas then listen back.* Is it in time? Are you happy with the rhythm? If you're not, delete the part and record it again.
8. When you are happy with the performance, *edit what you want to keep into a continuous part or sequence and delete what you don't want to keep.*

Fig. 6.2.7 – Edited hi hat information.

9. Proceed to the next track and continue building up your drum track until it contains a kick, snare and hi hat.

tip ▶ Try to record each drum sound onto a separate MIDI track in the sequencer, as this will make editing and quantizing easier.

Using Quantize

A blues rhythm should swing and have a lazy feel, so you may have to experiment using different Quantize values in order to achieve this. For example, if the kick drum pattern falls on beats 1 and 3, you could use a straight Quantize of 4 or 8. However, if the kick drum is more rhythmical, you could try a value of 12 to introduce a swing. The snare drum usually provides a strong emphasis on beats 2 and 4, so try using a Quantize of value 4. The hi hats can be straight or have a swing feel, so experiment with 8ths or 12ths. Some sequencers even have specific swing Quantize controls that allow you to determine the amount of swing that is added to a rhythm. For more information on using Quantize and swing, refer to Exercise 5.6.

Now build up a 12-bar arrangement by copying and pasting the drum tracks you have created. Remember to name the tracks you are using so it can be identified and to save your file to disk.

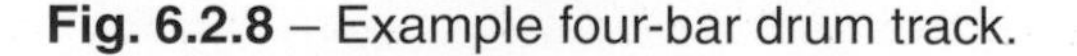

Fig. 6.2.8 – Example four-bar drum track.

producer says ▶

> Are your drums the same all the way through? Is this just a little bit boring? Create some more drum tracks and try overlaying some crash cymbals at the beginning of each 12-bar section or to highlight bars 9 and 10. Try using a ride cymbal instead of a hi hat to create variation and try adding some drum 'fills'.

Stage 2. Creating a bass line

Select a new MIDI track and assign it to a bass sound. Avoid using the same MIDI channel as the drum parts. Try and select a realistic bass sound, such as acoustic bass or fingered bass.

A blues bass line is often based around a repetitive riff or sequence of notes. Different notes can be used to form a scale or arpeggio style movement. A classic example of a 12-bar bass line in the key of C is shown in Fig 6.2.9. This sequence is based over two bars and uses the notes CEGAA#AGE. Notice how all the notes are the same length and how they are placed on each beat of the bar.

note ▶ A blues bass line should provide a strong rhythm and groove.

Fig. 6.2.9 – Typical blues bass sequence.

The example we used on the CD is shown in Fig. 6.2.10. The bass line is also based around the root note of C but alternates between the octave, fifth and seventh notes of the scale.

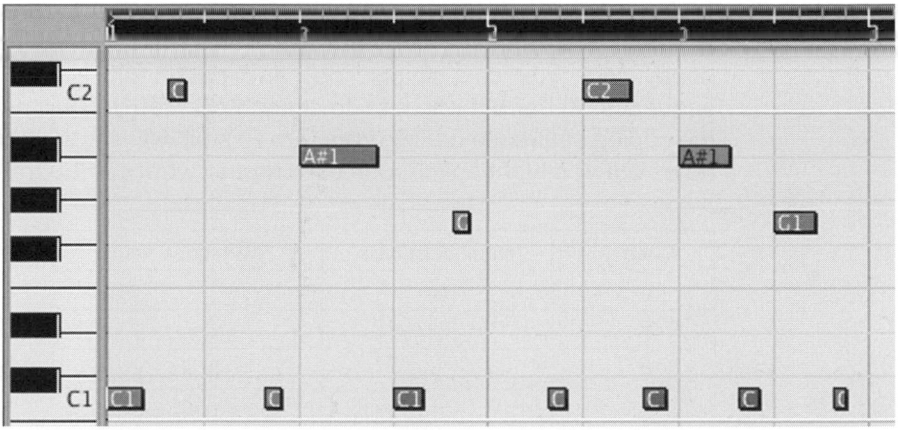

Fig. 6.2.10 – Blues bass line used on the CD.

Try and develop a simple two-bar phrase that will repeat over and over. Once you have developed an idea you will need to consider how the pitch of the phrase will need to change in order to create a 12-bar sequence. For example, if you start in C you will also need the same sequence in F and G.

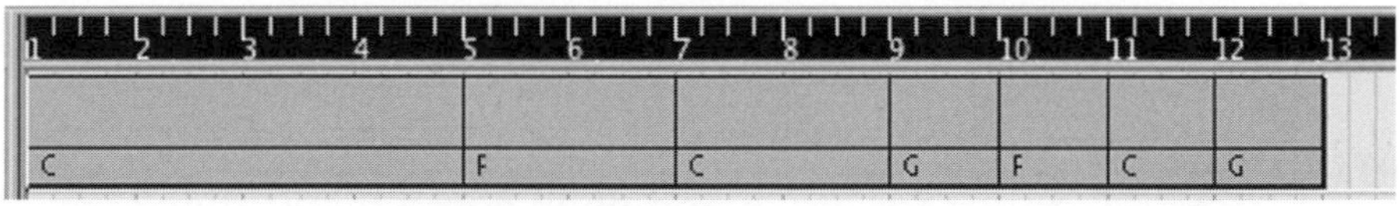

Fig. 6.2.11 – Typical structure of a 12-bar sequence.

note ▶ To create a classic *12-bar sequence* the pitch of the phrase will need to change at regular intervals.

tip ▶ Create a 12-bar cycle to practise playing your bass ideas along with the drum track.

Once you're clear in your mind what you're going to do, return to the beginning of the song and *drop into Record*. When you feel like you have recorded something worth keeping, edit and arrange it into a continuous part or sequence. Try quantizing the bass to 8ths or a swing value so it grooves with the drum track.

Using Transpose

With most 12-bar compositions you are often repeating the same sequence of notes over and over again but in different keys. It therefore may be easier to copy your initial phrase and then use Transpose to change the key. For example, if you have recorded a two-bar phrase starting in C that you are happy with, you can simply copy and paste this so it creates the next sequence. At bar 5 it will need to change pitch so you will need to transpose the copied sequence up +5 (from C to F) so it will play the same sequence of notes but at a higher pitch. Use this chart to work out the transpose values.

Chord	Number of bars	Transpose value
C	4	
F	2	From C+5
C	2	
G	1	From C+7
F	1	From C+5
C	1	
G	1	From C+7

Fig. 6.2.12 – Chart showing transpose values.

tip ▶

Transpose will allow you to easily repeat the same sequence or riff in three different keys.

producer says ▶

Stuck for ideas and need some inspiration?

You could always try step time recording or drawing in notes using a pencil. As a blues bass line often uses a set pattern of notes of the same length, you may find it easier to enter notes on a time-based grid.

At this stage you should have a track drum and bass line that last for 12 bars.

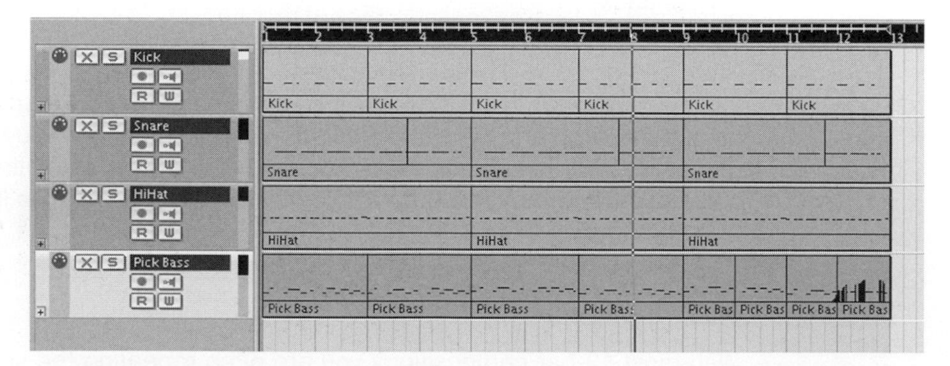

Fig. 6.2.13 – 12-bar bass and drum pattern.

Stage 3. Adding chords

The chords you will need to use will usually be determined by the notes being used in the bass line. If the bass line is based around the scale of C or the most commonly used note is a C then a C major chord will probably work. As the pitch of the bass sequence changes throughout the 12-bar sequence the chords will also need to change, so normally you will find that you will need to use three different chords. In our example we used the chords C, F and G.

note ► The chords will need to change throughout the 12-bar sequence to match pitch of the bass sequence.

The rhythm the chords create is also important as they can be played straight, broken or in a 'boogie woogie' style, so it's worth experimenting playing the chords using different rhythms to determine what works best. Listen to the track on the CD for an example.

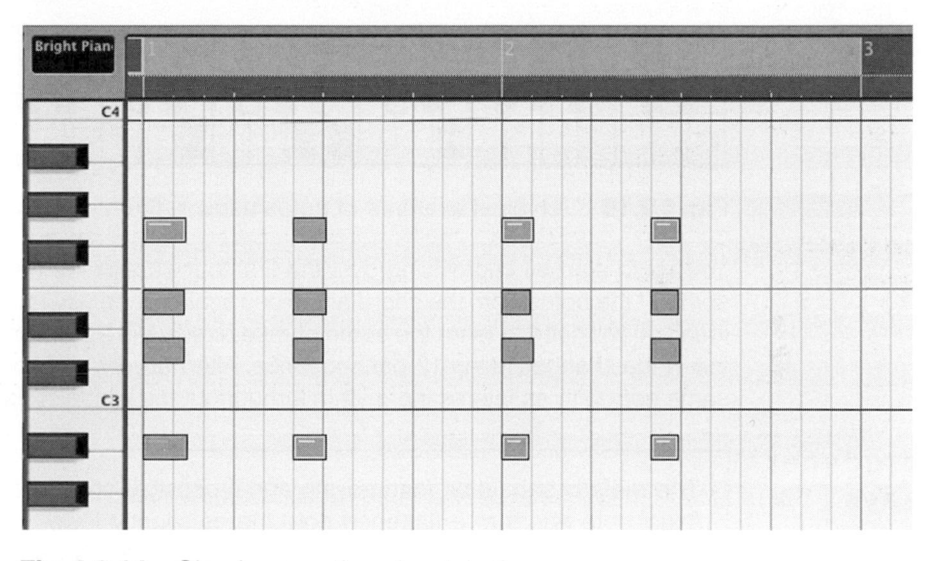

Fig. 6.2.14 – Simple repeating chord rhythm.

You can also experiment using different types of chords such as major or minor 7ths. Try adding a 7th note to a chord to give more character. For example, with the chord of C, try adding a B^b.

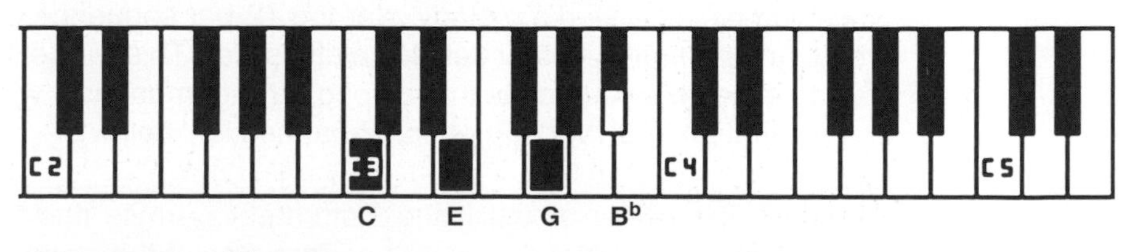

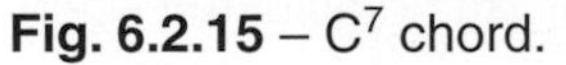

Fig. 6.2.15 – C^7 chord.

When you feel like you have recorded something worth keeping, edit the chord sequence into a 12-bar phrase.

tip ► If you're finding it difficult to play a complex chord rhythm, why not try slowing the sequencer's tempo down to around 90 BPM to do the recording and then speeding it back up to 120 for playback.

Stage 4. Creating a melody

Select a new MIDI track and assign it to a solo or lead instrument such as a saxophone or harmonica. Avoid using the same MIDI channel as the drums, bass and piano parts.

A blues melody is all about a riff or improvisation around a strong theme. Ideally the notes used for the melody should be taken from a blues scale. A typical example of a blues scale in C would use the notes C, E^b, F, F$^\#$, G, B^b, C.

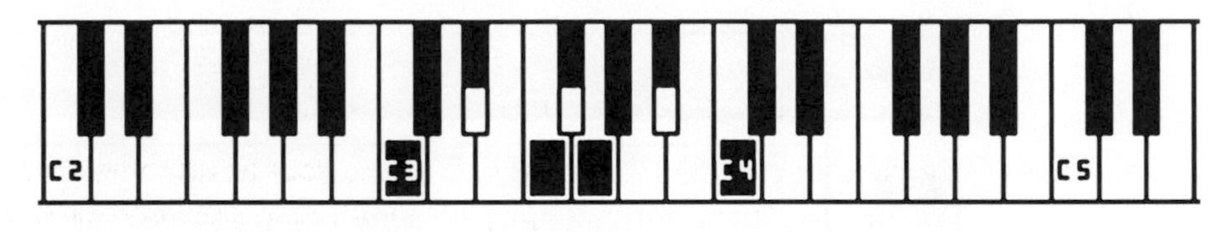

Fig. 6.2.16 – A typical example of notes used in the blues scale of C.

You could start by improvising over the first four bars of the rhythm track using some of the notes from the chord and blues scale. Once you have developed an idea you will need to learn the same phrase or scale in a different key in order to follow the changes in the 12-bar sequence. Alternatively, you could try using the same note with an interesting rhythm throughout the whole 12 bars.

producer says ► The melody should be fast moving and if possible contain a blue note. A 'blue' note refers to a flattened note that is slightly lower in pitch than it should be. On a MIDI keyboard this can be achieved by gently moving the 'pitch bend' wheel down to decrease a note's pitch. The pitch bend wheel is usually located on the left-hand side of a keyboard and allows you to slide a note up and down in pitch just like you can on a guitar (see Exercise 5.8).

Once you have created a melody over the 12-bar sequence you may want to create an alternative idea or develop a riff as well. This will be useful when you start to assemble your ideas into a song arrangement as it will help to create some variation (see 'Varying each 12-bar section' below).

tip ► Try sliding between notes if the instrument permits this to create 'blue notes'. Try using flattened 3rd, 5th or 7th notes of the scale.

Stage 5. Arranging the song

At this stage you should have recorded several different MIDI tracks. Use Copy and Paste to duplicate the first 12-bar section several more times to create an arrangement lasting around 2 minutes. As you're working in 12-bar chunks you should aim to create around *48 bars*.

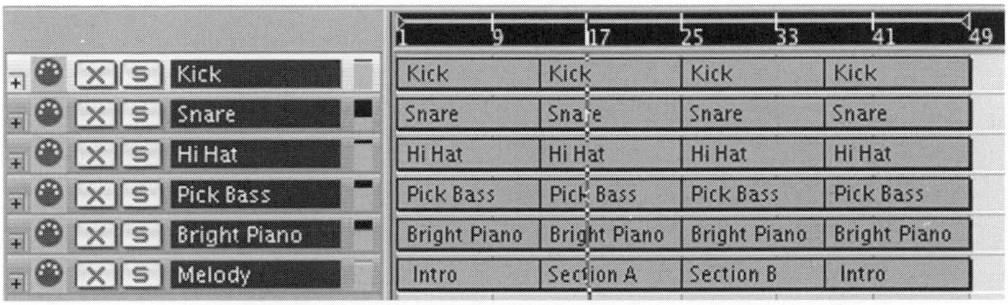

Fig. 6.2.17 – A 12-bar sequence copied four times.

Varying each 12-bar section

Each 12-bar section should vary to make your arrangement sound more interesting. You could, for example, try alternating between different solo instruments, so you hear a harmonica solo for the first 12 bars then a saxophone or piano solo for the next 12 bars. Alternatively, the first 12 bars could be based around a riff whilst the next section could be more random and varied or improvised. Note: in our example in Fig. 6.2.17 we use the first 12 bars as an introduction. Once you have created some variation, listen to your arrangement from the beginning to see how the sections work together.

tip ► Each 12-bar section should vary from the next so try alternating between different solo instruments.

Creating an ending

Decide how you want to end your composition. You could, for example make all the instruments come to a sudden stop or create a 'fade out' by gradually decreasing the volume of the whole piece of music until nothing is heard. Listen to our example on the CD.

Stage 6. Mixing and adding effects

As this is the final stage you need to listen carefully to the balance of the different instruments. Can you hear everything clearly? You may need to adjust the volume of the solo instruments, for example, so they cut through over the top of the rhythm track, or turn up the bass line so it drives the music. Also consider adding some reverb to the solo instruments to create more space and ambience around them.

You should now have completed a blues arrangement. Make sure you have saved your work and logged any settings you may need.

Project 6.3 Dance

Before attempting this project we recommend you complete Exercises 5.1–5.7, 5.9, 5.11 and 5.12 and *refer to Chapter 4 to brush up on anything you're not sure of.*

Introduction

Dance music usually has a fixed tempo of around 110–135 BPM and provides a strong regular beat that people can dance to. Most electronic sounding dance music is based around a drum pattern and uses regular repeating sequences and rhythms. This makes it an ideal style of music to create using a computer, as sections can be easily repeated and rhythms quantized to create a tight mechanical-type feel.

The sounds used are often 'synthetic' and sound artificial. The drums tend to sound 'electronic' and have a very short attack and release time which result in a very punchy feel. The rhythm section is often dominated by the kick drum which provides a strong regular pulse. Other sounds commonly used are luscious pads for chords and thin-sounding synth instruments for lead lines and melodies.

Another style of music that often falls into the dance music category is called the remix. This is essentially the same as dance music but contains sections or samples of audio that have been taken from another song. It is created by taking or sampling the main elements of a particular song and then building these elements into a new song. This allows you to interpret a song in any way you choose and be very creative. For example, you could take a vocal line or main phrase from an existing song and then build up a new drum and bass pattern to create a new song that sounds very different from the original.

The project ...

- **Stage 1.** Creating a drum track
- **Stage 2.** Creating a bass line
- **Stage 3.** Creating a chord sequence
- **Stage 4.** Creating a melody
- **Stage 5.** Arranging the song
- **Stage 6.** Mixing and adding effects.

Getting started

Load your MIDI sequencing software. Create a new project or choose Save As from the file menu. Navigate to a suitable location on the computer's hard drive where you want to save your new song or project. Give the song or project a name, e.g. Dance Project then press Save.

tip ▶ Always remember to name the file you are working on and to save it in a location on the hard drive where it can be easily located.

Stage 1. Creating a drum track

1. *Create four new MIDI tracks* or alternatively you can use any existing MIDI tracks. We will be creating more tracks later.

Fig. 6.3.1 – Four MIDI tracks assigned to a drum sound.

2. Now *assign each MIDI track to a drum sound* (drums are usually assigned to MIDI channel 10 as this is the GM standard). For more information on assigning sounds within the computer and virtual instruments, see Exercise 5.1.

producer says ▶ It's common practice to record each drum sound or percussion instrument onto a separate MIDI track in the sequencer. This gives you much better control when editing and quantizing individual sounds.

3. *Set the time signature to 4/4 and BPM to around 134.*

Fig. 6.3.2 – Tempo and time signature.

4. *Create a cycle* lasting one or four bars. Simple one-, two- and four-bar loops are common in dance music.

Fig. 6.3.3 – Four-bar cycle enabled and track 1 selected.

5. Select the first MIDI track and *enable the record ready switch.*

note ▶ The record ready switch enables MIDI from the controller keyboard into the sequencer.

6. *Using your MIDI controller locate a kick drum sound.* When using a GM drum kit you should find a different drum sound on each key of the keyboard. You should find the kick drum on key C1.

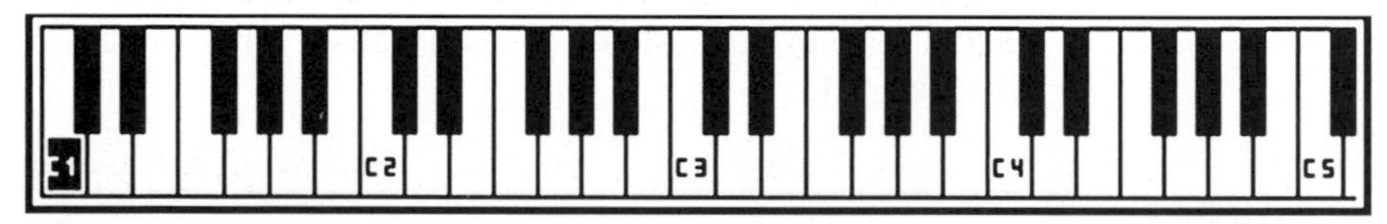

Fig. 6.3.4 – The kick drum is normally assigned to C1 on a keyboard.

note ▶ If you're using a non-GM sound source, make sure a drum or percussion set is chosen as the sound source.

We are going to start with a typical kick drum pattern used in dance music, called four to the floor. This means a kick drum will be placed on each beat of the bar. As the time signature is set to 4/4 there will be four kicks in each bar.

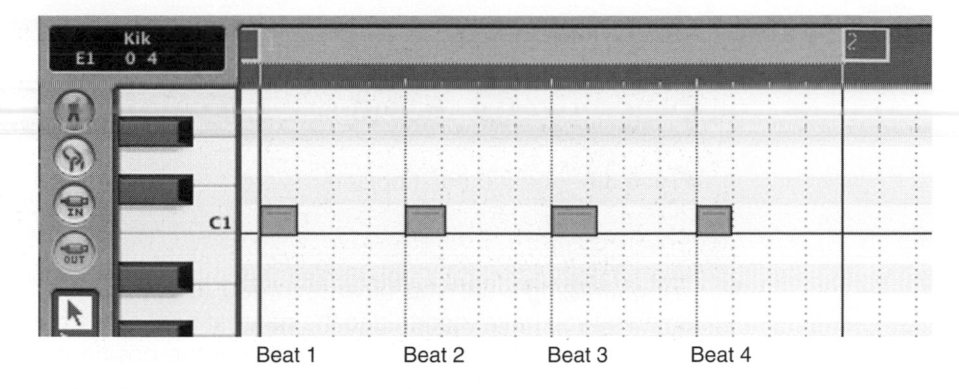

Fig. 6.3.5 – Four-beat kick drum pattern.

tip ▶ Try drawing the notes in with the pencil or use step time recording. See Exercises 5.11 and 5.12.

7. *Switch the metronome on and press Play on the sequencer.* You should now hear a regular pulse or beat at the chosen tempo to play along with. The metronome is usually set up to play four beats in a bar, so use this as a guide. Try playing the kick in absolute time as if you are a robot, four notes per bar, as with the click/metronome. When you're clear in your mind what you're going to do, stop the sequencer and return to the beginning of the song.
8. *Now drop into Record* and play the kick drum pattern into the sequencer. When you feel like you have recorded something worth keeping, *press Stop* and return to the beginning of the song.

note ▶ If you're recording while in Cycle, remember to stop playing once the cycle is complete.

9. *Press Play* to listen back to what you have recorded. Is it in time? Are you happy with it? If not, delete the part and re-record it again.

tip ▶ It's always a good idea to drop into Record, even when you are practising, as you might get lucky and record the perfect take.

10. Once you have got a recording you are satisfied with or a section worth keeping, separate the sections you want to keep by cutting/trimming the data with the scissors. Delete what you don't want to keep from the sequencer then arrange what you want to keep into a continuous part or sequence (glue together). Try to create a section that is exactly four bars in length, then move this sequence so it plays back from bar 1.

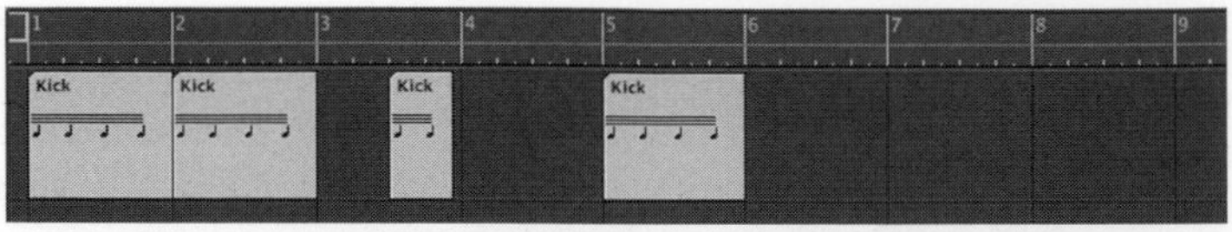

Fig. 6.3.6 – Edited MIDI objects.

11. *Try quantizing* what you have recorded to 1/4 notes. This will position anything you have recorded to the nearest beat of the bar. Even if the timing is poor this should help as it will limit the positions in a bar where a MIDI event can fall to just 4.

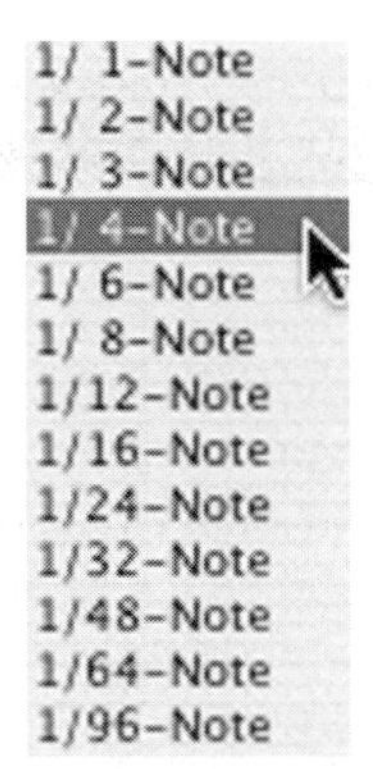

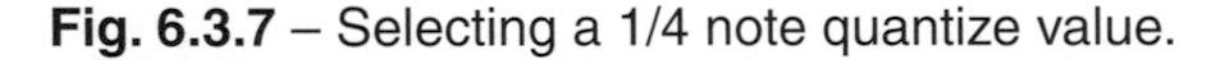
Fig. 6.3.7 – Selecting a 1/4 note quantize value.

tip ► Most parts of this composition should be quantized to help create a mechanical-type feel, as this is the most suitable setting for this style and genre of music.

12. If you have created a four-bar kick drum pattern you could try and introduce a slight variation at the end of every four bars by adding a double beat directly after the last kick drum.

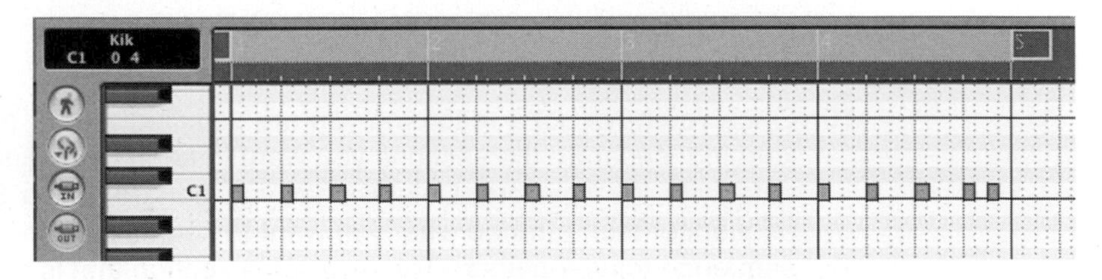

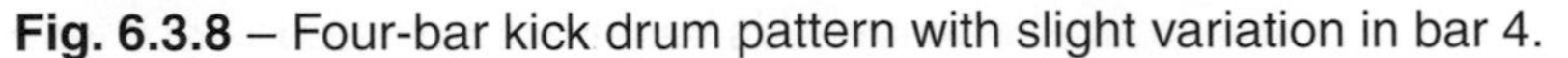
Fig. 6.3.8 – Four-bar kick drum pattern with slight variation in bar 4.

13. Once you have completed the kick drum pattern and created a four-bar sequence, *select a new MIDI track and locate a hi hat sound.* A hi hat can be open or closed and usually appears on two or more keys. If you are using a GM drum kit, the closed hi hat will be located on F#1 and open hi hat on A#1. Try and alternate between the open and closed hi hat to form a regular 1/8 pattern by placing the open directly after the closed in the same beat, e.g. closed – open – closed – open – closed – open – closed – open. Try counting to '8' in one bar and playing at the same time.

note ► When using a GM drum kit you should find a closed hi hat on F#1 and an open hi hat on A#1.

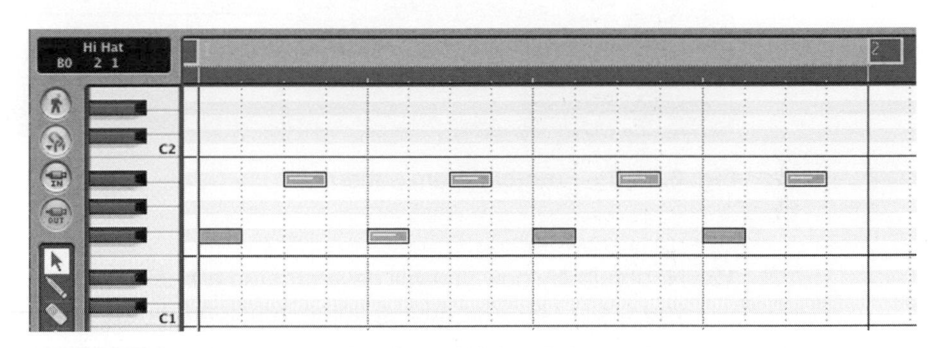

Fig. 6.3.9 – Example one-bar hi hat pattern.

14. *Now drop into Record* and record the hi hat pattern into the sequencer. When you feel like you have recorded something worth keeping, press Stop and return to the beginning of the song.

15. *Press Play* to listen back to what you have recorded. Is it in time? Are you happy with it? If not, delete the part and re-record it again. Try quantizing the hi hats to 8ths. This will position the hi hats to each 8th of the bar. Even if the timing is poor, this will help as it will limit the positions in a bar where a MIDI event can fall to just eight.

note ► Quantizing does not put your music in time, it just moves the notes to the nearest selected gridline. Therefore selecting the correct Quantize value is essential.

16. Once you have completed the hi hat pattern and created a four-bar sequence, *select a new track and find a hand clap sound.* A hand clap is meant to sound like a real hand clap, but they often sound electronic and synthetic. When using a GM drum kit you should find the hand clap on D#1.

17. *Record in the hand clap* on beats 2 and 4 and quantize to 1/4. You could use a snare drum sound instead of the hand clap if you wish.

tip ► Having trouble keeping up with the click? Well, remember you can slow down the sequencer to record your MIDI parts and then speed it up afterwards!

18. *Now edit and arrange your drum tracks.* Copy each drum part to create a four- or eight-bar pattern.

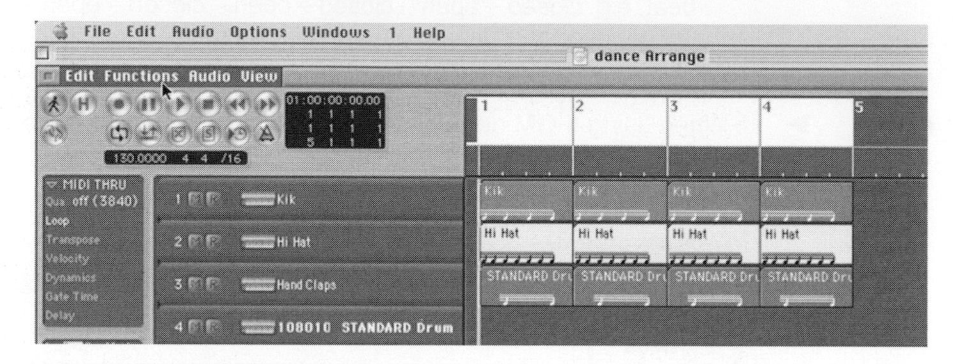

Fig. 6.3.10 – Four-bar drum pattern.

19. Hopefully by now you will have created a basic drum track. At this stage you could consider adding some additional drum sounds on a separate track to create some variation.

A crash cymbal could be used to emphasize the beginning of each 4 bar section or be used to create a crescendo by being placed on each beat of the bar (listen to our example). Alternatively, you could try using a reverse cymbal to introduce a new section or try overdubbing some extra kick drums to introduce a double beat or slight variation at the end of every 4 or 8 bars.

Creating a velocity snare roll

Dance music often features a snare drum roll that gradually rises in volume over several bars just before a new section begins. This production technique is known as a velocity snare roll and is commonly used to join different sections of a composition together and add tension and excitement to the music. It is usually created by entering a very fast robotic style snare drum roll into a sequencer using 8th or 16th notes and then editing the velocity of each note so the drum roll gradually gets louder towards the end of the bar.

Before you can use this technique yourself you will need to create a new MIDI drum track and locate a snare drum sound (a snare drum is usually located on key C1). You will then need to determine the duration of the snare roll and set up a cycle at this location in the song. Generally speaking you will find one, two or four bars work well. Now practice playing the snare drum quickly while observing the sequencer's metronome. Try hitting the drum quickly eight times in each bar. If you are finding this difficult you could try counting to eight as you play. When you are ready, record your roll into the sequencer and then quantize it using an 8th note value and listen to how it sounds.

At this stage you may want to view the notes you have played in Piano-roll-style edit window as this will make it easier to identify any mistakes or timing errors. You may also consider drawing in any additional notes using the pencil tool or entering the notes in step time.

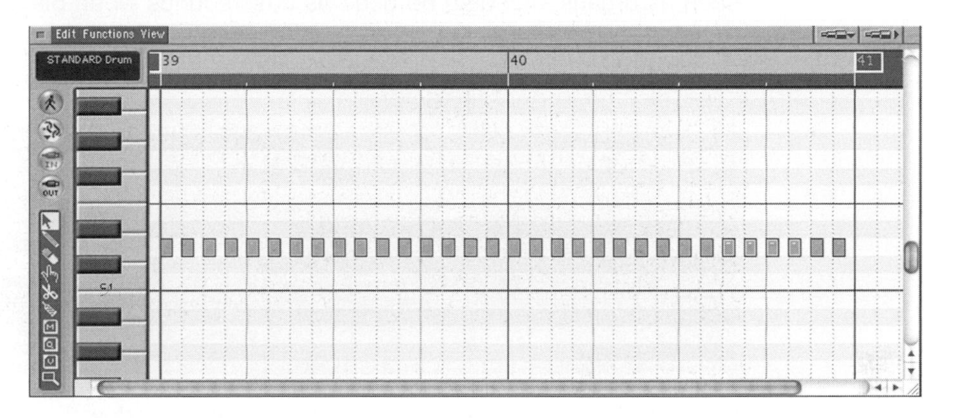

Fig. 6.3.11 – A fast snare drum roll. Notice how a drum is placed on each 16th of the bar.

Once you have recorded and edited a snare roll you are happy with, you then need to edit the velocity. Velocity is measured between 0 and 127, so in order to create a crescendo you will need to start low, around 10, and finish high at 127. Velocity can be edited and displayed in a number of different ways, such as in a list or graphically (see Exercise 5.7), so choose the simplest way for you. In our example we edited the velocity graphically using the pencil tool.

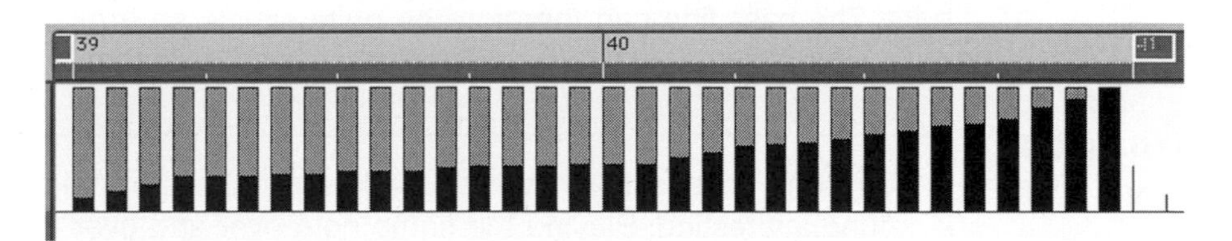

Fig. 6.3.12 – A velocity crescendo displayed graphically.

tip ▶ Now name the track you have been using and save your file to disk.

tip ▶ **Drum loops**

You may want to consider using a drum or percussion loop to add feel and groove to the rhythm. Audio loops are usually at a fixed tempo, so try and choose a loop that matches the tempo of your song.

Stage 2. Creating a bass line

Create a new track and assign it to a bass sound. Avoid using the same MIDI channel as the drum parts. Try looking for a sound which is deep and has a fast attack time and fast decay, i.e. it starts immediately and finishes once you let go of the key. Typically you will find that synthetic-type bass sounds work well in dance music. However, you may find that different types of sounds, such as organs, can also be used as bass sounds when played in the lower register of a keyboard. So, as always, experiment.

Fig. 6.3.13 – Bass track selected in the Arrange window.

note ▶

The bass and the kick drum need to work alongside each other to provide the low-frequency element of the track. If both sounds occupy a similar frequency range it will start to sound muddy and overcrowded in the low end.

Most dance tracks are based around repeating phrases lasting four or eight bars. The bass line can therefore be quite simple so try limiting yourself to only using four to five notes at a time and keep the rhythm simple!

producer says ▶

You don't necessarily have to use lots of notes to make a bass line sound interesting. Playing the same note over and over again is fine if the rhythm is interesting and you are creating a groove. Many popular dance records are based around simple bass lines that have great rhythms.

Find a few notes you like, for example A ,G ,D and C, and then try and create your own four-bar sequence. Practise playing your bass ideas along with the drum tracks to see how they fit together. If it doesn't sound interesting try playing the notes in a different order or in a different key. Once you have found a sequence of notes you want to use, experiment playing them back using different rhythms. You could, for example, try playing each note while counting to four so you place a different note on the beginning of each bar or speed up the sequence by only counting to two while holding each note, so you change pitch every two beats. In our example on the CD we use an off-beat 1/8 rhythm, similar to the ground bass exercise where each note falls on the 2, 4, 6, 8, of each bar.

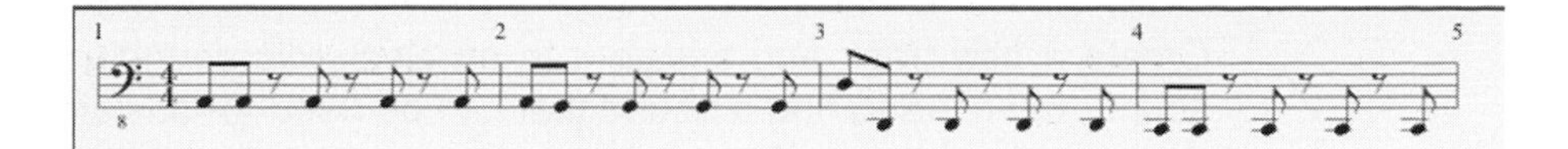

Fig. 6.3.14 – Example bass line.

tip ▶

> It is usually a good idea to record any ideas you may have into a sequencer even if they are not exactly what you want, as once they have been recorded you can edit the pitch and timing of each note.

Once you're clear in your mind what you're going to do, *return to the beginning of the song and drop into Record.* When you feel like you have recorded something worth keeping, edit and arrange it into a continuous part or sequence.

Fig. 6.3.15 – Editing the bass line.

tip ▶

> Always try and edit your ideas into two- or four-bar sections so they work as a loop.

producer says ▶

> Stuck for ideas? You could always try step time recording or drawing the bass notes into an edit window using the pencil tool (see Exercise 5.12). Alternatively, try slowing the sequencer's tempo down to record your bass line and then speeding it back up for playback.

Once you have edited your bass line, try quantizing it using 8ths or 16ths to give it a tight mechanical feel. Then *build up your song arrangement by copying and pasting* it to create a 16-bar sequence. Name the bass track and press Save to update your file to disk.

Stage 3. Creating a chord sequence

Create a new track and assign it to an electronic sounding string or choir sound. You need to find a sound that can be used for sustaining chords and that will fill out the mid range frequency of your composition, i.e. above the bass line but below the melody.

tip ▶ Avoid using the same MIDI channels as the drum and bass parts.

This part usually consists of several notes that are played at the same time to form a chord. As you have already recorded the bass line, you need to find a selection of chords that will work well with this and avoid chords that clash and sound discordant. A good starting point is to try and identify the root notes used in the bass line and then try using chords based around these notes (a root note is the main or most commonly used note). For example, if you have used the note A in your bass sequence, try playing a chord based around this note (e.g. A, C, E). Usually, you will find that chords change more slowly than the bass line (at the beginning of each bar for example), so holding down a chord over a faster changing bass line is common. Listen to our example on the CD.

Once you have found several chords that work with the bass line, try moving between them to create a chord sequence (e.g. A minor, G major, D minor, C major). This is only one example to help you create a chord sequence so, as always, experiment. You may find unusual combinations of chords work equally as well over the bass line.

Fig. 6.3.16 – Example chord sequence and bass line.

note ▶ The type of chords you choose will often determine the mood or feel of a composition. For example, major chords will usually create a happier feel than using minor chords.

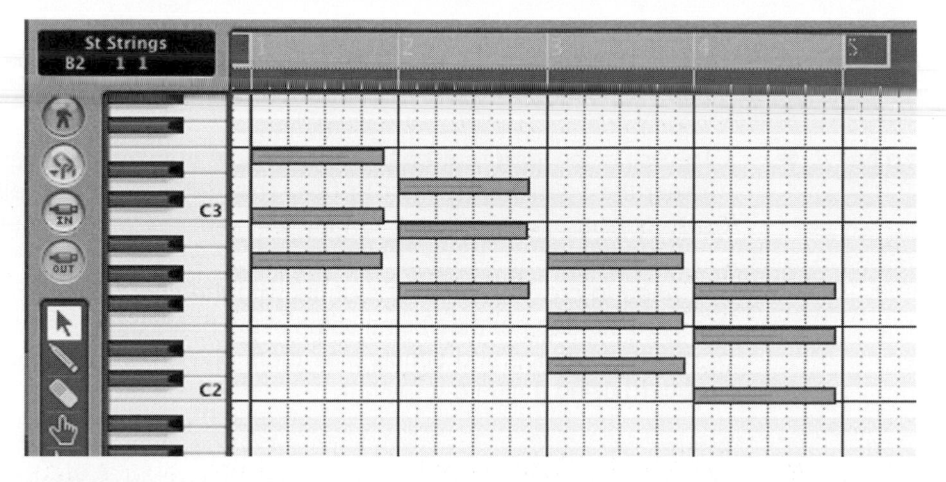

Fig. 6.3.17 – Example four-bar chord sequence.

Once you're clear in your mind what you're going to do, add the pad to the arrangement.

producer says ▶

> In order to create some variation and help define different sections of your composition, you could try and vary how your chord sequence is played every eight or 16 bars. For example, try playing back the same chord sequence using different inversions or transpose the chords so they play back one octave higher (see chords).

Once you're clear in your mind what you're going to do, add your chord sequence to the arrangement. Name the track you have used and press Save to update your file to disk.

Markers

At this stage you may want to try creating some markers to help you identify and locate to different sections of your arrangement more easily.

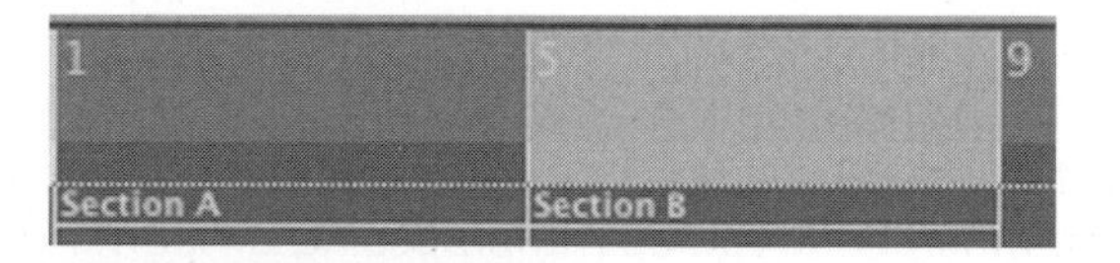

Fig. 6.3.18 – Markers defining different sections of an arrangement.

Stage 4. Creating a melody

Create a new MIDI track and assign it to a sound source. You need to find a sound that is suitable for playing a melody or 'top line'. In dance music this is normally a synthesized sound that is very angular and thin. These types of sounds are often called lead sounds and are usually quite bright sounding. This part needs to cut through, so when playing these sounds try using high notes on the keyboard above C4.

note ▶ Try using the program change command from within your software to select the sounds you want to use. This will ensure that all your sound selections are remembered as they will be saved with the song file.

You need to create a melody that fits with the chord structure and the bass line. As a starting point try using some of the notes being used in the bass line or chord sequence and simply play them one or two octaves higher on the keyboard using a different rhythm.

Fig. 6.3.19 – Example melody shown on the bottom line of the score. See how the bass line and chords support it.

tip ▶ A melody is the top line and often the main focus of a piece, so it needs to be interesting. Try adding some reverb or delay to give the sound more ambience so it floats above the rhythm track.

Practise playing your ideas along with the drums, bass and chord sequence to see how they work together. Once you have created a melody record it into the sequencer, edit and quantize it and then build it into the arrangement by copying and pasting.

producer says ▶ The last bit of advice on creating a melody is something which we reiterate throughout the book, i.e. the value of coming up with original ideas. No matter how much time you spend practising or learning about composing music, no one can write the music for you!

Stage 5. Arranging the song

At this stage you should now have several different MIDI tracks, so use Copy and Paste to create an arrangement lasting at least two minutes (around 50 bars).

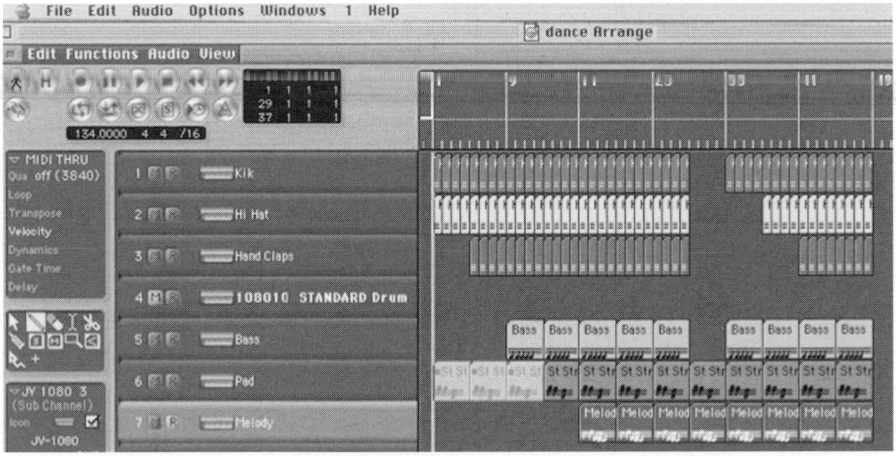

Fig. 6.3.20 – Typical song arrangement.

The arrangement needs to sustain interest over time. Listening to the same 8- or 16-bar loop will soon become boring. As the musical content is often very similar throughout most dance tracks, you will need to use your arranging skills to create some variation. One commonly used technique is to copy or repeat your main instruments several times, then experiment by muting out different instruments to create variation. Here are some examples for you to try:

- Introduce each instrument separately
- Introduce each drum track separately
- Experiment with different combinations of instruments playing together
- Mute out different instruments for four or eight bars
- Mute out the bass line for eight bars.

Your arrangement should be dynamic, so think how you can build it up at certain points and then drop it back down again. The possibilities are endless (listen to our example on the CD).

tip ► Try repeating all the parts several times and then use the mute tool to quickly try out different arrangements.

The breakdown

Most dance music includes a 'breakdown'. This is where the music suddenly drops down into a small section that only contains a few instruments before building up again into a crescendo. This can help create a sense of excitement in the music and add more dynamics to an arrangement. In our example the 'breakdown' section starts with the pad and melody playing for four bars, and then introduces the hi hats and bass line before bringing all the instruments back in. Listen to the example on the CD.

Creating an ending

Decide how you want to end your composition. You could, for example, make all the instruments come to a sudden stop or slowly mute out each instrument towards the end of the arrangement so you end up with only one instrument playing. Alternatively, you could create a 'fade out'. Try listening to examples of other pieces of music to see how they end.

Stage 6. Mixing and adding effects

This is the final stage so check if you can hear each instrument clearly and adjust any levels accordingly. Dance music is usually based around a drum pattern or loop, so make sure the rhythm is cutting through and that the kick drum is loud enough.

tip ▶ Be aware that adjusting the volume of the drum track may affect all the drum sounds. This is because all the drums are usually on the same MIDI channel. One solution is to adjust the velocity being sent to each individual drum sound.

Fig. 6.3.21 – GM MIDI mixer.

The volume of each MIDI instrument can usually be adjusted from within your sequencing software using the GM MIDI mixer or any parameter assigned to control change number 7 (see Exercise 5.9). MIDI volume operates between a parameter range of 0 and 127, so you need to ensure that you leave sufficient headroom to turn the instruments up and down in volume. Ideally you should aim to set the volume of each instrument between 80 and 100.

tip ▶ If all the sounds you are using are coming from one particular device, you will need to balance the volume level of each instrument internally within the module itself.

At this stage you may want to experiment and introduce some effects. Dance music is great for adding delays, as they can be timed exactly to each beat of the music and can add rhythm and distance to a sound. For example, using a stereo or ping-pong delay on a lead sound will make it appear to jump from left to right, whereas a gentle mono delay on the bass will create distance and atmosphere. Delays can be used on just about anything, so experiment. You may want to add reverb to the drums or lead sound to create some ambience and distance the sound, or use automation to create a sudden burst of reverb in certain places throughout the song (see 'Automation' section below). Chorus can be used to thicken a sound and create a rich texture, whereas phasers can be used on pad sounds to create a sweeping-type effect.

note ▶

Reverb and chorus are usually available as standard when using a GM-compatible sound device. You can determine the amount of effect that is added to each sound by using a GM MIDI mixer or by sending a control change message:

- Reverb responds to controller number 91
- Chorus responds to controller number 93.

Automation

As you listen through, you may want to make small changes throughout the arrangement, such as increase the volume of certain instruments during different sections. Automation allows you to record your parameter movements into the sequencer and then plays them back. Simple moves like volume can be recorded and then edited to perfection. You can then add additional moves to build up more complex automation moves. You can also use automation to create special effects such as automating the pan pot so the sound automatically moves from left to right (providing you're listening in stereo) or create a special effect by automating the reverb level in certain places throughout the song.

Saving your work

You should now have completed a dance arrangement, so make sure you have saved your work and logged any settings you may need.

Project 6.4 Reggae

Before attempting this project, we recommend you complete Exercises 5.1, 5.7, 5.9, 5.11 and 5.12. *You may also need to refer to Chapter 4 to brush up on anything you're not sure of.*

Introduction

Reggae music originated from the island of Jamaica and is a combination of traditional mento and American style R&B. It came into being in the late 1950s, when Jamaica was starting to develop its own record industry. Over the years it has influenced many different styles of music.

Reggae music features a heavy backbeat rhythm that often places the emphasis on the second and fourth beats of each bar (rather than the usual first and third). The rhythm is also characterized by regular chops on the off-beat which are usually played using a piano or guitar chord. The bass line is usually quite melodic as well as supportive and will often play short phrases that create a laid-back stop–start-type feel.

An offshoot of reggae is called Dub. This term is now commonly used to describe a remix in dance music but the term originated from Jamaica in the 1960s. A dub mix is an alternative version of a song where the lyrics are edited or removed and the music is stripped down or rearranged.

The project ...

- **Stage 1.** Creating a drum track
- **Stage 2.** Creating a bass line
- **Stage 3.** Adding a chord sequence
- **Stage 4.** Creating a melody
- **Stage 5.** Creating a new section
- **Stage 6.** Song arrangement
- **Stage 7.** Mixing and adding effects.

Getting started

Load your MIDI sequencing software. Create a new project or choose Save As from the file menu. Navigate to a suitable location on the computer's hard drive where you want to save your new song or project. Give the song or project a name, e.g. Reggae track, then press Save.

tip ► Always remember to name the file you are working on and to save it in a location where it can be easily located.

Stage 1. Creating a drum track

1. *Create four MIDI drum tracks* and label them kick, snare, hi hats and percussion.

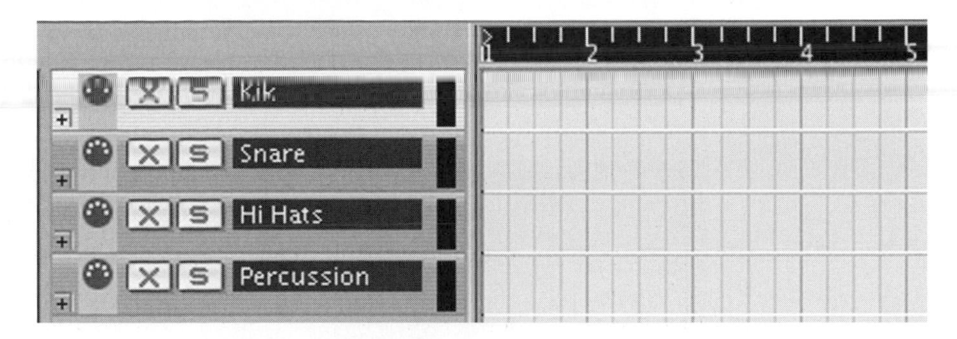

Fig. 6.4.1 – Four named MIDI drum tracks.

2. *Assign each new MIDI track to a drum or percussion sound.* Drums are usually assigned to MIDI channel 10, as this is the GM standard. Sounds may be generated internally within the sequencer or externally on a stand-alone sound module (see 'Sound sources' section on p. 37).

3. *Select the first MIDI track and enable the record ready switch.* The record ready switch enables MIDI from the keyboard into the sequencer.

Fig. 6.4.2 – Track enabled.

4. Now *locate the sounds you want to use* on the keyboard. Try and find a kick, snare and hi hat sound (when using GM drum sounds you should find a different drum sound on each key).

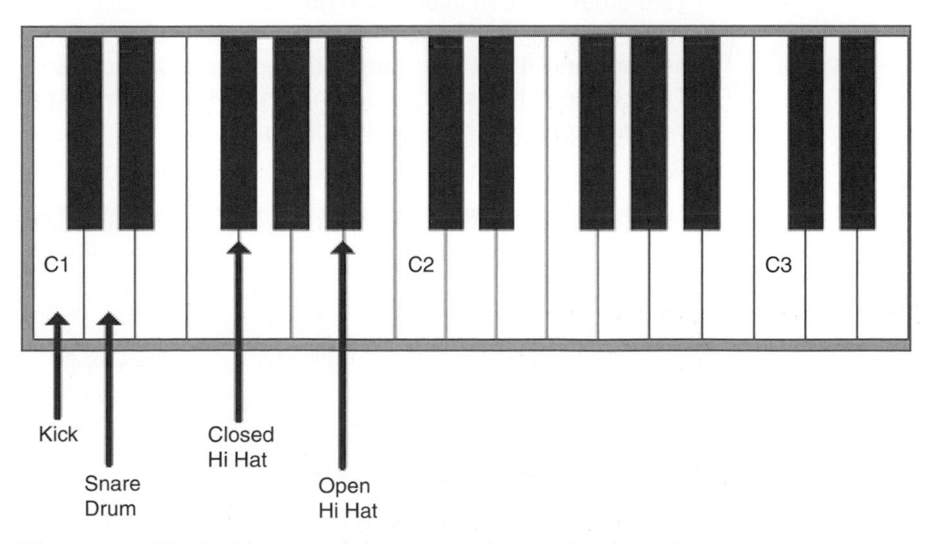

Fig. 6.4.3 – Typical layout of drum sounds on a keyboard.

5. *Set the tempo* between 65 and 95 BPM. The music doesn't want to sound rushed or too fast. Slowing down the tempo will create a more relaxed feel. Set the time signature to 4/4.

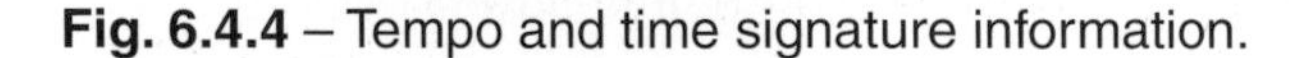

Fig. 6.4.4 – Tempo and time signature information.

6. *Switch on the metronome and press Play on the sequencer.* The metronome will provide a regular pulse or beat at the chosen tempo to play along with.
7. *Now experiment using the kick, snare and hi hat sounds.* You need to create a slow groove that doesn't sound rushed. You could start by trying to create a simple rhythm using either the kick or hi hat sound or try and develop a groove by playing several different drum sounds at the same time. Either way, practise playing in time with the metronome until you have developed an idea.

tip ► Sometimes you may find it easier to create a rhythm by playing several different drums at the same time. Remember, there are no rules, so just experiment.

In our example we placed a kick drum on beats 1 and 3 within each bar. This created a sense of space as the tempo is quite low. (Alternatively you could try placing the kick on beats 2 and 4 instead). The snare drum is on beats 2 and 4 but alternates between two different snare drum sounds. The closed hi hat is based around an off-beat 8th note pattern with a slight variation towards the end of each bar.

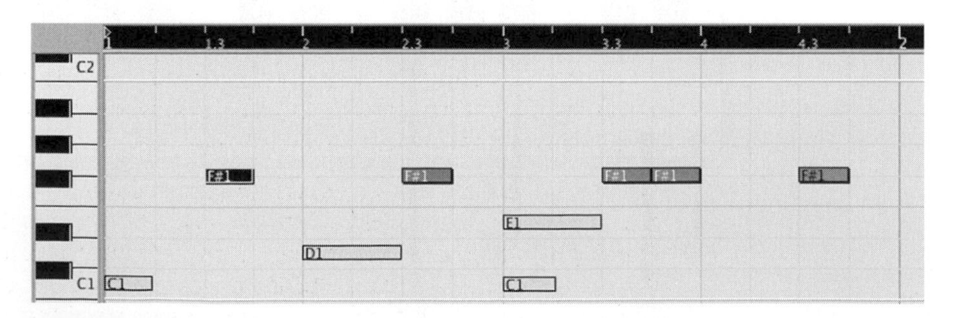

Fig. 6.4.5 – Example drum pattern displayed in the Piano-roll-style edit window.

tip ► Try placing a snare drum on beats 2 and 4 to provide a constant beat to build the kick and hi hat on.

8. Once you're clear in your mind what you're going to do, stop the sequencer and return to the beginning of the song. *Drop into Record* and remember to wait for the count-in before you start playing.

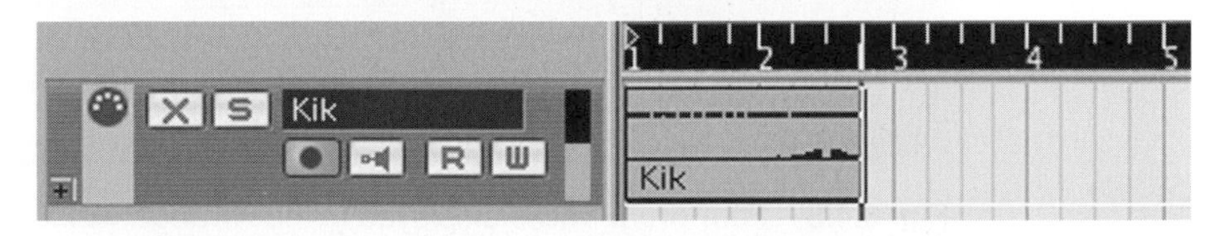

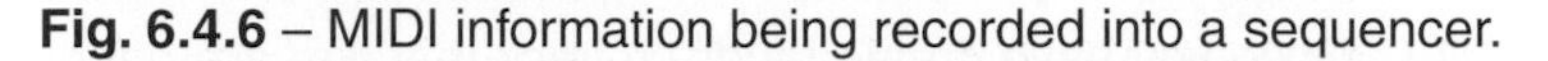

Fig. 6.4.6 – MIDI information being recorded into a sequencer.

9. *Play until you have recorded a good selection of ideas.* Try not to stop recording if you make any mistakes. Just keep playing, as you can edit small mistakes out later. When you feel like you have recorded something worth keeping, *press Stop and return to the beginning of the song.*

10. *Now listen back to what you have just recorded.* Is it in time? Are you happy with the rhythm? If you're not, delete it and record it again. When you are happy with the performance, edit your idea into a continuous one-, two- or four-bar sequence.

Fig. 6.4.7 – Kick drum information edited into a four-bar pattern.

tip ► Try and edit your ideas into one-, two- or four-bar sections.

11. Try *Quantizing* what you have recorded as this will help correct the timing and provide a more solid base to build other instruments on. In our example on the CD, we quantized the drums to 16ths.

12. *Now build up several more drum tracks* using the same method to create a more complex rhythm. Once you have a solid kick, snare and hi hat pattern, try adding some percussion sounds, such as shakers, tambourines, steel drum or timbales. Remember, building up a drum track using individual sounds requires practice so try not to make each of the individual drum parts too busy, as you may start to loose the laid-back reggae feel and it may start to sound like a Latino track.

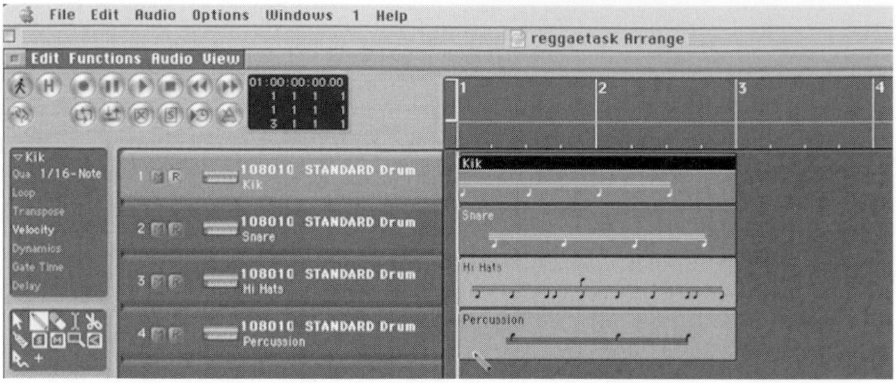

Fig. 6.4.8 – Four drum tracks displayed in the Arrange window.

13. Once you have created a drum track that you are happy with, use Copy and Paste to extend it into a four- or eight-bar sequence. This will provide a rhythm bed to build the other instruments on.

producer says ▶

Making your drum tracks more interesting

Are your drums the same all the way through? They will be if you've just copied and pasted them. Is this just a little bit boring? Well, why not use the matrix editor to vary the kick or hi hat pattern at the end of each four-bar section or try dropping out some of the individual drum or percussion instruments every so often. Alternatively, create some more tracks and add some 'fills' or embellishments or overlay some bongos, toms or cymbals. Be creative and experiment with your drum and percussion sounds.

14. Before starting the next stage, *press Save to update your file.*

Stage 2. Creating a bass line

1. *Create a new MIDI track* and assign it to a sound source.

Fig. 6.4.9 – Bass track selected in the Arrange window.

note ▶ Avoid using the same MIDI channel as the drum parts.

2. *Select a bass sound.* Types of instruments you may wish to try:
   - Fingered bass (GM034)
   - Picked bass (GM035).

tip ▶ Try using the program change command from within your software to select the bass sound you want to use.

3. A reggae bass line is usually quite melodic and has a stop–start-type feel. To get you started you could try using the individual notes from a major or minor chord and play them like a slow arpeggio or melody in the lower register of the keyboard. Once you have found a sequence of notes you want to use, experiment playing them back using different rhythms and timings and try and create some space between the notes. You could try playing an occasional double or quadruple time phrase and then try stopping for two beats. This would create a stop–start feel (double or quadruple timings are notes played very fast.) Listen to our example on the CD.

tip ▶ Try playing a melodic bass line with lots of movement, just like playing a slow melody in the bass register.

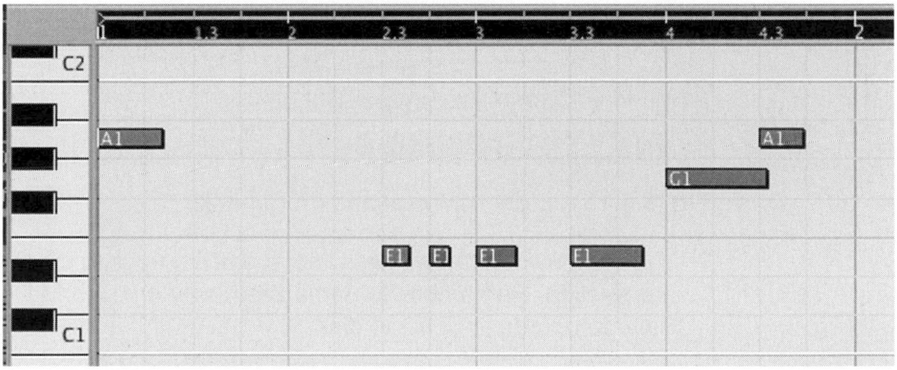

Fig. 6.4.10 – Example bass line displayed graphically.

It is common for the kick drum and bass line to play together on the first beat of the bar. However, as reggae has a more relaxed feel you could try missing the first beat of the bar and start the bass line on beat 2.

tip ▶ Try playing a short phrase of notes then stopping to leave some space before continuing with the next phrase.

4. Try playing your bass line along with the drum tracks to see how they fit together. Once you're clear in your mind what you're going to do, stop the sequencer, return to the beginning of the song and drop into Record.

tip ▶ It is usually a good idea to drop the sequencer into Record even whilst practicing, as inspiration can strike at any time and you may record something worth keeping.

5. Once you have recorded something worth keeping, stop the sequencer and return to the beginning of the song. If there are lots of mistakes when you listen back, delete what you don't want to keep and then arrange what's left into a continuous part or sequence. Alternatively, if what you played is pretty close, you could use an edit window to make any small adjustments to the timing and pitch of the notes. Ideally you should try and edit your bass ideas into a four- or eight-bar phrase so they work as a loop.

tip ▶ Sometimes keeping one small section and repeating it will inspire you to build up a more complex bass sequence.

6. Consider quantizing what you have just played as this may help improve the timing and help lock the song together (see Exercise 5.6).

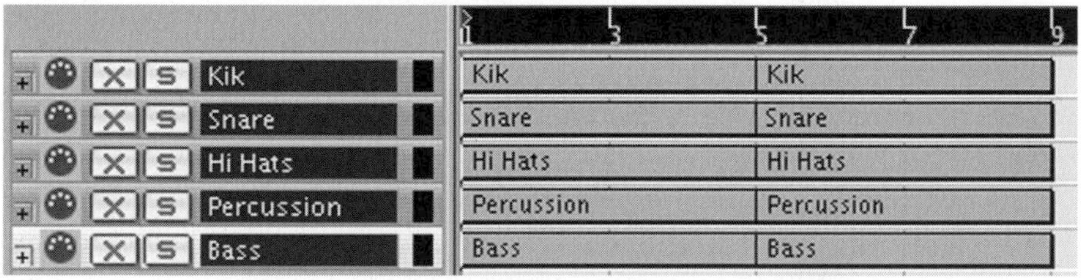

Fig. 6.4.11 – Eight-bar bass and drum track.

7. You can now repeat the bass line and drum track to create a four- or eight-bar section. Remember to name the newly created MIDI track and press Save to update your file to disk.

Stage 3. Adding a chord sequence

1. Create a new MIDI track and assign it to a sound source. You need to locate a piano or guitar sound that can be used to create an off-beat chord rhythm over the bass and drum track.

   Types of instruments you may wish to try:

   - Piano General MIDI sound 001
   - Organ General MIDI sound 017
   - Guitar General MIDI sound 027.

2. As you have already recorded the bass line, you need to find chords that will work well with this and avoid chords that clash and sound discordant. A good starting point is to try and identify the root notes used in the bass line and then try using chords based around these notes. Once you have found a chord that works try finding another chord and then move between them to create a chord sequence. Usually, you will find that chords change more slowly than the bass line (at the beginning or each bar or every two beats). In our example in Fig. 6.4.12 we change between two different chords each bar:
   - The chord of A minor (A, C, E) for bar 1
   - The chord of G major (G, B, D) for bar 2.

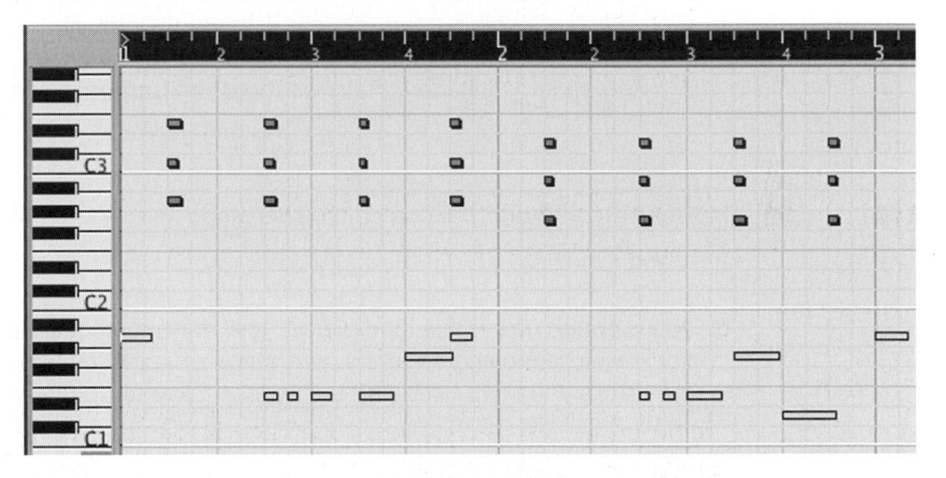

Fig. 6.4.12 – Example chord sequence and bass line displayed in Piano-roll-style edit window.

tip ► The bass part will usually have a root note that will help you determine the overall pitch and the chords to use.

3. The rhythm the chords create in reggae music is really important as the chords often provide an integral part of the rhythm. It is common to use an off-beat staccato-type rhythm where each chord is placed between each beat of a bar. When attempting to play this type of rhythm it may help if you try counting '1 and 2 and 3 and 4 and…' and play the chord on the 'and'.

Fig. 6.4.13 – Example chord rhythm.

When playing this type of rhythm on a multi-pitched instrument such as a piano, you should try and play each chord as short as possible and remember to change the notes being used as the chord sequence progresses.

tip ▶ Try creating chord sequences that form two- or four-bar loops.

4. *Now record your ideas into the sequencer.* When you feel like you have recorded something good, or worth keeping, edit and arrange it into a continuous part or sequence, then build this idea into an arrangement by copying and pasting, so it matches the drum and bass line.

tip ▶ Always try and edit your ideas into two-, four- or eight-bar sections.

5. You should now have completed the rhythm section of your reggae track. Before moving on to the next stage ensure that you have named all the tracks you have used and then press Save to update your song file.

note ▶ If you already have an idea for an alternative chord sequence or bass line, jump to Stage 5.

Stage 4. Creating a melody

1. Create a new MIDI track and assign it to a sound source. You need to find a sound that can be used to play a melody over the drums, bass and chords. Types of instruments you may wish to try are:
   - Saxophone General MIDI Sound 066
   - Fantasia. General MIDI Sound 088
2. You need to create a melody that fits with the chord structure and the bass line. As a starting point try using some of the notes being used in the bass line or chord sequence and simply play them one or two octaves higher on the keyboard using a different rhythm. In our example (see Fig. 6.4.14) we have created a melody using some of the individual notes from the chords A minor and G major.

Fig. 6.4.14 – Melody, chords and bass line displayed in the Score edit window.

Practise playing your ideas along with the drums, bass and chord sequence to see how they work together. Once you have created a melody, record it into the sequencer, edit and quantize it and then build it into the arrangement by copying and pasting.

producer says ▶

> A melody is the top line and often the main focus, so it needs to be interesting. Think about how you can make it unique – introduce some originality into your music.

Before starting the next stage, *press Save to update your song file.*

Stage 5. Creating a new section

Once you have created a section of music you are satisfied with you may want to consider creating an additional section that uses a different bass line, chord sequence and melody. This essentially means developing another idea that is different from the one you already have but in a similar style. However, any new sections that you create will need to blend with the existing section but be different enough to provide some variation.

producer says ▶

> If each section has its own chord sequence, then it's like having several different songs in the same arrangement. The key to making this work is ensuring that the different sections blend together.

1. Before trying to create a new section of music give yourself a head start by copying all the drum and percussion parts from your first section. This will give you a rhythm bed to develop a new idea over.

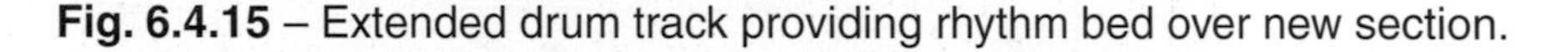

Fig. 6.4.15 – Extended drum track providing rhythm bed over new section.

tip ► When creating a new section copy all the drum tracks to provide a rhythm bed.

2. Usually, it's best to work in four- or eight-bar sections, so set a cycle over the new section. Alternatively you want to set up a cycle that allows you to listen to the existing section before you start playing as this will allow you to hear how the different sections will blend together.
3. Experiment using the bass or chord sound to create a new idea. This can be simpler or more complex than your first idea – there are no rules. You could try playing the existing chord sequence in a different order or use totally different chords. In our Reggae example on the CD we created a new section by repeating the first section and then transposing it up (+2) by two semitones.
4. When you're clear in your mind what you're going to do, drop into Record. When you feel like you have recorded something worth keeping, listen back to see how it blends with the previous section. If it doesn't, delete it and try something new.
5. When you have created an additional bass line or chord sequence you will then need to build up this section using the other instruments.

tip ►

Using Transpose

An easy way of creating musical variation or a B section is to use Transpose. First, make a copy of the A section, then transpose all the copied instruments up or down to create a B section. Listen to our example on the CD.

Optional stage

Create a third new section or C section. This, again, could be a new set of chords and bass line or a transposed version of a previous section.

Stage 6. Song arrangement

At this stage you will need to decide how to arrange the different sections of your composition. You could, for example, try playing back each different section in the order that it was created. However, you may find that you want to feature one particular section more than the others so you will need to create a different structure that will incorporate this. In our example we created three different sections and then arranged them in the order listed below.

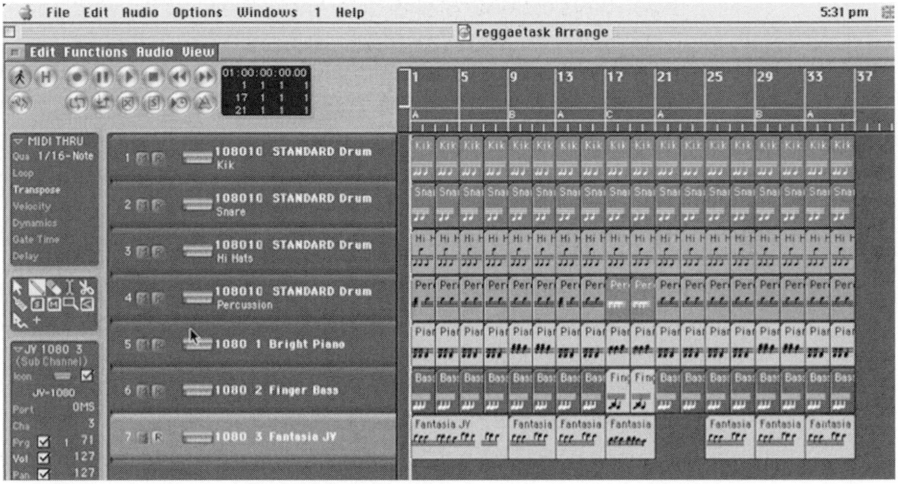

Fig. 6.4.16 – Typical song structure.

- A section for 8 bars
- B section for 4 bars
- A section for 4 bars
- C section for 4 bars
- A section for 8 bars
- B section for 4 bars
- A section for 4 bars
- End.

Creating an ending

Decide how you want to end your composition. You could, for example, make all the instruments come to a sudden stop or create a 'fade out' by gradually decreasing the overall volume until nothing is heard.

Creating markers

At this stage you may want to try creating some markers to help you identify and locate to different sections of your arrangement more easily.

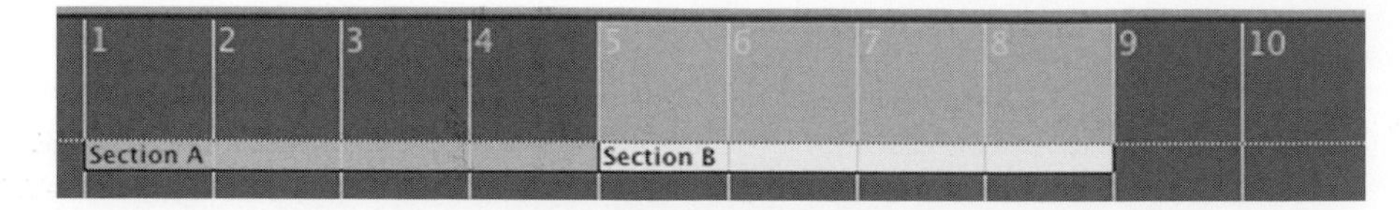

Fig. 6.4.17 – Markers defining different sections.

producer says ▶

Adding a melody across the entire arrangement

If you have created several different sections within your arrangement, you will need to create a number of different melodies or themes to complement each section. Reggae often uses small themes or riffs, so think about how you can achieve this. Perhaps try a simple spacious melody for your first A section before developing this into a busier catchy chorus-type riff for the B section.

Stage 7. Mixing and adding effects

This is the final stage of the process, so check that you can hear each instrument clearly and adjust any levels accordingly. Reggae tends to feature the bass line and drum pattern, so make sure the rhythm is cutting through. Also check you can hear the off-beat chord pattern, as this often plays an integral part in making the rhythm track groove.

The volume of each MIDI instrument can usually be adjusted from within your sequencing software using the GM MIDI mixer or any parameter assigned to control change number 7 (see Exercise 5.9). MIDI volume operates between a parameter range of 0 and 127, so you need to ensure that you leave sufficient headroom to turn each instrument up and down in volume. Ideally you should aim to set the volume of each instrument between 80 and 100. Also make sure that any external MIDI devices you are using are set up to receive control change messages, as it is possible to filter them out.

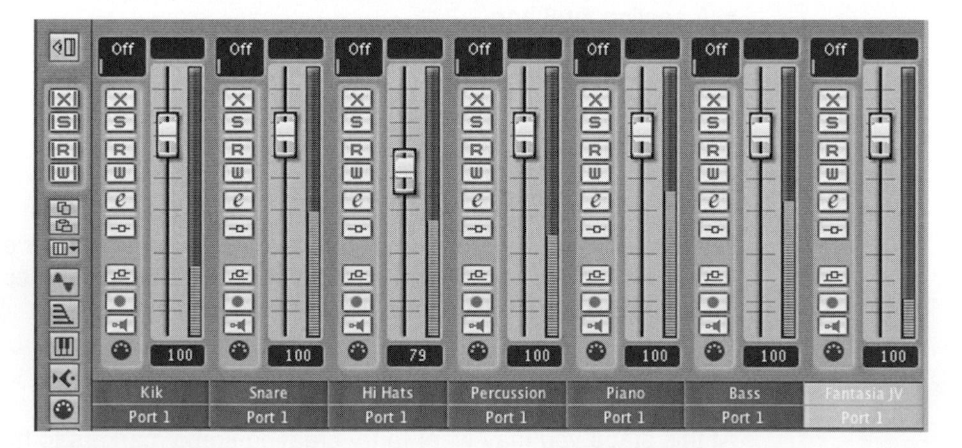

Fig. 6.4.18 – MIDI mixer.

Adding effects

As reggae has quite a slow beat, you may want to add some reverb to the drums or percussion sounds to create some ambience around them. You could even try adding a touch of reverb to the off-beat chord pattern to create a more ambient sound. A medium amount of chorus could be used to thicken the bass sound and create a rich texture.

Reverb and chorus are usually available as standard when using a GM-compatible sound device. You can determine the amount of effect that is added to each sound by using a GM MIDI mixer or by sending a control change message.

- Reverb responds to controller number 91
- Chorus responds to controller number 93.

Automation

As you listen through, you may want to make small changes throughout the arrangement, such as increase the volume of certain instruments in different sections. Automation allows you to record any parameter movements you make into the sequencer and then plays them back, so it's like having someone sitting next to you helping you move the parameters. You could, for example, try automating the reverb send level to create a special effect by introducing a sudden burst of reverb for just one snare hit, or try sending the hi hat or the off-beat chord into a delay so it echoes over a particular section.

You should now have completed a reggae arrangement, so make sure you have saved your work and logged any settings you may need.

7
Audio

The recording process

The actual process of recording music can be broken down into several different stages regardless of the type of equipment being used. Before you start it is essential that you understand what needs to be achieved within each stage and allocate enough time to it. Note that, depending on the type of recording method you choose, you may find several stages naturally combine together.

- Preparation
- Recording
- Overdubbing
- Editing
- Mixing
- Mastering
- CD.

Preparation

You first need to establish whether the music or performance is at a suitable standard to actually be recorded. This can be a personal decision, but can usually be achieved by listening to a rehearsal of the performance or by listening to a demo tape. You then need to establish what type of instruments you will be recording and how many instruments will be playing at the same time. This will help you plan how each instrument will be recorded and how to utilize the equipment and resources you have available to you. For example, a live band or vocalist will require a microphone to translate the sound they produce on to a recording device, whereas electronic sound sources such as keyboards, sound modules or sequenced MIDI instruments can usually be recorded without using a microphone.

Live stereo or multitrack recording

You then need to decide the type of recording method you are going to use – live stereo or multitrack. One of the simplest and most economical ways to make a recording is to capture a live performance in real time directly to a stereo recorder. This is known as a live stereo recording and is where all the different instruments and sounds are recorded and mixed at the same time. The recording and mixing stages are combined and there are no editing or overdubbing stages. The other method of recording is called multitracking. This is unlike a stereo live recording as you have the option of recording each individual sound or instrument independently, so you don't have to record all the instruments at the same time and a recording can be gradually built up in stages.

tip ▶ Always try and do as much preparation as you can before actually starting the recording session. That way, you will have time to plan the most suitable method to make the recording and prepare the equipment you plan to use.

Recording

This stage is where initial recording takes place. Each sound or instrument will need connecting or miking so it can be routed to a recording device. Suitable levels will also need to be set going to the recording device. The main goal of a recording should be to capture the full frequency range of the performance onto the recording device, without hiss or distortion.

note ▶ If you are using a multitrack recorder you should aim to record each sound or instrument on a separate track, as this will give you more flexibility.

How many tracks do you need?

In order to have complete control over a recording you ideally need to record each sound or instrument on a separate audio track. However, this isn't always possible or practical, as you may be restricted by the number of tracks you have available or by the amount of inputs on your mixer or soundcard. You will therefore need to plan ahead carefully and consider whether you have enough tracks available to record each instrument on a separate track or if you will need to combine certain instruments or sounds together. You don't want to find yourself running out of audio tracks during the recording session.

Imagine you only have eight tracks available to make a recording of a band. If you allocate a separate audio track to each drum sound you will quickly run out of tracks and be unable to overdub any additional instruments, such as vocals and guitar solos. One solution would be to combine the individual drum sounds to a single audio track, therefore leaving plenty of tracks available for overdubs.

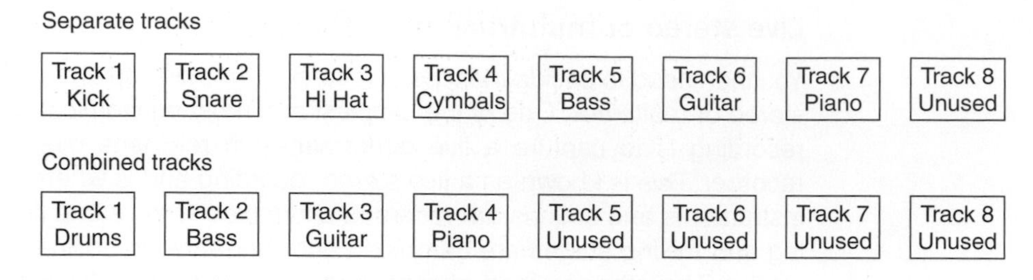

Fig. 7.1 – Example of how tracks can be combined.

Deciding which sounds or instruments to combine together can often be a difficult decision, as once they are merged you will have little flexibility to adjust the volume or process the combined sounds independently.

Stereo sounds

You also need to consider whether any of the sounds you plan to record will benefit from being recorded in stereo (see 'Stereo sounds' on p. 260). Remember, stereo sounds may require two audio tracks, so only do this if it is absolutely necessary and if it's actually improving the sound.

Overdubbing

If you are using a multitrack recorder you will be able to add additional instruments to an existing recording by overdubbing. Note: stereo recording devices, such as minidisks and stereo cassette recorders, only record in stereo and don't allow you to overdub. This process allows you to record new or replace existing instruments while listening back to previously recorded tracks. This can be particularly useful if it isn't possible to record all the different instruments at the same time or if you wish to repair a mistake on one track without affecting the other tracks. It can also be used to create a perfect performance by building up a recording in small sections or capturing several different recordings of the same instrument and then choosing the best performance (see Exercise 11.6).

note ▶ Overdubbing is a process that allows you to add additional sounds or instruments to an existing recording.

Editing

This stage is optional depending on the style of music you are recording and the quality of the performance. However, you will normally find there are a few adjustments and changes that can be made to generally improve a recording. The complexity of the editing can vary from simply removing a noise or mistake from a recording to completely rearranging all the instruments so they play back in a different order.

The options you will have available to actually edit the audio will vary depending on the type of recording equipment that is being used. You will find most of the examples given are relevant to both stand-alone digital multitracks and computer-based systems.

- Move the audio to a new location
- Copy and repeat the same section over and over
- Remove unwanted sections
- Change the arrangement of a piece of music so the sections play back in a different order
- Try out a different arrangement of a song
- Join different songs together
- Create a composite by picking out the best bits to create a perfect recording.

Mixing

Once you have finished recording and overdubbing, and have completed any editing, you are ready to mix. The idea is to create the best possible blend of all the individual sounds and instruments that have been recorded by adjusting the individual levels of each sound/instrument. Note: mixing will not be required if you have made a stereo live recording.

This is a very creative stage, where time can be spent experimenting blending individual instruments together. You also have the option of using equalization (EQ) to tonally change or enhance a sound by cutting or boosting certain frequencies or exploring different effects and processors (see Exercise 11.7). The complexity of the mix will vary depending on how many individual tracks or instruments have been recorded and what you are looking to achieve from the mix (see Exercise 11.9). Generally though, the more options you have the longer the mixing process will take.

Creating a CD

Once you have a mix that you are happy with, you will need to create a stereo master. This is a stereo recording of the mix that can be used to make a CD (see Exercise 11.10). Before this recording is transferred onto CD it will often pass through another stage called mastering. This stage is designed to tonally adjust and tweak the overall sound of the mix, so when it is played back it should sound comparable to any other commercially available CD. Note that professional mastering is a highly specialized skill.

Different types of recording environments and set-ups

The actual environment where you make a recording can vary tremendously. The main considerations are usually based on the studio's physical size, the type of equipment available, location and budget. If you're planning to record a 50-piece symphony orchestra, for example, you will need a recording area large enough to accommodate all the musicians and all the relevant recording equipment, including several different microphones. If you are simply recording a vocalist or guitarist, you will need less physical space and could use a less sophisticated recording set-up.

You also need to consider your requirements at each different stage of the recording process. You may, for example, need a large recording area and lots of microphones to make an initial recording, but then may find it cheaper and more practical to add any overdubs using a smaller recording set-up. You could then consider doing any small edits and adjustments using simply a computer with audio-editing software and then complete the final mixdown in a professional studio. Note that moving a recording project between different studios will only be practical if the recording equipment being used in each studio is compatible with the other (see 'Different types of multitrack' on p. 239). As you can see, there are many variables and choices when considering whether a recording environment will be suitable. We are therefore going to take a look at some of the most commonly used examples.

Professional recording studio

Professional recording studios are designed to provide an ideal working environment when recording and mixing music. Although they can vary in size and design, they are usually based around two separate rooms:

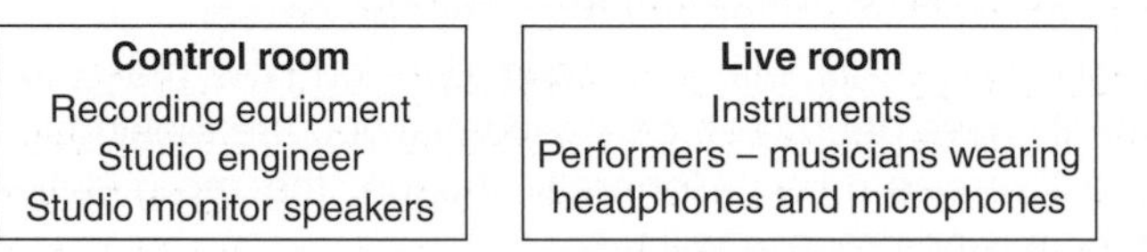

Control room
Recording equipment
Studio engineer
Studio monitor speakers
Live room
Instruments
Performers – musicians wearing headphones and microphones

Fig. 7.2 – Typical layout of a professional recording studio.

The control room

The control room is where the recording equipment is situated, such as the mixing desk and studio monitor speakers. This room is designed to be used by the recording engineer and producer to accurately listen to any sound being produced. It can also be used by all the people who are not performing. A well-designed control room will provide total sound isolation from the live room, allowing the studio engineer to make detailed judgements on the studio monitor speakers without being influenced by sound coming from the musicians in the live room.

The mixing desk will allow the engineer and producer to control the volume of each individual sound. They will be able to communicate with the musicians in the live room by using a small microphone that will pick up their normal talking voice. This is called talkback and is usually enabled by holding down a small button on the mixing desk. This allows the producer to talk to and moti vate the performers.

producer says ►

The control room should also be acoustically designed so any sound produced from the studio monitor speakers is clear and uncoloured by the surroundings. This is so that the engineer can make decisions without hearing sounds attributed to the room rather than to the speakers! Note: there is no point in trying to make detailed judgements on expensive and powerful speakers in a room that has bad acoustics, as the sound will bounce all over the place and confuse the listener.

The live room

The live room is normally just an empty room that provides the performers/musicians with their own space to set up and perform while making a recording. Live rooms are usually soundproofed to isolate any sound made by the musicians from the recording engineer in the control room. This is an essential part of the monitoring process, especially when using microphones, as it allows the recording engineer sitting in the control room to only hear the sound that is being picked up by the microphone, without being influenced by the sound coming from the live room or the actual instruments themselves. Note that a similar result can be achieved by wearing headphones.

A live room in a professional recording studio will often be acoustically designed to either reflect or absorb sound in a certain way. For example, a stone wall or a very high ceiling will create a live vibrant sound that could be useful when miking percussion and drum kits, whereas a room with lots of fabric or curtains will dampen the sound and make it more intermittent and less ambient. All of these design features can have a drastic effect on the sound being recorded.

note ►

Sound isolation between each room is essential, as it allows the recording engineer sitting in the control room to make detailed judgements on the studio monitor speakers without being influenced by the sound coming from the actual instruments themselves.

Project studio/home studio

A *portastudio* is a self-contained box that contains a mixing desk and a multitrack recorder. Portastudios can be analogue or digital and use tape or record directly to disk. As the names suggest they are usually portable, which makes them ideal for recording ideas quickly.

Virtual studio

A virtual studio is a piece of audio software such as Cubase, Logic, Digital Performer, Sonar, that is run on a computer. It allows you to recreate much of what exists in a real studio inside a computer.

Studio personnel

The process of making and recording music can often involve several different people.

Producer

A producer is responsible for overseeing and delivering a finished recording. Throughout the recording process they will make decisions on what is suitable and generally motivate people to get the best out of them. Some producers will distance themselves from actually operating or using any of the recording equipment themselves and purely make decisions based on what they are hearing. Other producers will be more hands on and may also be composers/arrangers and engineers themselves.

Recording engineer

Engineers are responsible for operating the studio equipment and getting the best out of it. They also need to be able to interpret any ideas made by the producer. Their skills usually include: operating a mixing desk and multitrack recorder, using effects and processors and different microphone techniques. As much of today's technology is based around computers, they may also be capable of using an audio sequencing program, such as Pro Tools or Logic, and editing audio. As well as technical skills, an engineer will also need the ability to listen and make judgements based on what they are hearing. This is an art in itself and only comes from experience. The engineer, along with the producer, is responsible for the final mix.

Assistant engineer

An assistant engineer is there to help and assist the recording engineer. Their duties can vary from making the tea and helping set up equipment to actually doing some of the jobs of the recording engineer.

Programmer

Traditionally a programmer was a person who specialized in using synthesizers and sound modules (boxes full of sounds). They were responsible for understanding and operating the equipment during a recording session, and for creating and editing any sounds that were required. As technology has progressed and computers have become more widely available, the term programmer also refers to someone who operates and records information into a MIDI/audio sequencer. A producer will often hire a programmer to help bring their vision to life, as a programmer will often bring their own unique style and set of sounds to a project.

Pro Tools engineer

Pro Tools is a complete recording and editing package that runs on a computer. If a studio or producer wants to use a Pro Tools system they may consider employing a dedicated person to operate it, as not everyone knows how to use it properly.

Hangers on!

During the recording process you may find other people in the studio who are not directly involved with the recording process. These can be friends of the artist, managers and representatives from the record label known as A&R (Artist and Repertoire). Although their agendas and views may be the same, you need to make sure that any guests are well behaved and do not distract or slow the recording process down.

Audio equipment

The actual process of recording audio has evolved over the years into a sophisticated state-of-the-art process using digital technology and computers. This advancement in technology has closed the gap between so-called professional and semiprofessional recording equipment by making it more affordable and accessible to everyone.

It is not uncommon for a professional recording to be completed simply using a computer-based studio or for schools and colleges to be using the same software as a professional recording studio.

Fig. 7.3 – The control room of a professional recording studio.

producer says ▶

> Using the latest state-of-the-art equipment does not necessarily guarantee a good recording. The key to a good recording is usually understanding how to get the best out of the equipment you are using.

In this chapter we are going to explore the main components of a typical recording set-up and give you an overview of what each component does. This should be helpful regardless of the type of equipment that you are using, as once you become familiar with the basic signal flow between each element, you will be able to apply the same principles to any recording system. Any terms used that you don't understand, please refer to the Glossary.

Multitrack recorders

This is the device you actually record onto. It will allow you to capture and store sound and then play it back. One of the main advantages of using a multitrack recorder is its ability to record individual sounds and instruments onto their own separate tracks. Generally, this will give you more control and creative flexibility when making a recording, as each individual track can be processed and erased independently of any others.

note ► A multitrack recorder will allow you to record, store, play back and erase sound.

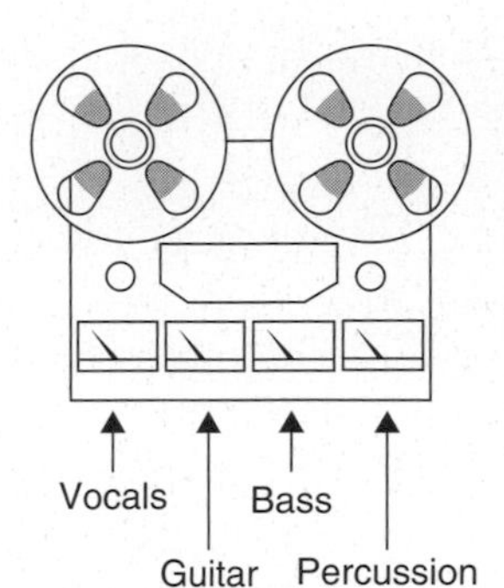

Fig. 7.4 – Multitrack recorder.

A multitrack recorder allows individual sounds and instruments to be recorded onto separate tracks. These are usually sent to the inputs of the multitrack from the bus outputs of a mixing desk (see Figure 7.10). This type of recorder is generally more suitable for studio recordings.

Overdubbing

A process called overdubbing can be used to add additional instruments to an existing recording while listening back to previously recorded tracks. This means all the instruments don't have to be recorded at the same time and complex recordings can be built up in small sections by layering different instruments on top of each other (see Exercise 11.6).

producer says ► A multitrack recorder allows individual sounds and instruments to be recorded on their own separate track and complex recordings to be built up in stages using a process called overdubbing.

Different types of multitrack

Most multitracks are either stand-alone devices (separate boxes) that are designed to be connected and used alongside a mixing desk or sophisticated software programs that run on a computer. They can either be digital or analogue and record to tape or disk. Here are some examples.

Tape-based multitracks

Multitracks that require a tape in order to make a recording can either be digital or analogue. They are usually designed to be connected and used with a mixing desk. A common example is the Alesis ADAT, which is an eight-track digital recorder that uses an SVHS cassette or the Tascam DA-88.

Hard-disk recorders

Digital multitracks can record directly to disk, removing the necessity for tape. This can prove to be a very efficient way to make a recording, as any location can be accessed instantly without having to wait for a tape to fast forward or

Fig. 7.5 – Tascam DA88

rewind. It also opens up many possibilities for manipulating and editing sound once it has been recorded. Stand-alone hard-disk recorders are usually designed to be connected and used with a mixing desk.

Computer-based recorders (digital audio workstations)

A computer can be used as a recording device providing it is capable of running recording software such as Cubase, Pro Tools or Logic (see Chapter 1). In order to provide the necessary connections to record and play back audio, you will need to connect an audio interface to the computer. These are normally connected using a USB or Firewire cable or by inserting a card into the back of the computer. Note that any additional hardware connected to a computer will need to be compatible with the software.

- *Audio software* is a program that allows a computer to record audio
- *Audio interface* is a separate device that provides audio connections.

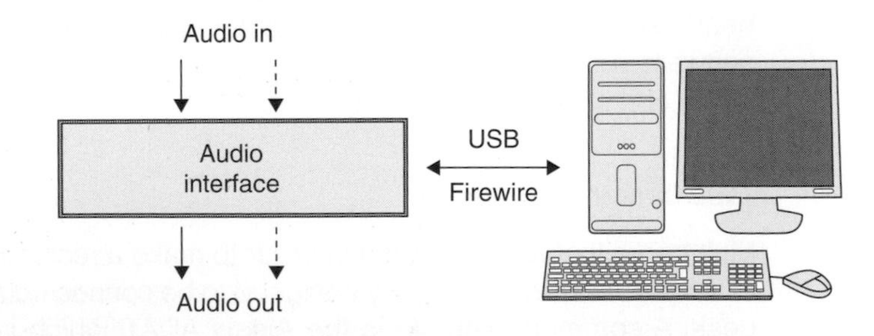

Fig. 7.6 – Computer and audio interface.

Most audio interfaces offer a minimum of two inputs and two outputs. This is adequate for making a live stereo recording or recording up to two separate instruments at the the same time. However, if you're planning to make multitrack recording you may require more inputs to allow you to record several sounds or instruments on different tracks at the same time.

note ▶

Combining MIDI and audio

One of the most powerful features of audio programs such as Pro Tools, Logic and Cubase is the ability to use audio and MIDI at the same time. This allows you to combine audio files with MIDI information within the same arrangement. For example, one audio track could play back a guitar or a vocal, while the next track could be a piano MIDI track.

Portastudios

These are self-contained boxes that contain a mixing desk and a multitrack recorder. This makes them ideal for recording ideas quickly. Portastudios can be analogue or digital and use tape or record directly to disk. They are usually portable, which makes them ideal for location recording.

producer says ▶

How many tracks do you need?

In order to have complete control over a recording, you ideally need to record each sound or instrument onto a separate audio track. However, this isn't always possible or practical, as you may be restricted by the number of tracks a multitrack has available or by the amount of inputs on multitrack or soundcard. Generally, the more tracks you have available, the more flexibility you will have. Popular formats range anywhere from eight, 16, 24 tracks or more. Note: when using a computer the amount of tracks you will have available will usually be determined by the software itself and the power of the computer.

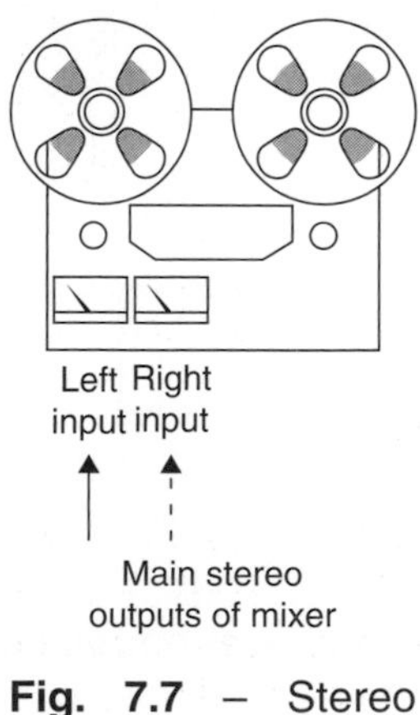

Fig. 7.7 – Stereo recorder.

Stereo recorders

A stereo recorder (or two-track) is similar to a multitrack in the sense that it allows you to capture and store sound and then play it back. However, it will only allow you to record and play back two tracks at the same time. When you drop into Record, both tracks are recorded together as if they were just one track. This means you will be unable to overdub or add any additional sounds or instruments to a recording, as dropping into Record would simply overwrite what was there before. This type of recorder is more suitable for making live stereo recordings and for storing completed stereo mixes (a stereo mix is the final blend of all the individual sounds and instruments together).

A stereo recorder is usually connected to the main stereo outputs of a mixing desk. This allows a blend of all the sounds or instruments being used to be recorded onto both tracks in stereo (see Fig. 7.10).

Microphones

A microphone is a device that will allow you to translate any sound that you can hear into a signal that can be recorded. You will need to use one if you are planning on recording vocals or any acoustic instrument such as an acoustic guitar, violin or a drum kit, or if you plan to make a live stereo recording.

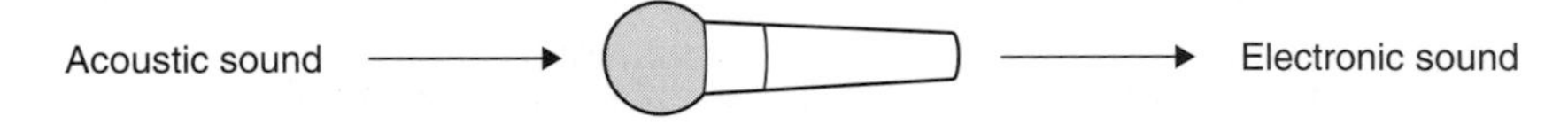

Fig. 7.8 – Microphone signal flow diagram.

note ► In order to record vocals or an acoustic instrument, you will need to use a microphone.

Connections and signal flow

When making a recording, a microphone is often the first stage of the signal path, as it collects any sound produced and translates it into a small electrical current that can then be connected and used by other studio equipment, such as a mixing desk or recording device.

note ► Electronic instruments that provide their own output connection, such as electronic keyboards and sound modules, will not require a microphone, as the sound they produce can be directly connected to a mixing desk before being passed on to a multitrack.

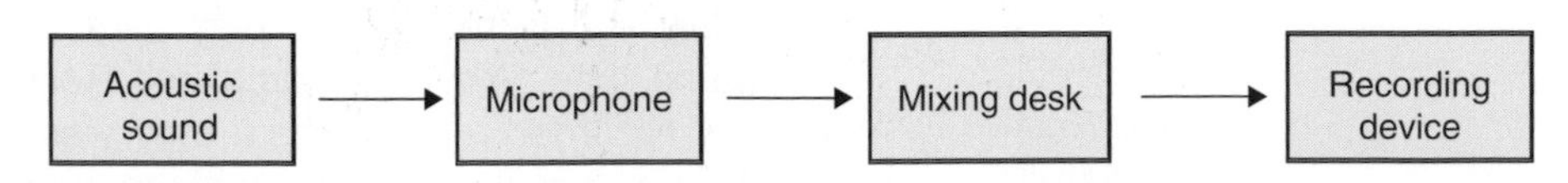

Fig. 7.9 – Signal flow to a recording device.

Microphones come in all different shapes and sizes and you will find some will translate sound better than others.

note ► When making a live stereo recording you will need to use two microphones.

The mixing desk

A mixing desk is often the main focal point of a recording studio, as all the sounds and instruments being used will pass through at some stage. It can also be seen as a hub for connecting other types of studio equipment, such as microphones, multitracks and monitor speakers. In its simplest form a mixing desk allows you to combine several different sounds and instruments at the same time and choose how they are blended together. However, most mixing desks can do a whole lot more, such as:

- Control the volume of each sound or instrument
- Route signals to various destinations
- Send signals to and from a multitrack and stereo recorder
- Send signals to and from effects units, like reverbs and delays
- Tonally adjust a sound by using equalization
- Send signals to headphones and studio monitor speakers.

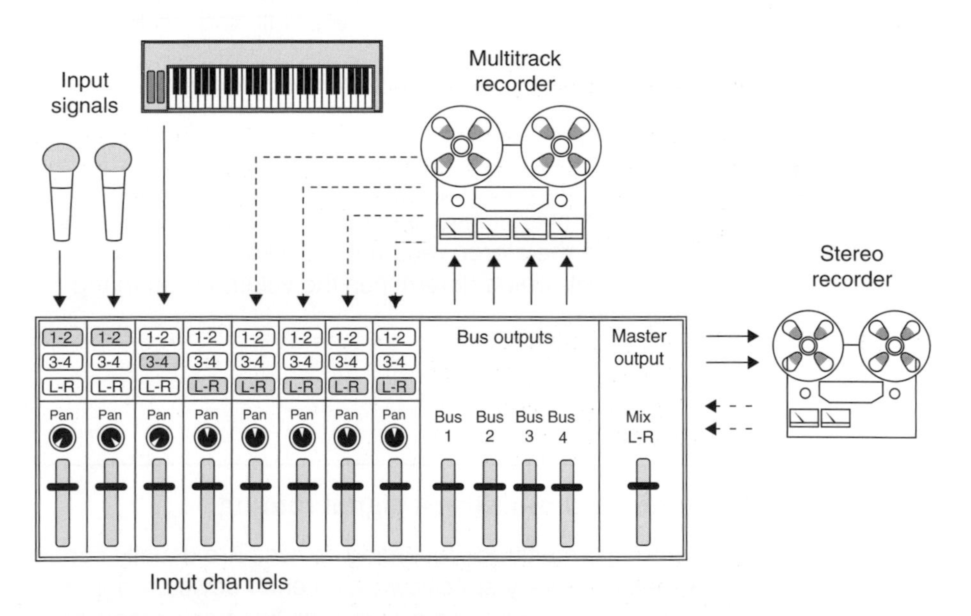

Fig. 7.10 – Different types of equipment connected to a mixing desk.

Identifying each section

Mixing desks come in all different shapes and sizes and can be stand-alone ('separate devices') and exist in software within a computer system. The actual layout of a mixing desk can vary tremendously depending on whether it is digital or analogue and hardware or software. However, you will find that most mixing desks are based around three main sections:

- *Input channels*. Input channels can be used to connect and listen to various different sound sources, such as electronic keyboards and microphones. The number of input channels will usually determine how many different sounds can be used and heard at the same time. Input channels can also be used to monitor the returns from a multitrack, although some mixing desks may provide dedicated tape monitor returns instead. These are additional inputs that allow you to listen to a multitrack without having to use the input channels.
- *Bus outputs (group outputs)*. Most mixing desks provide ways of sending signals to different destinations, such as multitracks. If a mixing desk has four buses, for example, then it will allow you to send a signal to four inputs on a multitrack recorder.

tip ▶

A mixing desk can often look confusing at first. However, once you can identify the different sections, you will find all the input channels are the same.

- *Master section*. This section is where the master fader is located. This fader controls the overall level of all the other audio channels and the level that gets sent to the stereo recorder. It's also where you'll find the control room monitor level, which determines the volume level that gets sent to the amp and speakers.

producer says ▶

Stereo and multitrack signal routing

Stereo and multitrack recording devices are usually connected to a mixing desk in a very specific way, as each device requires a different type of signal. The inputs to a stereo or two-track recorder should be connected to the main mix or stereo outputs of the mixing desk. This output sends a blend of all the sounds or instruments that can be heard to the recorder. The inputs to a multitrack recorder, on the other hand, should be connected to the mixer's bus outputs, so sounds or instruments can be isolated and routed to different tracks on the multitrack.

Studio monitor speakers and headphones

Before starting a recording project you will need to consider how you are going to monitor the recording process. This is normally done by using a pair of studio monitor speakers. In a professional recording studio, the studio monitor speakers are spaced around 2 metres apart and angled towards the centre of the mixing desk, so the person operating the recording equipment can hear the sound coming from both speakers equally (see 'Stereo sound' on p. 260). Professional recording studios often have several pairs of studio monitor speakers to simulate how a recording will sound when played loudly at a club or quietly on the radio. The smaller near-field monitors are usually placed on the actual mixing desk itself, whereas the larger studio monitor speakers are often built in to the wall.

Fig. 7.11 – Different types of studio monitor speaker.

note ► A pair of studio monitor speakers allows you to accurately hear any sounds or instruments during the recording process.

Speaker connections

Most speakers will require an amplifier to power them. However, some speakers have amplifiers built in and are known as powered monitors. The quality and performance of an amplifier is as important as the sound of the speaker itself, so we would always recommend using an amplifier that is specifically designed for powering studio monitor speakers. You also need to ensure that good-quality cables are used and that the correct connections are

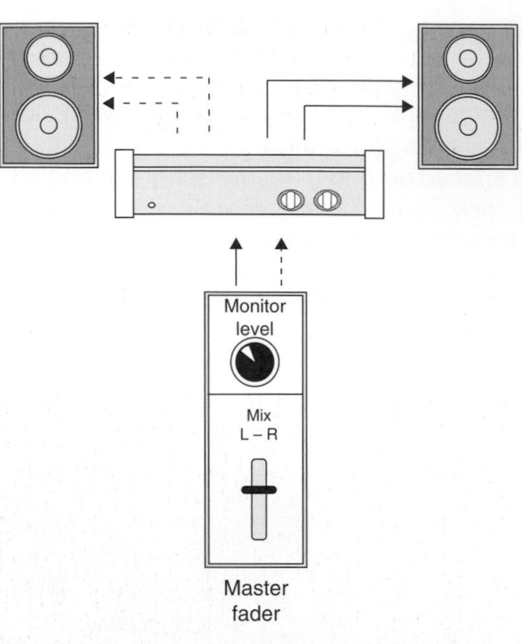

Fig. 7.12 – Signal flow from a mixing desk to a pair of studio monitor speakers.

made between the amplifier and speakers, i.e. the cable used on the red terminal on the left speaker should be connected to the red terminal of the amplifier. This will help avoid any problems such as *phase cancellation*.

In the example in Fig. 7.12 the stereo control room outputs from a mixing desk have been connected to the left and right inputs of an amplifier. This is the most logical connection to make if you are using a stand-alone mixing desk, as it allows you to control the level being sent to the amp and speakers from the mixing desk.

note ►

The control room monitor level allows the listener to control the volume level that gets sent to the speakers from the mixer.

producer says ►

Based on what you hear with a pair of studio monitor speakers, you will make decisions and judgements, so in a way an accurate pair of monitor speakers is more important than the quality of the recording equipment itself. A common problem is a situation where a recording will appear to sound OK when played back on the studio monitor speakers but will sound completely different when played back elsewhere. This is a far from ideal situation, as you need to trust what you are hearing and make sure the sound you are hearing while making a recording will translate when played elsewhere. We would therefore always recommend setting up the most accurate monitoring system you can, as this is often an overlooked investment.

Headphones

Headphones may be used as an alternative to using studio monitor speakers if the performers/microphones are sharing the same room as the equipment and recording engineer, or if you are making a live stereo recording. Headphones can usually be connected to the centre section of a mixing desk or sometimes directly to the stereo recorder when making a live stereo recording. This is a far from ideal situation during a recording, as it may be difficult to judge what is actually being recorded since you would be hearing both the live sound in the room and the recorded sound in the headphones. However, sometimes this is unavoidable.

Effects

An effect is a general term used to describe a device that can be used to enhance or change a sound or instrument in some way during the recording

Fig. 7.13 – Photograph of 19-inch equipment rack.

process. This can be useful for a variety of different reasons – for example, you may want a sound or instrument to appear as though it is playing in a large hall or echo away into the distance, or it may simply make a sound more interesting when an effect is added – there are no rules. Effects can be added at any stage of the recording process and are usually connected to a mixing desk via an auxiliary send, so you can choose how much effect is added to each sound or instrument.

External effects units

An external effects unit is a separate piece of equipment that usually fits into a 19-inch equipment rack. These stand-alone boxes can produce a variety of different effects, sometimes at the same time. They are normally connected to a mixing desk, so you can choose how much effect is added to each sound or instrument. Professional recording studios will usually have several different types of effects.

Plug-ins

If you are using a computer system with a software mixing desk it is likely that you will be able to use a plug-in. This is an additional piece of software that can be used inside a computer to produce an effect. Most audio programs come with a good selection of plug-ins built in.

Advantages of using a plug-in

- No need to use a stand-alone effects unit
- Audio does not have to leave the computer to be processed
- You can use several of the same plug-ins at the same time
- All parameters can be controlled on screen
- All parameters can be automated
- All parameters settings are saved with the song.

Different types of effects

Some of the most commonly used effects are reverb, chorus and delay; each one will produce its own unique sound when added to a sound or instrument. Note that when an effect is added you will hear both the original sound and the effect. Here are some examples.

- *Reverb* is probably one of the most commonly used effects, as it can be used to simulate the sound of a real environment such as a room or hall. When added it can create the illusion of space around a sound by making a sound appear more distant, which can help create distance and perspective. It can also be used to add a final gloss to a sound.
- *Delay* is a time-based effect that can be used to create several repeats of sound. This is achieved by delaying the sound by a set amount of time so

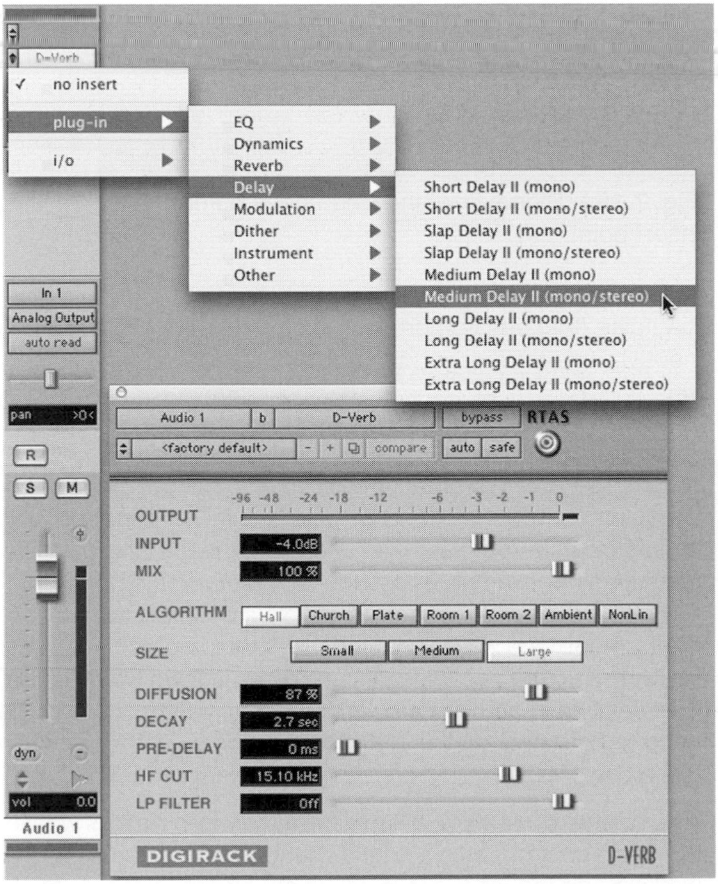

Fig. 7.14 – A software reverb plug-in being inserted in Pro Tools.

you hear the original sound followed by a delayed version – the repeat. It can be used to create the illusion of a sound echoing away into the distance.

- *Chorus* can be used to add a fuller texture to a sound. This is achieved by using a small delay and a slight variation in pitch. It can create the illusion of two of the same instruments playing together in unison.

Processors

Other devices, such as compressors and noise gates, can also be used to enhance or change a sound or instrument in some way during the recording process. These are referred to as processors as they are usually connected in a different way from effects, so 100% of the audio signal actually passes through them. Processors are usually connected to a mixing desk via an insert point or can be directly inserted into a slot on a software mixer.

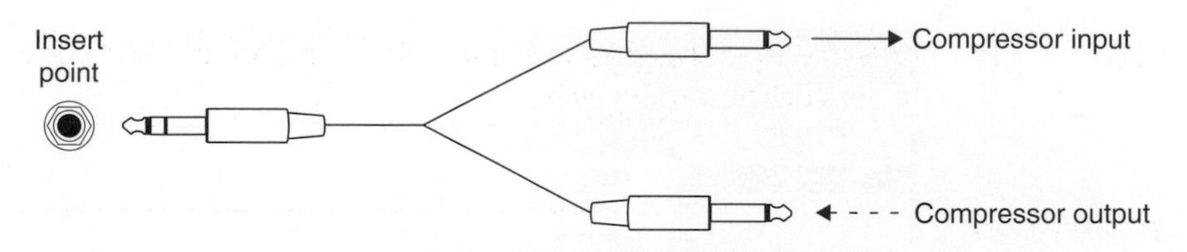

Fig. 7.15 – Signal flow showing how a compression can be connected to an insert point.

note ► A noise gate and compressor should be connected via an insert point and not via an auxiliary send and return.

Compressors

A compressor is like an automatic volume control that can be used to even out the volume level of a sound or instrument. The benefit of doing this is that you should be able to hear the sound more clearly and consistently throughout a recording. For example, most sounds naturally vary in volume and you may find any really loud sections seem to jump out while the quieter sections are more difficult to hear. One way to compensate for this would be to move the volume fader up when a sound became quiet and move it down when the sound became loud. This solution is not very practical and may be difficult to achieve, as you would need to anticipate all the changes in volume. Therefore, it is more common to use a compressor.

A compressor will allow you to determine when it starts working by adjusting the threshold control. Once a sound passes this threshold the compressor will start working and reduce the level of the sound automatically by the ratio set. A gain reduction display allows you to see how hard the compressor is working. The attack control allows you to determine how quickly the compression starts once the signal crosses the threshold, whereas the release time determines the time it takes for the compression to return back to normal when the signal drops below the threshold.

Fig. 7.16 – Compressor controls.

Once a sound has been compressed the distance between the loudest and softest sections is reduced, making the overall signal more equal. This can be extremely useful when making a recording, as it allows you to increase the overall recording level without having to worry about any sudden rises in volume.

tip ▶ Gently compressing a signal while recording will help maintain a more constant recording level.

producer says ▶

Compressors can be used as a level control or artistically as an effect to make things sound bigger or tougher. They are commonly used on vocals and solo instruments, and can also be used to control the overall volume of a stereo mix.

Noise gates

A noise gate is like an automatic on/off switch and will open and close depending on the strength in level of a sound. When a signal is loud enough it will open the gate and the sound will be heard. If a signal is below the threshold set, the gate will remain closed and the signal will not be heard. This can be very useful, as unwanted sounds such as noise and tape hiss are usually at a lower level than the actual sound itself, so providing the threshold of the gate is set correctly you will be able to ensure no sound is heard when an instrument stops playing.

Fig. 7.17 – A two-channel Drawmer noise gate.

The *threshold* control sets the level at which the gate will decide whether to open or close. When a signal is above the threshold, the gate will be open and sound will be heard. When it is below the threshold, the gate will be closed. There is usually a visual display to help you see when the gate is opening and closing. The attack control allows you to determine how quickly the gate opens once the signal crosses the threshold, whereas the release time determines the time it takes for the gate to close when the signal drops below the threshold. An additional control called hold allows you to determine how long the gate will remain open once the signal drops below the threshold.

8
Digital Recording

In this chapter we are going to explain the process of transferring sound into digital audio and discuss some of the options you will encounter along the way. This should be helpful regardless of the type of recording equipment that you are using, as most of the parameters explained are common to all digital recording devices.

Basic sound theory

When we hear sound we don't normally think about how it is actually being produced. However, as soon as you consider capturing and recording it, you will find it helpful to understand the basic principles behind it.

Sound is created by vibrations in the air. The air itself doesn't actually move but the process causes the air molecules to squash together and then retract. The rate of this to-ing and fro-ing determines the distance between the peaks and troughs of the sound waves that are created. By measuring the number of peaks and troughs that occur every second, we can calculate the frequency of a sound wave. This is the number of complete cycles each second. For example, a sound wave that completes three complete cycles in a second would have a frequency of 3 Hertz or Hz. A sound wave that completes 100 cycles in a second would have a frequency of 100 Hz and so on. Basically, the more cycles per second, the higher the frequency. However, in order for our ears to hear a frequency it must fall somewhere between 20 and 20 000 Hz. Note: Hertz is a term used to express the number of cycles per second.

Amplitude

Another characteristic of any sound is its volume. This is referred to as its amplitude and in the diagram in Fig. 8.1 is portrayed by the height of the waveform. The louder the sound, the higher the waveform.

Visualizing sound

All sounds have two basic ingredients: *amplitude* and *frequency*. If we could actually see sound, we would think of it differently.

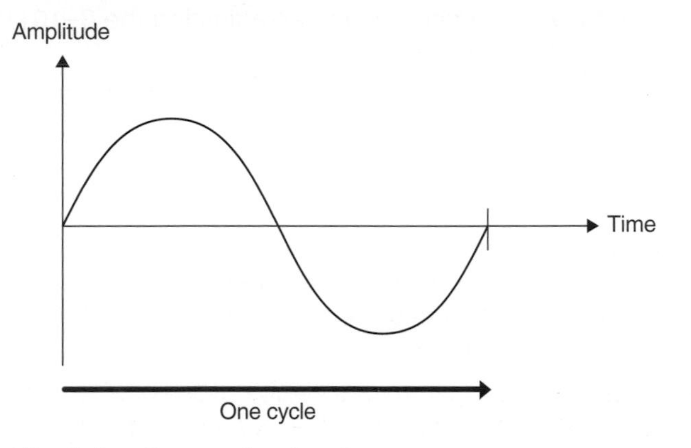

Fig. 8.1 – One cycle of a sine wave.

Amplitude refers to how loud we perceive a sound to be and will determine the height of the waveform. The louder the sound, the higher the waveform.

Frequency refers to the number of cycles in the wave per second. The rate at which these variations take place determines the pitch or frequency of the sound that we hear.

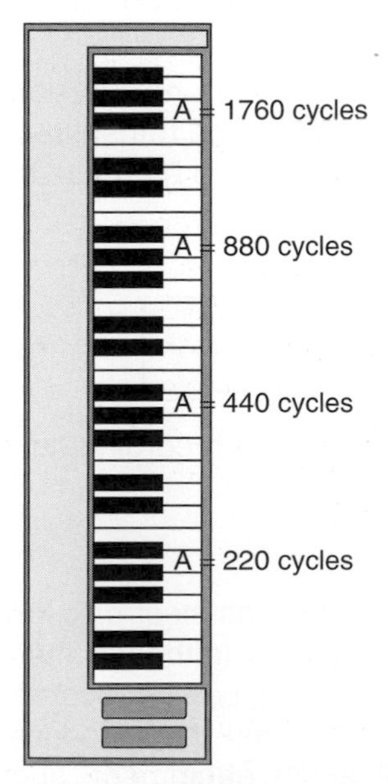

Frequency and pitch

It's also worth noting that there is a direct relationship between frequency and pitch. Frequency is the term used to express the number of cycles in a sound wave each second, whereas pitch is used to determine the notes on a musical instrument such as a piano or guitar. However, each note is simply a set frequency. For example, a frequency of 440 Hz would produce the note A. This means a change in pitch is also a variation in frequency, and the higher the pitch, the higher the frequency. Note: when an instrument drifts out of tune, the pitches of the notes are simply producing the wrong frequency.

Fig. 8.2 – Keyboard displaying relationship between frequency and pitch.

producer says ▶

> Understanding how different frequencies can change a sound or instrument is particularly useful when using an equalizer on a mixing desk. For example, cutting or boosting frequencies around 60–120 Hz will affect the bass or low frequencies, whereas adjusting the 8–10 kHz region will affect the high frequencies.

Recording sound

The process of capturing and recording sound will vary depending on the type of recording equipment being used and whether or not a microphone will be required. It is important to understand the route each sound or instrument will take to a recording device.

Acoustic sounds

Sounds or instruments such as acoustic guitars, violins, drum kits, and vocals will require a microphone to capture and translate any sound that is produced into a format that can be recorded and stored (see Chapter 9).

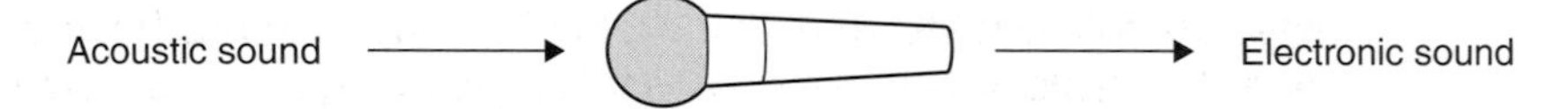

Fig. 8.3 – Microphone translating acoustic sound into an electronic sound.

The sound will be captured by the microphone and translated into a small voltage that will travel down an audio cable and be transferred into the mixing desk. This will allow any adjustments in volume to be made before sending it to the recording device.

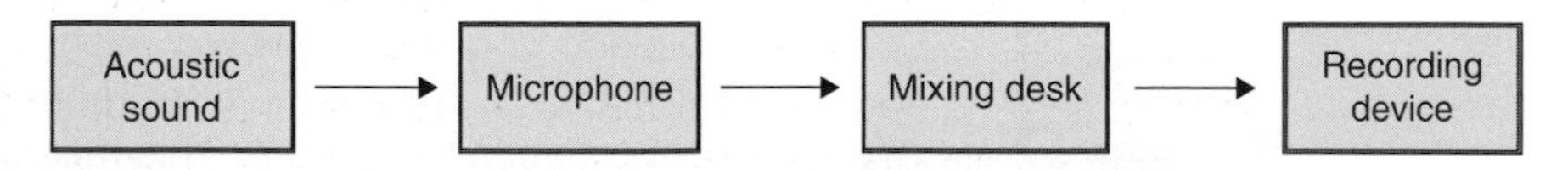

Fig. 8.4 – Signal flow of an acoustic sound to a recording device.

Electronic sounds

Powered electronic musical instruments such as electronic keyboards, sound modules, and even bass and electric guitars (although these will need an additional piece of equipment called a DI box), can be recorded without using a microphone. This is because the sound that they produce can be transferred directly into another piece of audio equipment, such as a mixing desk or recording device, by simply connecting it with an audio cable.

note ▶ Understanding the route each sound or instrument will take to a recording device is particularly important.

Fig. 8.5 – Signal flow of an electronic sound to a recording device.

Analogue or digital

Once sound has been recorded it is referred to as audio. The actual medium used for storing it can either be analogue or digital. Both provide a suitable medium for storing audio. However, for the purpose of this book we will focus on using digital recording devices.

Analogue

When you record something as analogue it usually means using magnetic tape such as a cassette or video tape. The analogue signal is captured on the tape by aligning tiny magnetic particles on the tape to represent the signal (note: these are too small to see). This method of recording has been used for many years to create many classic-sounding recordings, but is now being superseded by digital technology, mainly because it's cheaper and generally offers a lot more flexibility. Some of the downsides of recording to analogue tape are tape hiss and the time it takes to locate to different positions on the tape. Also, editing is more complex, as you are unable to see the audio as a waveform.

producer says ▶

Degrading sound

Every time a sound is transferred from one device to another it can lose quality or be slightly changed in some way during the process. Additionally, the technology itself can add attributes that can change the sound. This is particularly true of analogue equipment such as multitracks and mixing desks.

Digital audio

Digital audio is generally more flexible than analogue and has revolutionized the way people can work with audio by making it more affordable and accessible to everyone. It can be stored in various formats, such as tape or hard disk, and allows you to achieve things that would have been impossible to do several years ago, at a fraction of the cost. Once a sound has been recorded digitally it can be manipulated with ease. Most digital recorders allow instant access to any location within a recording and there are usually lots of editing functions, making it easy to copy and rearrange the audio. You can also see the audio as a waveform, so it can be edited visually. However, digital audio does not guarantee high quality, so you need to be aware of the mechanics behind a digital system in order to achieve the best results from it.

Displaying audio as a waveform

Once audio is recorded or transferred into a DAW (digital audio workstation), you should be able to see the audio as a waveform. This is a major advantage over analogue or tape-based systems, as you can actually see where the audio starts and stops. You can also instantly access any location in the audio with no rewind time like with tape. It also makes editing a lot easier.

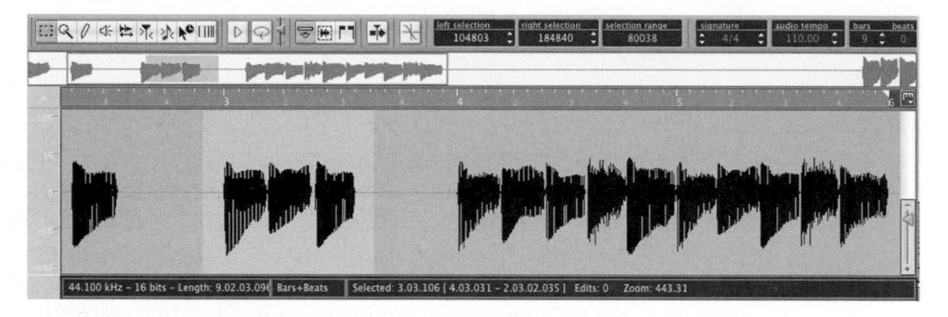

Fig. 8.6 – Digital audio displayed as a waveform.

Digital audio waveform

The waveform displays the amplitude (volume) of the audio over time, so it's easy to see where the loud and quiet section are (the higher the waveform, the louder the sound).

note ► When sound is recorded digitally into an audio sequencer, it can be displayed as a waveform.

Sample rate

When a digital recording device is used to record an analogue sound from a microphone or electronic keyboard, the sound is converted and broken down into lots of pieces called samples. Each sample then represents the sound at a particular moment in time. Generally, the more samples, the better the sound quality will be. On a CD, for example, 44 100 pieces of sound are used to represent each second of sound. The reason you need so many pieces is to measure the sound as accurately as possible.

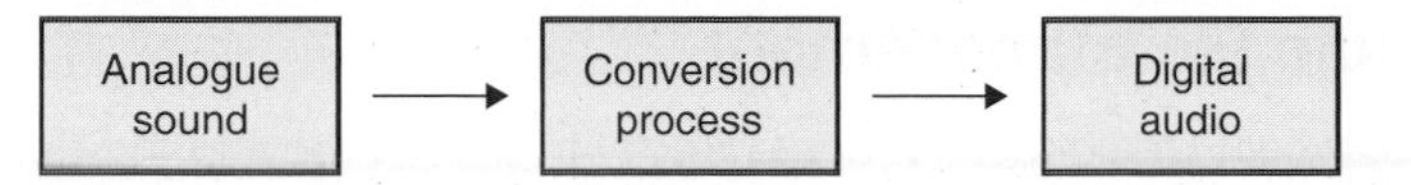

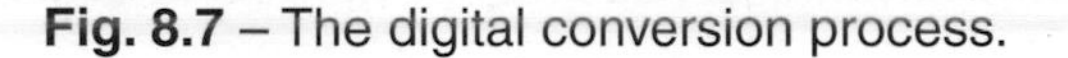

Fig. 8.7 – The digital conversion process.

note ▶ The higher the sampling rate, the more accurately an audio signal will be converted into a digital format.

Selecting a sample rate

All digital recording equipment has its own sampling rate, so before starting a recording it is essential that this is set correctly. Commonly used sampling rates are:

✓ Sample Rate 44100
Sample Rate 48000
Sample Rate 88200
Sample Rate 96000
Sample Rate 176000
Sample Rate 192000

- 44.1 kHz – for CD
- 48 kHz – often used for broadcast and on DAT machines
- 96 kHz – for very high-quality recordings and DVD audio.

Fig. 8.8 – Selecting a sample rate of 44 100.

tip ▶ Always ensure the correct sampling rate has been set before you start recording, as changing it mid-session would result in the audio playing back at a different speed.

Bit rate

The other important factor along with the sample rate when recording sound digitally is the bit rate. This determines how accurately the volume level or amplitude of a sound is calculated. As a signal enters a digital device, its volume is calculated against a scale of measurement, which is just like measuring the height of a sound with a ruler. With a 16-bit system (which is the bit rate used on a CD), the height of the waveform is measured against a ruler that has 65 536 positions. Most digital recorders will allow you to choose between 16-bit or 24-bit. The downside to using higher bit rates is that it will use more hard disk space. A typical example is given below:

- To store one minute of stereo, 16-bit audio requires 10 Mb of memory
- To store one minute of stereo, 24-bit audio requires 15 Mb of memory.

note ▶ Using higher sample and bit rates will use more hard disk space.

Audio file management

Most digital multitracks record directly to disk, removing the necessity for tape. This gives you many advantages, such as being able to access any location instantly without having to wait for a tape to fast forward or rewind. It also opens up many possibilities for manipulating and editing sound once it has been recorded. You will, however, need to manage all the audio files and information relating to a particular recording.

The audio record path

Before making a recording it is essential to know where the audio is going to be stored. Most systems require you to create a project folder before you start, so all the files relating to a particular project are stored in one location. We recommend you work this way, as it makes it easier to back up or transfer your project and you will be less likely to lose a vital piece of information. On systems that don't require you to create a project folder, you will need to set the audio record path manually, i.e. where the audio will get recorded to on the hard disk.

note ► It is essential to know where the audio is going to be stored before pressing Record.

As your audio recording progresses, lots of different files will be created, such as song files, audio files, fade files and even samples. Each time you press Record a new audio file is created, so you may find you have several small pieces of audio that are being used to make up one complete track. It is therefore a good idea to get organized at the start of a project and make sure all the files are named correctly and stored in the same location. You will also need to remember to save your session or song file independently of the audio files, as this contains all the settings relating to all the audio files.

- *Song files* store the settings and don't contain any audio
- *Audio files* are useless without a song file.

note ► Each time you press Record, a new audio file will be created.

Hard disk performance

The speed and capacity of a hard drive will directly affect the performance of a digital recorder. In order to use a hard drive successfully you need to ensure that it has enough space to store the information you wish to record. If you run out of hard disk the recorder will simply stop recording. A three-minute stereo file will need around 30 megabytes of disk space when recorded at 16-bit and 44.1 kHz, so you need to plan carefully as each track will require additional space. It's also worth taking into account that recording silence takes up just as much space as recording something useful on a digital system.

note ▶ When recording audio onto a hard disk you will need to ensure it has enough space to store the audio files that will be created.

You also need to ensure that the hard drive is fast enough to play back the amount of audio tracks you plan to use. The faster the drive spins, the quicker it can access the information and the more audio tracks you will be able to use at the same time. The slower the drive spins, the less audio tracks you will be able to use at the same time. Most internal hard disks will work fine for basic audio use. However, a hard drive that rotates at 7200 RPM is usually recommended for professional audio use.

Before starting a project

- How much recording time will you need?
- Have you enough spare hard disk space?
- Is the drive fast enough to play back audio tracks?

note ▶ If you run out of hard disk space, a digital recording device will simply stop recording.

Digital audio file formats

Digital audio can be stored in many different file formats. However, not all formats offer the same sound quality and some formats, such as MP3 and mini-disk, are 'compressed' and reduce the sound quality in order to make the file size smaller. We therefore always recommend that you use an uncompressed audio file format, such as AIF, SD2 or WAV, when making a recording, as you can always convert or mix to a compressed audio format later.

- AIFF (Audio Interchange File Format) is a popular Mac- and PC-compatible file audio format.
- WAV or WAVE (broadcast wav) files are a popular format for PCs.
- Sound Designer 2 (SD2) is a Mac-only file format. Originally used for Pro Tools, it time stamps the file so it always has a positional reference.
- MP3 (MPEG Layer 3) is a compressed audio file format popular for Internet use, as the file size is smaller.
- Apple's AAC is a compressed audio file format used for the distribution of music for consumers via the Internet.

Sample Accurate Automation: SDII (SoundDesigner II)
Recording File Type: ✓ AIFF
Pass Keyboard Events to Plug-ins: WAVE

Fig. 8.9 – Varying the amplitude waveform with a ruler.

Stereo sound

When we listen to sound we have the ability to determine the direction it is coming from. When a sound appears to come from one particular direction, we hear it slightly differently in each ear. For example, a sound made directly in front of you would usually appear equal in both ears, whereas a sound made to the left-hand side of you would appear slightly louder in your left ear. The ability to perceive sound differently in each ear means we have the ability to hear in stereo. To hear sound in stereo you will either need a pair of studio monitor speakers or a pair of headphones.

note ► When a sound appears to arrive from two directions at the same time, it can be termed stereo.

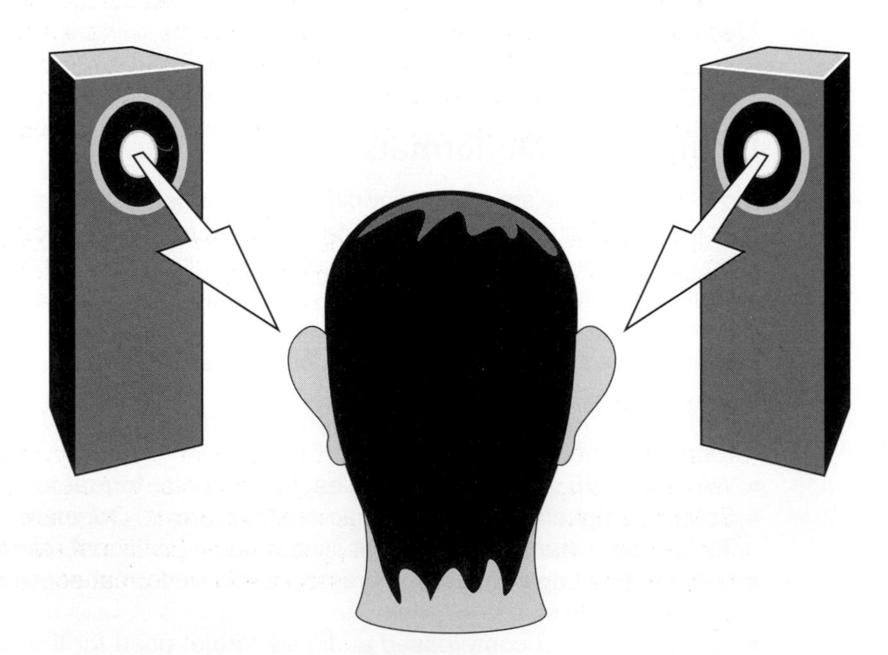

Fig. 8.10 – Listening in stereo.

A stereo sound consists of two separate elements (one for the left ear and one for the right ear) that arrive at the listener from different directions to make up the overall sound.

Pan control

Each individual channel on a mixing desk will have a *pan control.* This is a rotary dial that allows you to choose where you want to position a sound or instrument within a stereo field. Note that this will only work if you are actually listening to the sound in stereo. Moving the pan control will move the position of the sound from left to right or ear to ear, speaker to speaker. Setting the pan pot to the centre (12 o'clock position) will send equal amounts of the sound to both the left and the right, making the sound appear in the middle (see 'Pan control', p. 276).

note ▶ Using the pan control does not make a sound stereo, it simply adjusts its position within the stereo field.

Mono or stereo

When making a recording you will need to decide whether to record a sound or instrument using a single audio track in mono or using two audio tracks in stereo. This decision is usually based on whether the sound or instrument has a stereo capability and whether it actually sounds better in stereo than in mono. You will find that most instruments, such as vocals, guitars and bass, can be recorded in mono using an audio single track. If you're planning to make a live stereo recording or are using sounds that have a wide stereo image, such as synthetic string pads, you will need two tracks available on a multitrack or stereo recorder.

note ▶ Stereo sounds will require two separate audio tracks: one track for the left signal and one track for the right signal.

Stereo file formats

Computer-based digital multitracks usually allow stereo sounds to be combined into a single audio file. This single file contains the individual information for two separate tracks. This generally makes it a lot easier to manage – for example, if you edit the Left file, you automatically edit the Right file, as if they were one complete unit. This audio file format is called interleaved.

9
Microphones

A microphone will allow you to translate any sound that you can hear into a signal that can be recorded and used by other studio equipment, such as a mixing desk or recording device. Electronic instruments that provide their own output connection will not require a microphone, as the sound they produce can be directly connected to other studio equipment, such as a mixing desk (see Exercise 11.2).

note ► In order to record vocals or an acoustic instrument, you will need to use a microphone.

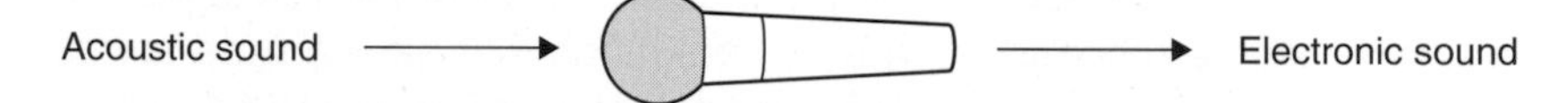

Fig. 9.1 – Microphone signal flow diagram.

Instruments that are typically recorded using a microphone are:

- Acoustic guitar
- Strings
- Brass or woodwind
- Drums
- Percussion
- Piano
- Vocals.

Different types of microphones

Microphones generally fall into two different categories: dynamic or condenser.

Dynamic microphones

Dynamic microphones are usually quite robust and designed for live use and close miking (close miking means placing the microphones close to the sound source). This makes them suitable for miking loud sound sources such as drums and guitar amplifiers. The Shure SM58 is a classic example of a good dynamic microphone.

Condenser microphones

Condenser microphones offer a good frequency response and are capable of picking up a wider range of the sound than dynamic microphones. This makes them more suitable for use in a recording studio. They can be used to record a wide variety of different sounds and instruments, but are generally more suitable for miking solo instruments and vocals. They can also be used for ambient miking and stereo live recordings (ambient miking means placing the microphone further away from the sound source). Condenser microphones are usually designed to be placed on a microphone stand as they are usually quite delicate.

Fig. 9.2 – Condenses microphone set up to record vocals.

Phantom power

If you are using a condenser microphone you will need to supply phantom power in order for the microphone to work. This is normally achieved by enabling a +48 V *phantom power* button on the device the microphone is connected to (see Exercise 11.2). Alternatively, power may be supplied from a dedicated stand-alone box or in some cases the microphone will be able to use an internal battery.

note ▶ If you are using a condenser microphone you will need to supply it with +48 V phantom power in order for it to work.

Microphone polar patterns

All microphones have what is termed a 'polar pattern'. This refers to the area around the microphone that picks up sound. This should be taken into consideration when positioning a microphone as it will determine the direction the microphone will be most sensitive to.

For example, a microphone using a cardioid polar pattern will be most sensitive to sound entering the front of the microphone and less sensitive to sound coming from behind it. This creates a heart-shaped recording area at the front of the microphone and is a popular choice as it only captures sound directly in front of the microphone.

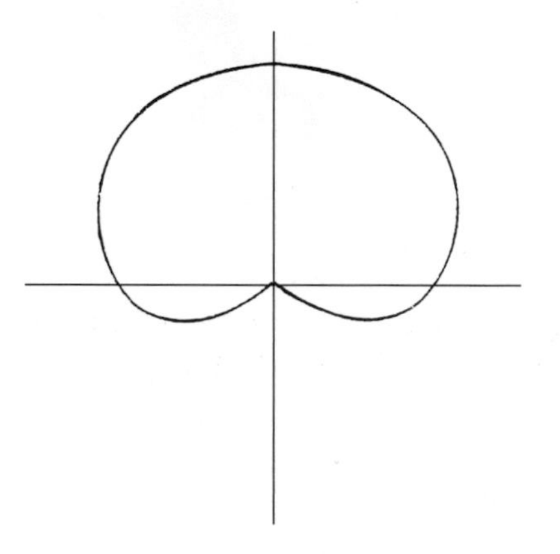

Fig. 9.3 – Cardioid (heart-shaped) microphone polar pattern. Picks up sound directly in front of microphone.

note ▶ A polar pattern is the recording area around the microphone.

Other types of polar patterns include omni, where sound is picked up equally all around the microphone, and figure-of-eight where the front and back of the microphone collect sound equally. Generally you will find that most microphones have a fixed polar pattern, however some condenser microphones (usually the expensive ones) allow you to switch their polar pattern.

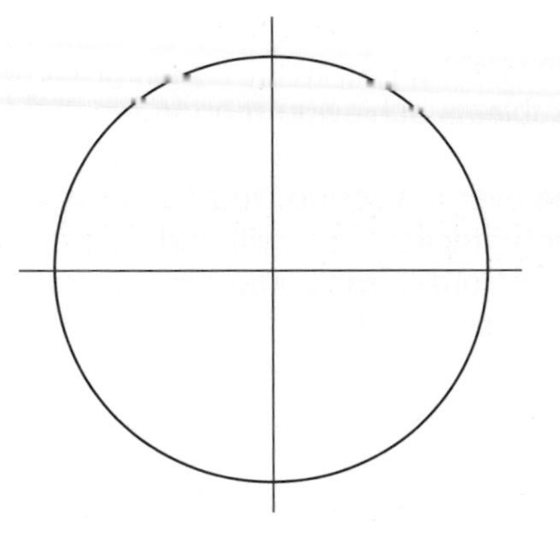

Fig. 9.4 – Omnidirectional (omni) microphone polar pattern. Picks up sound equally all around the microphone.

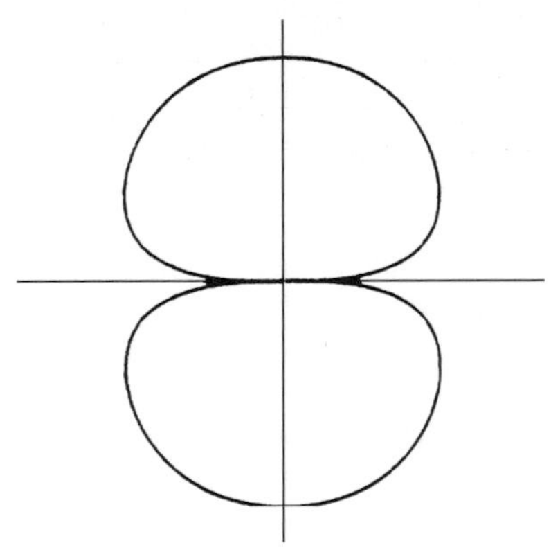

Fig. 9.5 – Bidirectional (figure-of-eight) microphone polar pattern. Picks up sound equally at the front and back of the microphone.

Microphone placement

When using a microphone your objective should be to only capture the sound you want to use. This can be more difficult than it sounds, as there may be several sound sources close together or the microphone may have a wide pick-up area.

You will also need to decide if you are going to close mike an instrument or use a more distant ambient miking technique. Close miking means placing the microphones close to the sound source. Ambient miking means placing the microphone further away from the sound source. Dynamic microphones are generally less suitable for ambient miking than condensers, as they only have a limited high-frequency response and start to lack sensitivity when moved beyond the range of one foot (30 cm) or more away from the sound source.

producer says ▶

> Microphone placement is a real art, so don't expect great results straight away. If you're having problems achieving a good result, try varying the distance of the microphone from the sound source or adjusting the height of the microphone over the sound. You can even try pointing the microphone in a different direction. You will find that moving a microphone a small amount can considerably affect the way in which it picks up the sound. As always, experiment.

Using several microphones at the same time

When using several microphones at the same time, you should try and achieve the best separation you can between each microphone, as even the smallest amount of leakage will become noticeable on a recording. If a microphone is picking up two sounds at the same time it will be impossible to separate them at a later stage, so always try and position the microphones so they only record what they are pointing at. Professional recording studios often use screens and isolation booths to help improve separation between different instruments.

tip ▶

> When miking an instrument you should always try to point the mike directly at the sound source so it rejects sound coming from other directions.

A typical example of where several different microphones can be used at the same time is when recording a drum kit. This is where microphones are carefully positioned to collect sound from different areas of the drum kit.

Fig. 9.6 – A drum kit is typically recorded using several different microphones.

tip ▶

> When using a microphone with a cardioid pick-up pattern, try pointing the back of the microphone at the sound you don't want to record.

Stereo microphone techniques

You also need to consider if any of the sounds you plan to record will benefit from being recorded in stereo (see 'Stereo sound' section in Chapter 8). Remember, stereo sounds require two audio tracks, so only do this if it is absolutely necessary and if it's actually improving the sound. You will find there are several different ways you can position two microphones to make a stereo recording. Here are some commonly used examples.

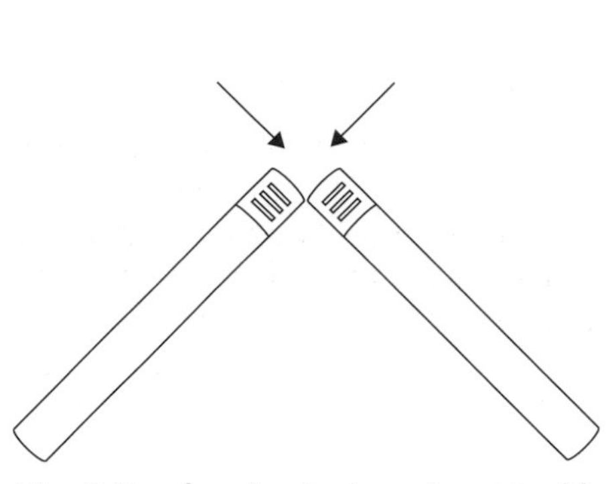

Fig. 9.7 – A pair of microphones with their capsules positioned very close together.

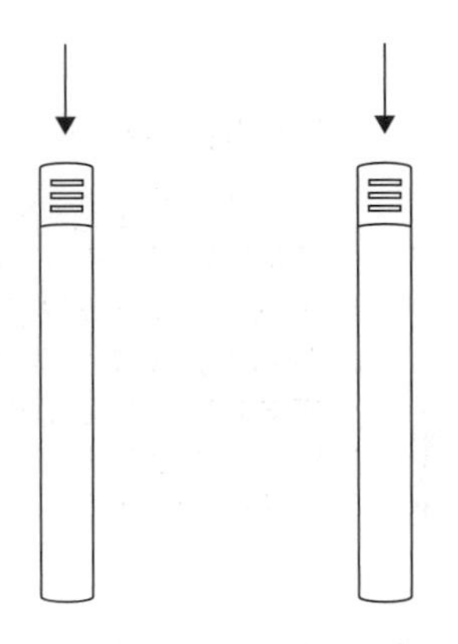

Fig. 9.8 – A pair of microphones at a set distance from each other.

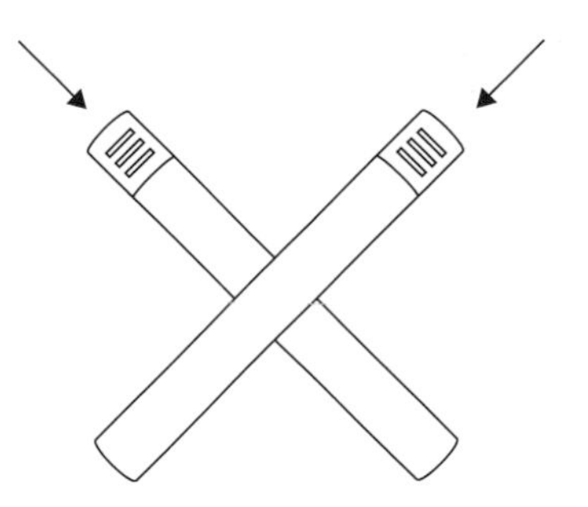

Fig. 9.9 – A pair of microphones crossed at a 90° angle.

note ►

If you are planning to make a live stereo recording you will need to use two microphones.

Positioning *stereo microphones* in a room can be a very difficult job, as you have to consider:

- The sound of the actual room itself
- How to make best use of the stereo field

- How ambient you want the sound to be
- The distance between each microphone
- The balance between the *original sound* and the *sound of the room*.

producer says ▶

> If you are making a stereo recording, always try and use a pair of matching microphones, i.e. two identical microphones that are of same model and make.

10
The Mixing Desk

A mixing desk is usually the main focal point of any recording project, so it's essential that you become familiar with its operation. In this chapter we are

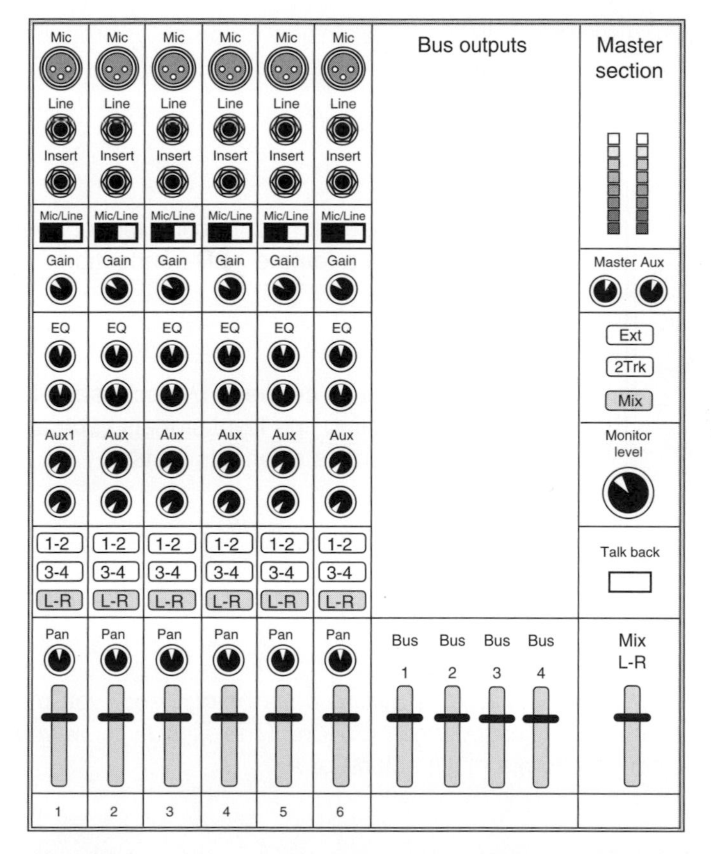

Fig. 10.1 – The actual layout of a mixing desk can vary tremendously depending on whether it is digital or analogue, or hardware or software. In this example, the mixer is divided into three main sections: input channels, bus outputs and master section.

going to explore the many different controls you would normally expect to find on a traditional stand-alone mixing desk. You will, however, find most of these controls relevant to any type of mixing desk, whether it is stand-alone or incorporated into a computer software program.

Input channels

Input channels can be used to connect and listen to various different sound sources, such as electronic keyboards and microphones. The number of input channels will usually determine how many different sounds can be used and heard at the same time. The first stage of a channel is where you actually connect the instruments or devices you plan to use. Most mixing desks will usually offer you a choice of different inputs, allowing you to choose the most suitable for the device you plan to connect.

note ▶ If you're using a software mixer inside a computer program, connections will be made to the audio interface.

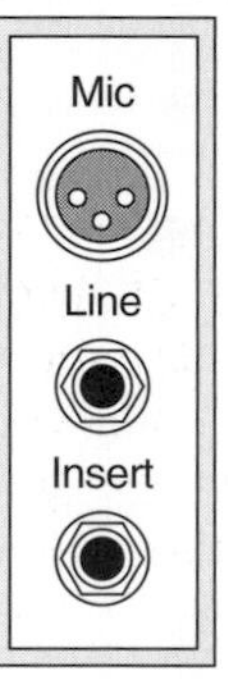

XLR inputs

These can be used for connecting microphones or DI boxes. Note: this will be a female XLR connection.

Line inputs

These allow you to connect 1/4-inch jack plugs and can be used for connecting line level signals such as electronic keyboards, synthesizers or effects units.

Fig. 10.2 – Input channel connections.

note ▶ **Tape monitors**

Depending on the type of mixing desk being used, you may have some additional inputs called tape monitors. These inputs allow you to return signals coming from a multitrack recorder.

Insert point

This is a special type of connection which is designed to simplify the process of connecting external devices, such as noise gates and compressors, to a mixing desk. It works by providing an output to allow signals to be sent directly to the input of an external device and an input to return signals from the output of an external device. Once a device has been 'inserted' it effectively becomes part of the channel. If you are using a digital mixing desk you may be able to insert a software plug-in.

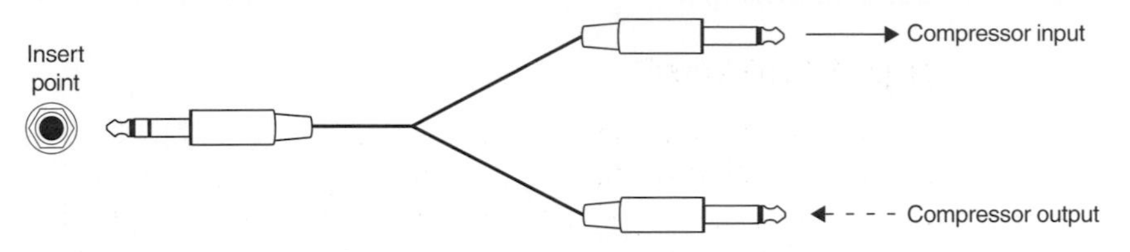

Fig. 10.3 – Signal flow diagram showing how a compressor can be connected using an insert point.

It's common for a mixing desk to only provide one connection for an insert point. This can appear quite confusing at first, as you would expect to see two connections – a send and a return. If this is the case you will need to use a special type of cable that uses a stereo jack plug to separate the two signals.

note ▶ An insert point allows you to send and return a signal to and from a device, such as a compressor or noise gate.

The Mic/Line switch

This switch will allow you to choose which input signal enters the channel. When set to Mic it will allow any signals made via an XLR connection to enter the channel. When set to Line it will allow any 1/4-inch jack signal to enter the channel. Regardless of the type of connection you have made to the channel, you will still need to choose which input you want to use.

Fig. 10.4 – Mic/Line switch set to Mic.

note ▶ The Mic/Line switch allows you to choose which signal enters the channel.

Phantom power

Located near the XLR input you may find a (+48 volts) phantom power switch. This is a small voltage that can be sent out of the XLR input of a mixing desk in order to power condenser microphones and DI boxes. It can usually be switched on or off for each individual channel or switched globally for all the microphone inputs. The reason it is called phantom power is because a small voltage travels invisibly down a microphone cable with the audio signal. If it didn't, you would need to connect a separate cable to power the microphone.

Pad

The input section may also include a pad switch. This switch allows you to reduce the amount of signal entering the input of the mixer by a set amount. It is useful for turning down very loud signals so they can enter the mixer without overloading it.

Gain

The gain control will allow you to adjust the level of the input signal, which can be either Mic or Line. The idea is to use the gain control to get the strongest signal possible into the channel without noise or distortion. This can sometimes be difficult to achieve, as each sound source will vary in level and most levels naturally fluctuate.

- If you turn up the gain too high, you will get distortion
- If you set the gain too low, you will not get a very strong signal and may start to hear noise.

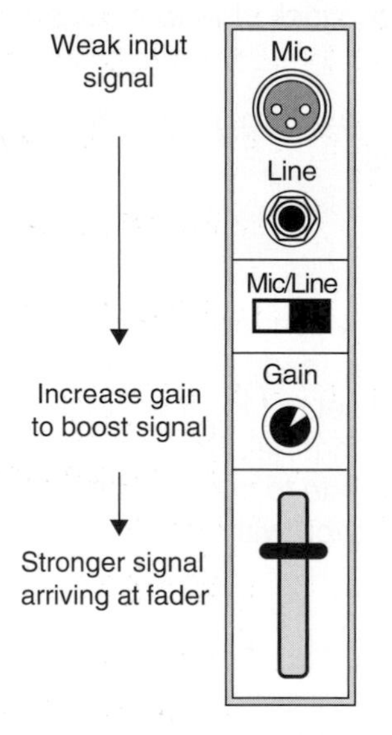

Fig. 10.5 – Diagram showing how the level of an input signal can be increased using the gain control.

note ► If you set the gain too high and the channel starts distorting, moving the channel fader down will not get rid of the distortion, it will simply turn down the volume of the distortion. You will need to turn the gain control down!

Equalization

Most mixers have equalization (EQ) built into each separate channel. This allows you to make different tonal adjustments to each individual channel. In its simplest form, EQ allows you to cut or boost the bass or treble of a sound, so if you want a brighter sound, boost the high frequencies; for a basser sound boost the low

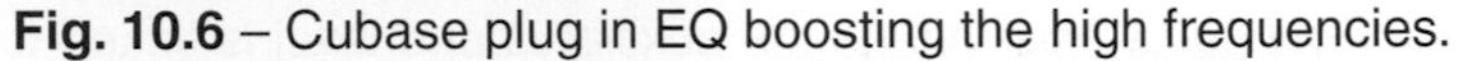

Fig. 10.6 – Cubase plug in EQ boosting the high frequencies.

frequencies. However, there are many different types of equalizer available that offer more choice and flexibility by allowing you to choose the frequency which you want to cut or boost. Most equalizers found on a mixing desk are split into separate frequency bands that allow you to cut or boost the selected frequency.

- *Shelving equalizers* are the simplest type of EQ – they allow you to turn up or down all the bass or all the treble
- *Parametric equalizers* allow you to make detailed adjustments to frequency and level
- *Graphic equalizers* allow you to control the levels of many different frequencies at the same time
- *High pass filters* (HPF) allow you to remove low frequencies such as the bass without affecting the high frequencies (see Fig. 10.7)
- *Low pass filters* (LPF) allow you to remove high frequencies such as hiss without affecting the low frequencies (see Fig. 10.8) .

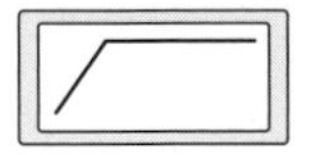

Fig. 10.7 – HPF symbol.

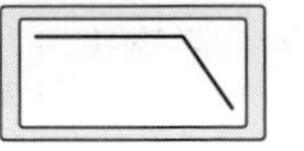

Fig. 10.8 – LPF symbol.

Auxiliary and cue sends

Each channel on a mixing desk will normally have access to an auxiliary send. A dial will allow you to choose how much signal is sent out from each channel to an effects unit to add reverb or delay, or to a pair of headphones to provide a foldback mix. Most mixing desks usually have two or more Aux sends, making it possible to send signals to several different effects at the same time.

note ► When using a software mixer you may have to create and route each auxiliary send before it can be used.

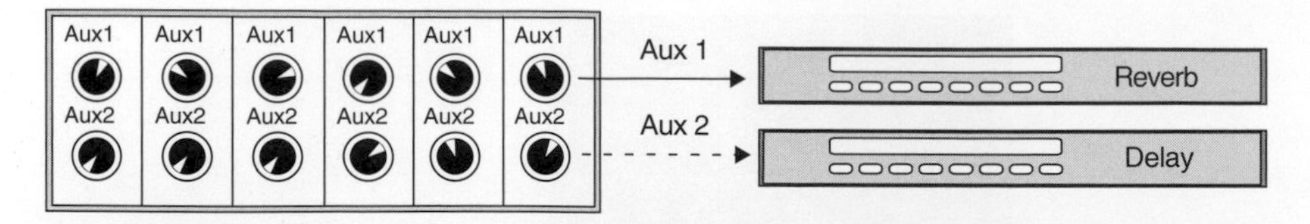

Fig. 10.9 – Each auxiliary send can be used to send signals to different devices. In this example Aux1 is being used to add reverb while Aux2 is being used to add delay.

Pre or post

You will normally find an additional switch or option to choose how each auxiliary send will work in relation to its corresponding channel fader.

- *Post send.* This means the auxiliary send level is linked to the level of the channel fader. For example, as you move the channel fader down, the auxiliary send level will also move dow n. This mode should be selected when using an auxiliary to send a signal to an effects unit, such as reverb or delay, and is also commonly used during mixdown.
- *Pre send.* The auxiliary send level becomes independent and is not linked to the channel fader level in any way. This is more commonly used when creating a 'foldback mix', as it allows you to set up two independent mixes – one on the faders and one using the auxiliary sends.

Fig. 10.10 – The level of each auxiliary send can usually be configured to work in two different ways in relation to the channel fader.

Channel routing buttons

Routing buttons allow you to choose where to send the signal contained in the channel. Normally, each channel is routed to the left and right stereo mix bus, so you can hear it via the main stereo output. However, if you want to record a signal to a multitrack, you may need to send the signal to a bus in order for it to reach the multitrack's input. The number of destinations you can route a signal to will vary depending on the flexibility of the mixer. If you are using an eight-bus mixer, for example, then you will have eight other possible destinations to send the signal to.

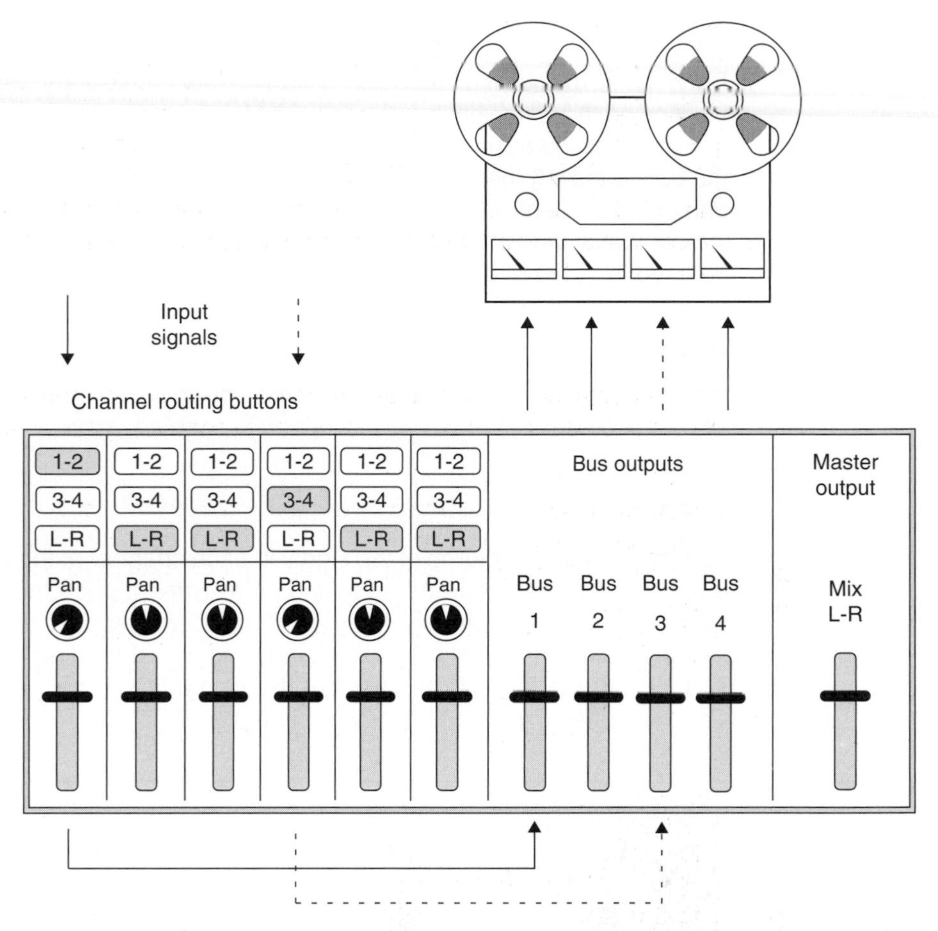

Fig. 10.11 – Signal flow diagram showing how input signals can be routed to a multitrack recorder.

note ► Routing buttons allow you to send a signal to different destinations, such as the master left and right stereo bus or to a multitrack recorder.

Routing and pan

Sometimes one button will route a signal to two different destinations, as some routing buttons work in pairs, 1–2, 3–4, 5–6, 7–8. You can determine the destination by using the pan control – pan left for odd numbers and right for even numbers. Left goes to 1, right goes to 2, etc.

Fig. 10.12 – Diagram showing different pan positions.

Pan control

Each individual channel will have a *pan control*. This rotary dial allows you to position each sound within a stereo field. Moving the pan pot will move the position of the sound from left to right or ear to ear, speaker to speaker. If you move the pan to the left then the sound will move over to the left speaker. Setting the pan pot to the centre (12 o'clock position) will send equal amounts of the sound to both the left and the right, making the sound appear in the middle. It can also be used in conjunction with the channel routing buttons to select between odd and even channels and to monitor a stereo input signal in stereo.

note ▶ You will only hear the pan pot working correctly if you are actually monitoring in stereo.

Channel faders

Faders control the volume of the signal in each channel of the mixing desk, so if you want to hear more of the sound or instrument push the fader up. If you want to hear less of the sound, move the fader down. It's also worth noting that the gain control will affect the amount of signal feeding into the channel. If you push the fader to the top but the signal is still not loud enough, you could increase the gain to give you more signal.

producer says ▶ Setting two fader levels the same doesn't necessarily mean their volume levels will be the same. The strength of the signal entering a channel is determined by the gain. Note that if a channel contains noise or is distorted, the fader will only turn up or down the noise or distortion, not remove it.

Solo and mute buttons

A channel mute or cut switch allows you to switch each channel on or off. This means the signal will still be there but you are just not hearing it. Pressing solo allows you to listen to the selected channel on its own; however, there are several different modes of solo.

Different types of solo

- *In place solo* – cuts all the other channels and you hear the channel at its current fader level and pan position.
- *PFL (pre-fade listen)* – this brings the signal directly to a separate monitor amplifier and ignores the fader and pan settings. This is useful for setting the correct amount of gain.
- *AFL* – this option is similar to PFL but the fader level is also taken into account.

- *Solo safe* – this allows you to exclude a track from being soloed. A good example would be to use solo safe on your reverb returns so you could still hear the reverb while in solo.

note ► Pressing solo will allow you to hear one channel at a time.

Channel metering

Channel meters show individual signal levels for each channel. Some mixing desks may have an individual meter to show the signal strength for each channel, while others may simply have a master meter to show the overall signal strength for the left and right stereo bus. If you are using a mixing desk that only has a stereo LED meter, you will need to solo a channel to visually view the strength of the input signal.

Bus outputs

The bus output section of a mixing desk is normally used to send signals out of a mixing desk to a multitrack recorder. It works in conjunction with the channel routing buttons, allowing you to choose which signals arrive there. For example, if a channel signal is routed to bus 1 using one of the channel routing buttons, it will arrive at bus 1. You would then be able to use the bus 1 output fader to control the level of the signal being sent out from the mixing desk.

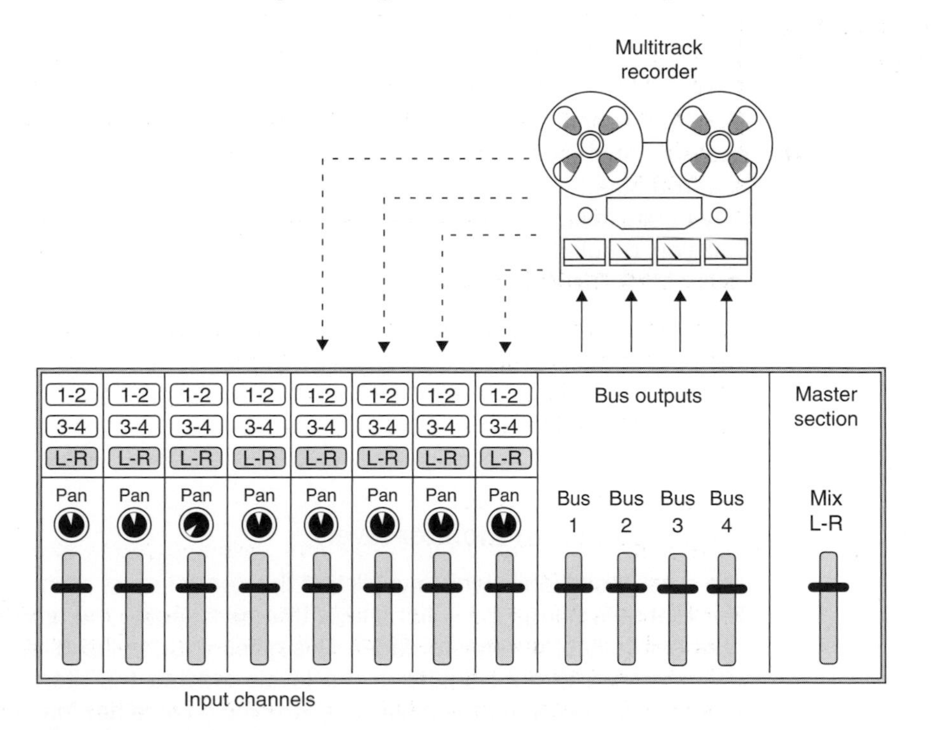

Fig. 10.13 – Mixer-to-multitrack connections.

If you are using a mixing desk and a stand-alone multitrack recorder you will be able to send signals out from the mixer in to the multitrack. This will require a physical connection between the two devices. Each connection will need to be correctly configured so the bus outputs of the mixing desk correspond to a multitrack's inputs.

Master section

The master section is one of the most important sections of a mixing desk as it contains several controls that allow you to adjust the overall volume level leaving the mixer, and your listening level.

The master fader

This is where the master fader is located and where all channels that are routed to the stereo bus will arrive. This is a stereo fader that controls the overall signal level leaving the desk. Moving the master fader will adjust the level being sent out of the mixer to the amp and speakers, and the stereo output levels being sent to a two-track or stereo recorder. It should normally be set around 3/4 of the way up or at the marked zero or unity gain position and can be used to create a fade out at the end of a mix.

Overloading the stereo bus

All channels that are routed to the stereo bus will arrive here and the level of each channel fader will contribute to the overall level arriving at the stereo bus. You should therefore take care not to overload the master fader, as this may cause 'distortion' or 'clipping' on the final mix. We therefore recommend that you regularly observe the master meters that show the overall output level of the desk.

tip ▶ Care must be taken to not overload the stereo bus, as this may cause distortion on the final mix.

The control room monitor level

The master section is where you'll find the control room monitor level control. This dial determines the volume level that gets sent to the amp and speakers independently of the master fader. This gives you the flexibility to adjust your listening level without adjusting the levels leaving the stereo mix bus and should always be used to adjust your listening level rather than using the master fader. Remember, your listening levels are different to the recording levels.

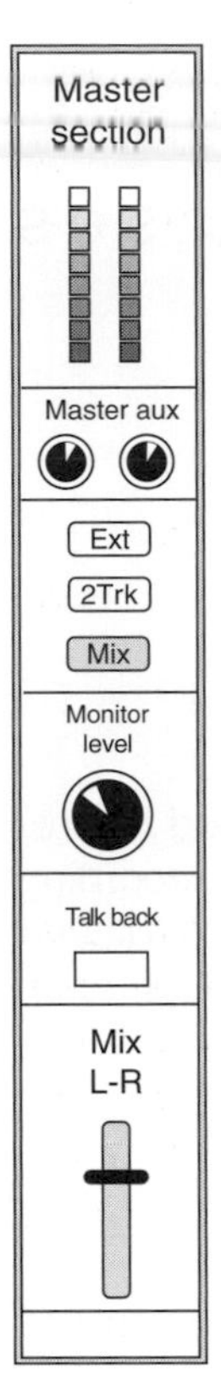

Fig. 10.14 – Master section.

Control room monitor switches

The master section may provide a set of buttons so you can choose which signal you will hear through the control monitor speakers. The vast majority of the time you will want to hear the output of the stereo bus; however, this gives you the option of hearing other external stereo sound sources such as a CD player, or stereo recording device such as a DAT or cassette, through the studio monitor speakers without having to connect them to the input channels. These buttons are often labelled Mix, Tape1, Tape2 and External.

Talkback switch

This switches on or off a small microphone that is usually built in to the master section console so you can communicate with the performers if they are wearing headphones.

Headphone output

This will allow the engineer to check how a mix sounds on a pair of headphones or to check the foldback mix (what the performers are hearing in their headphones).

Master auxiliary sends

These control the overall send level for each auxiliary. This is useful for turning up or down the overall foldback level or the overall send level to an external effects device.

Auxiliary returns

These are designed so you can connect the outputs of any external effects units. They are basically scaled-down input channels that only consist of a level and pan control. The advantage of using auxiliary return is that you don't have to waste an input channel in order to hear your external effects.

Alt speaker switch

Some studios have more than one pair of studio monitor speakers. This switch allows you to choose which set of speakers you want to listen to. Small speakers placed near the desk are known as near fields, while larger speakers are known as main monitors.

11 Audio Exercises

This chapter contains a series of audio exercises that are designed to prepare you for the main audio projects. The exercises break down the recording process into manageable chunks so you can focus in detail on one particular aspect at a time. We therefore recommend you work your way through each exercise before attempting the main audio projects.

Exercise 11.1 Connecting a microphone

Before attempting this exercise we recommend you read the 'Audio equipment' section in Chapter 7, along with Chapters 9 and 10.

This exercise will cover the following:

- Different types of connection
- Switching between mic and line
- Using phantom power
- Recording directly into a computer
- Recording in stereo
- Using the phase reversal switch.

The exercise ...

If you are planning to record or capture any analogue sound, such as vocals, acoustic guitar, piano, drums, or make a live stereo recording, then you are going to need a microphone. In this exercise we are going to explore the different options you will have when connecting a microphone. Note: throughout this exercise you should not be hearing any sound from the microphone.

Before you start

Turn your monitoring system down. It's always a good idea to turn your monitoring system down before making a new connection, as this will help protect your monitoring system from any sudden bursts of sound. If you are using a mixing desk you should turn down the control room monitor level.

Select the type of microphone you plan to use. You will first need to select the type of microphone you wish to use for the recording, e.g. dynamic or condenser, and make sure the appropriate polar pattern is selected (see Chapter 9).

Microphone placement

Microphones usually require microphone stands, unless the performer is planning to hold the microphone. Note that certain types of dynamic microphones are designed to be hand-held. When using a microphone stand, make sure the stand offers good support for the microphone and that you tighten the stand firmly to prevent the mic from changing position. A *mic clip* will also be needed to attach the mic to the stand. Make sure you have the correct type of mic clip to firmly support the mic, as some mics require their own special clips. In the example in Fig. 11.1.1 a condenser microphone is being supported using an elastic cradle to help provide better isolation from any floor vibrations and a pop shield has been placed in front of the mic.

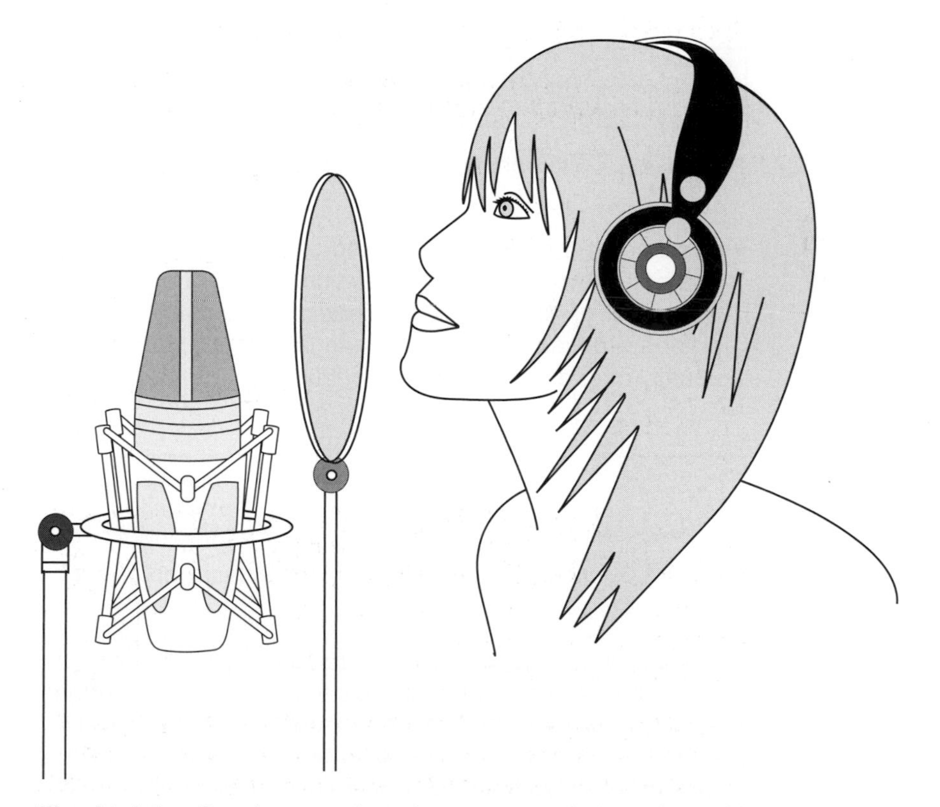

Fig. 11.1.1 – Condenser microphone set up to record vocals.

Before making a connection to a microphone it is usually preferable to roughly position the microphone where it is going to be used, as you can always make finer adjustments later. This makes it easier to determine the length of microphone cable needed to connect it.

Different types of recording equipment

The device you actually connect a microphone to may vary depending on the type of recording equipment that you are using (see 'Different types of recording environments and set-ups' section in Chapter 7). Here are some commonly used examples:

- *Mixing desk.* A mixing desk with XLR/mic inputs is probably one of the most commonly used devices that microphones are connected to, as this usually provides the most convenient way to route the microphone signal to a recording device. When using a microphone in the live room of a recording studio, you will probably have to connect it to the mixing desk via a mic box. This simply provides an extension of a console's mic inputs. When connecting a microphone to a software mixer inside a computer, you will need to connect the microphone directly to the computer's audio interface.

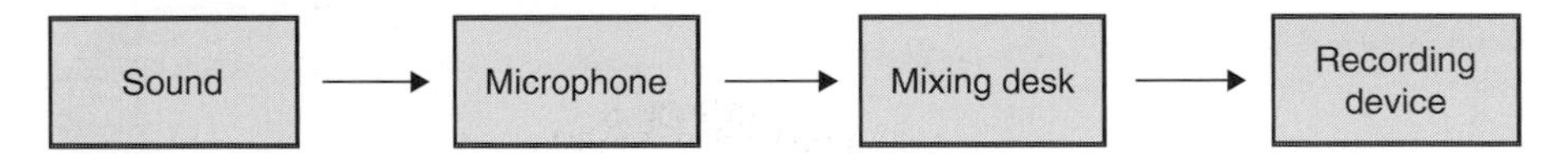

Fig. 11.1.2 – Microphone connected to a mixing desk.

- *Mic amp or channel strip.* A stand-alone mic amp or channel strip offers an alternative way of connecting a microphone. They can be used alongside or instead of a mixing desk to connect a mic directly to a multitrack or computer soundcard. Note: most stand-alone mic amps or channel strips will usually offer a higher-quality signal path than most mixing desks.

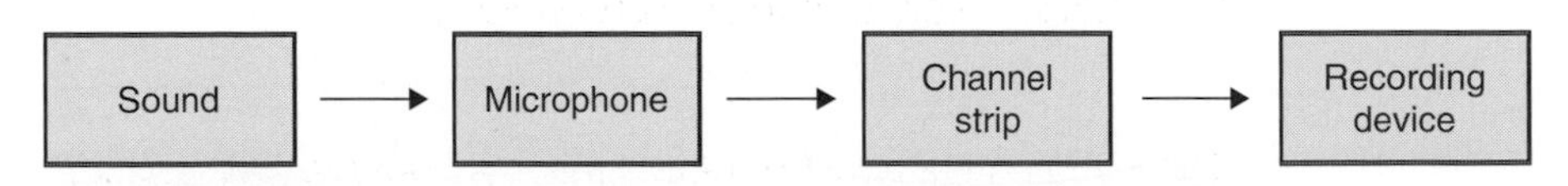

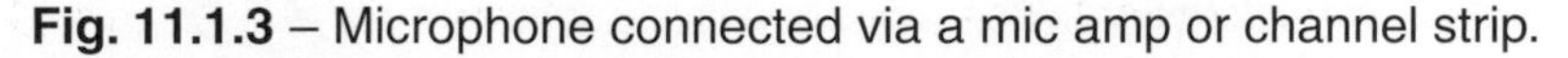

Fig. 11.1.3 – Microphone connected via a mic amp or channel strip.

- *Computer soundcard or audio interface.* If you are using a computer with an audio interface it may be possible to connect a microphone directly to it and avoid using a mixing desk or mic amp altogether. This will only be possible if the interface itself has an XLR/mic connection and you are able to monitor the signal without audio latency. It may therefore be more practical to connect a microphone to a mixing desk or channel strip before sending a signal to a computer, as it will give you more control over the signal and help you monitor it if latency is a problem.

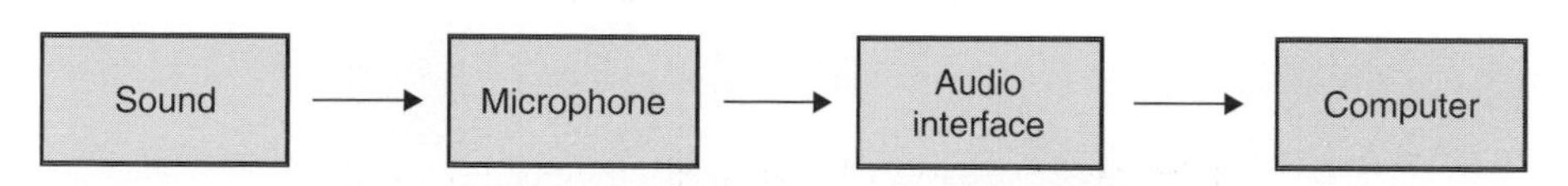

Fig. 11.1.4 – Microphone connected directly to a computer soundcard.

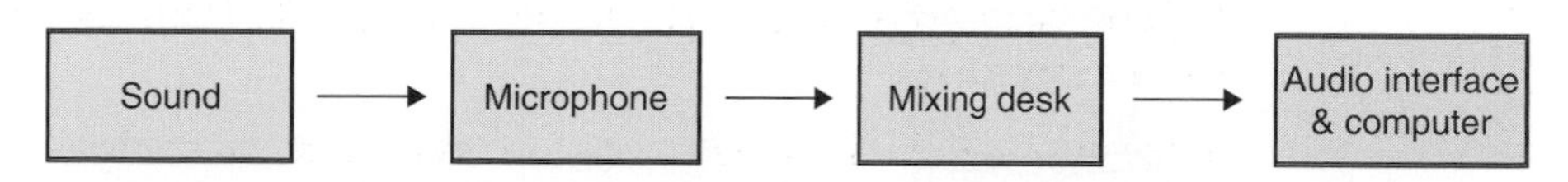

Fig. 11.1.5 – Microphone connected via a mixer to a computer soundcard.

Making a connection

A microphone is normally connected using an XLR cable. Make sure the cable you are planning to use has a female XLR on one end and a male XLR on the other (the female XLR end goes into the microphone).

Fig 11.1.6 – Male to female XLR cable.

If you don't have an XLR input available, see the next section, 'Different types of connection'. The options you will have available to actually connect the microphone to will vary. Here are some examples:

- XLR/mic input on a mixing desk
- XLR/mic input on a stand-alone mic amp or channel strip
- XLR/mic input on a computer soundcard or audio interface
- XLR/mic input on a recording studio mic box
- 1/4-inch jack input on a mixing desk (see next section, 'Alternative ways of connecting a microphone').

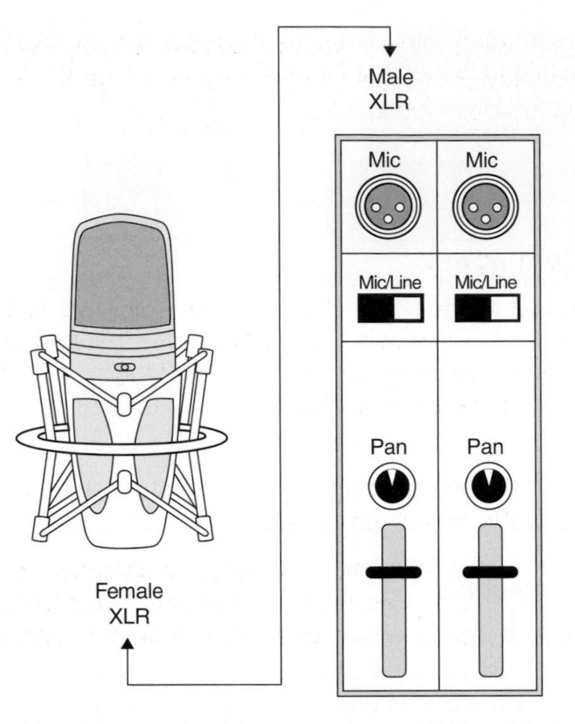

Fig. 11.1.7 – In this example an XLR cable has been used to connect the microphone to the XLR input on the mixing console.

tip ► Make sure the microphone cable is long enough to reach between the microphone and recording equipment. If you have a spare XLR cable it is possible to join two cables together to extend the cable.

Alternative ways of connecting a microphone

If you don't have an XLR input on the equipment you plan to connect the mic to, or the microphone has a cable already attached, you will need to connect via the line input instead of the mic input. We don't recommend this method of connection, as it often introduces more noise into the signal path and you may struggle to get sufficient signal to make the microphone signal loud enough. However, sometimes it's unavoidable.

Switching between mic and line

If a microphone has been connected to a mixing desk or other device using the XLR/mic input, you may have to press a switch to allow the mic input to be used. This is normally labelled Mic or Line and allows you to choose which signal enters the device (see Fig 11.1.7).

note ▶ The Mic/Line switch allows you to choose which signal enters the channel. If you have connected a microphone via the XLR/mic input, then you will need to select Mic input.

Using phantom power

If you are using a condenser microphone you will need to supply phantom power in order for the microphone to work (see 'Phantom power' section in Chapter 9). This is normally achieved by enabling a +48 V button on the device the microphone is connected to.

Recording directly into a computer

If you have connected a microphone directly to a computer via an audio interface then you will need to select the input you are using from within the software. This is usually found in the audio mixer page of the software.

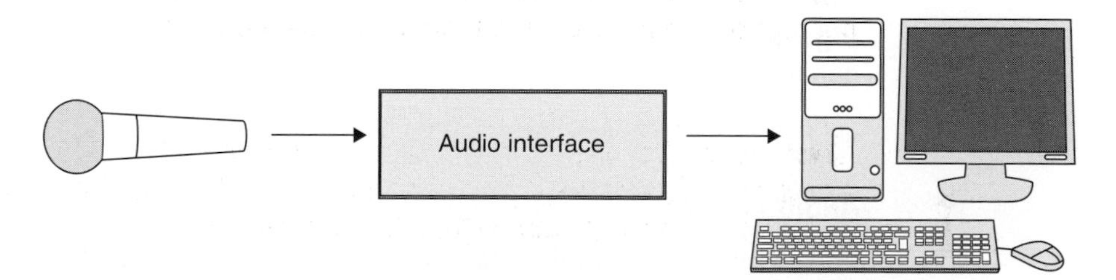

Fig. 11.1.8 – Microphone connected directly to an audio interface.

note ▶ If you're planning to record directly into a computer, it's now possible to purchase microphones with a USB connection. This removes the necessity of having an audio interface with an XLR input.

Switching between mic or line

You may be able to select between mic or line on the audio interface, or this may have to be done from within the software. It's worth noting that most audio interfaces provide line inputs, while others may offer an XLR connection, but only on some channels.

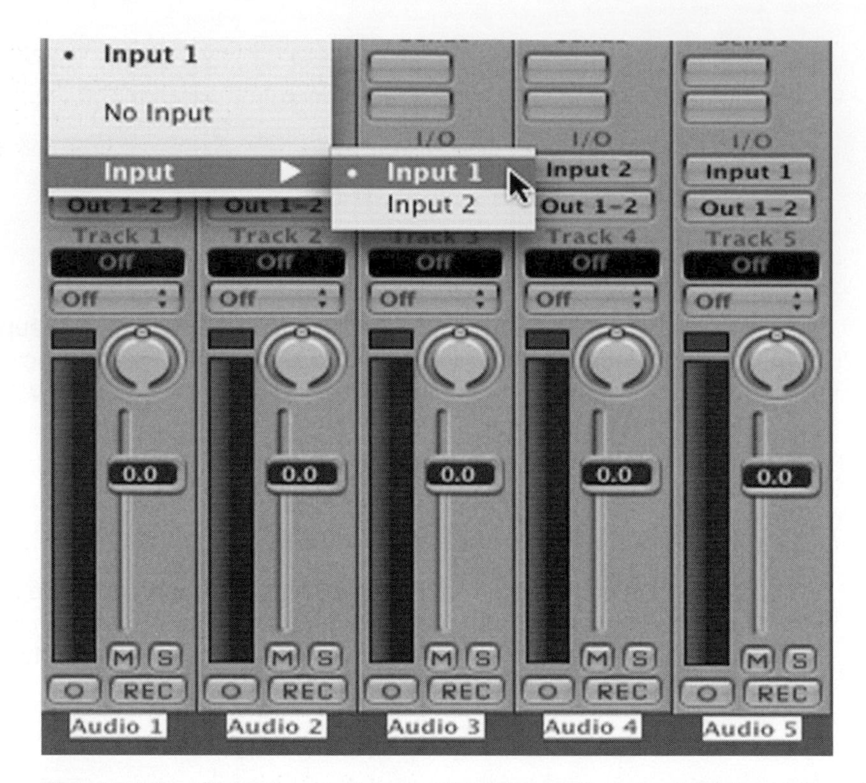

Fig. 11.1.9 – Selecting an input will tell the software which input the signal is coming in on.

Audio latency

You need to consider whether latency will be a problem if you are planning to monitor a signal passing through a computer. Latency is a delay between the computer's input and output. This will make the sound you are listening to appear later than it really is, so it sounds out of time (see 'Audio latency' section on p. 312).

Recording in stereo

In order to capture a sound in stereo you will need to use two microphones at the same time. Ideally, you should use two identical microphones as this will help balance the overall sound collected from each.

note ► Two microphones will be required to record sound in stereo.

When using two microphones to capture the same sound, each microphone will pick up the sound slightly differently. This can create an interesting effect, making some sounds appear bigger and wider. Note: you will only hear this if you are actually monitoring in stereo and the outputs from the sound source are directed to separate speakers.

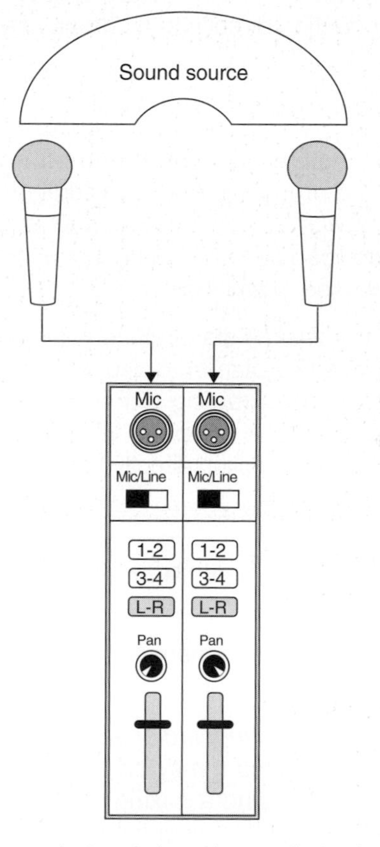

Fig. 11.1.10 – In this example two microphones have been connected to the XLR inputs of two separate channels of a mixing desk. Note the pan settings for each channel.

tip ▶ If you are monitoring a stereo signal using two channels of a mixing desk, make sure you pan each channel hard left and right in order to hear it in stereo.

Exercise 11.2 Connecting a sound source

Before attempting this exercise we recommend you read the 'Audio equipment' section in Chapter 7, along with Chapters 9 and 10.

This exercise will cover the following:

- Making a connection
- Switching between mic and line
- Connecting a stereo sound source
- Connecting an electric guitar via a DI box
- Connecting several instruments at the same time
- Connecting directly to a computer via an audio interface.

Introduction

Before making a recording you will need to connect any external instruments or sound-producing devices you plan to use. These will include electronic keyboards with built-in sounds, sound modules, samplers and electric guitars. The device you will actually connect them to will vary. Here are some commonly used examples:

- *Mixing desk*. A stand-alone mixing desk is a good option for connecting any devices you plan to record, as it will usually provide several different inputs and a way of routing signals to a recording device. When using a software mixer inside a computer without an external mixing desk, you will need to make a connection directly to the computer's audio interface (see Fig. 11.2.2).

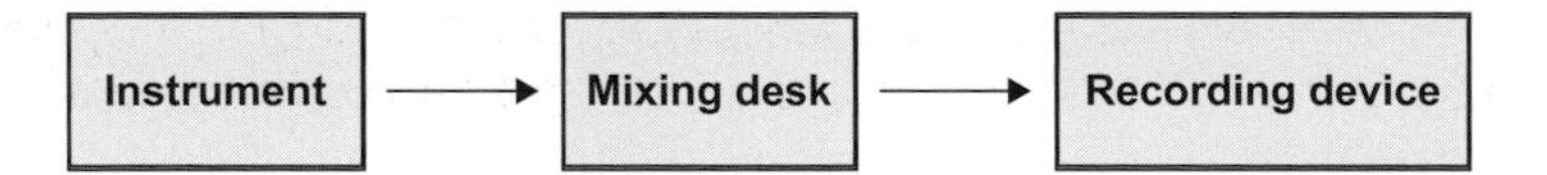

Fig. 11.2.1 – Instrument connected to a mixing desk.

- *Computer soundcard or audio interface*. If you are using a computer with an audio interface it may be possible to connect a sound source directly to it and avoid using a mixing desk altogether. However, it may be more practical to connect a sound source to a mixing desk before sending a signal to a computer, as this may give you more control over the signal level and help you monitor it if latency is a problem.

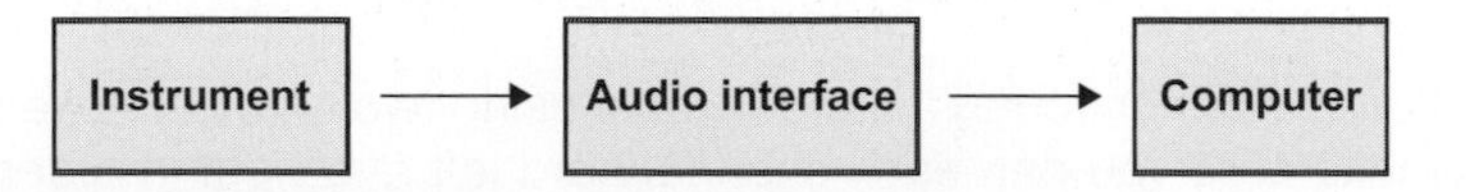

Fig. 11.2.2 – Instrument connected directly to an audio interface.

- *Digital connections*. It also may be possible to connect a sound source using a digital connection. This will only be possible if the sound source itself actually has a digital output and the device you plan to connect it to has a digital input. You will also need to make sure that both digital connections are compatible with each other.

note ► Some sound-producing devices now come equipped with a USB connection. This makes it possible to make a direct audio connection with a computer (see USB section in Chapter 3).

Choosing a cable

Most powered electronic musical instruments such as keyboards and sound modules can be connected directly to a mixing desk or computer soundcard using a 1/4-inch jack cable. This type of cable has a 1/4-inch jack plug on each end.

Fig. 11.2.3 – Jack cable.

The exercise ...

This exercise will explain how to connect various different electronic sound sources, such as electronic keyboards and sound modules, and electric guitars using a DI box. Note: throughout this exercise you should not be hearing any sound.

Before you start

Turn your monitoring system down. It's always a good idea to turn your monitoring system down before making a new connection, as this will help protect your monitoring system from any sudden bursts of sound.

note ► When using a mixing desk you should turn down the control room monitoring level. This is the overall level from the mixing desk that goes to the amp and speakers.

Making a connection

A sound source may have several different outputs, such as L and R or output 1 and 2. If you are planning to use the device in mono, choose either the left output or the output marked mono and connect this to the line input on the mixing desk (for stereo sounds, see 'Connecting a stereo sound source' section below).

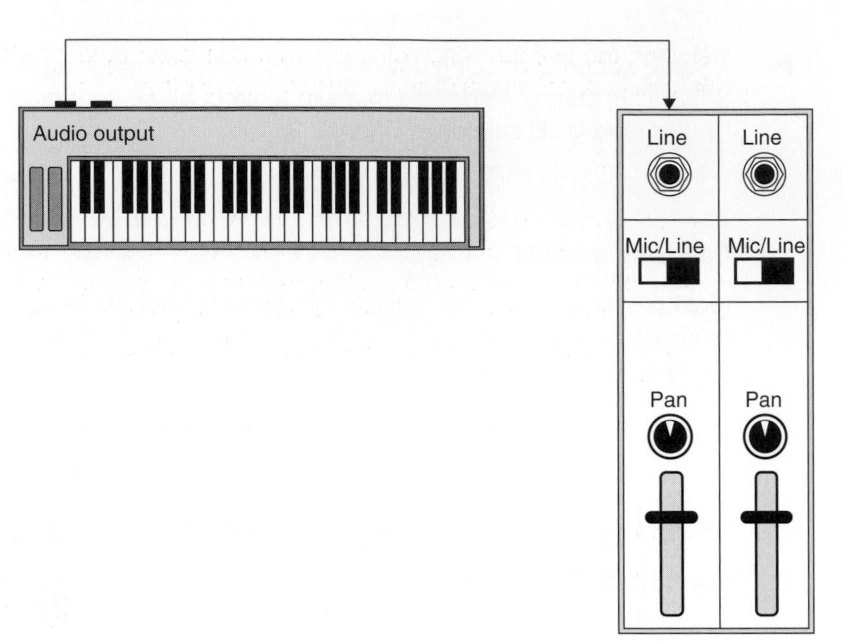

Fig. 11.2.4 – In this example the left mono audio output of the keyboard has been connected to the line input on the mixing desk using a 1/4-inch jack cable.

Switching between mic and line

If you have the option to switch between mic or line on the mixing desk's input channel, then choose line. This will allow the 1/4-inch jack signal to enter the channel. Note that selecting mic will allow any signal connected to the XLR input to enter the channel.

Fig. 11.2.5 – Mic/Line switch set to Line.

note ▶ The Mic/Line switch allows you to choose which signal enters the channel.

Alternative types of connection

Most mixing desks and soundcards will provide 1/4-inch line inputs. You may, however, encounter a phono-style connection, as some equipment may provide this instead of using a 1/4-inch jack plug. If you need to translate between the two formats, you can use a converter plug or a cable that has a 1/4-inch jack on one end and a phono on the other.

Connecting a stereo sound source

In order to use a device in stereo you will need to make two connections: one for the left output and one for the right output. Stereo sounds produce a slightly different sound from each output and can create an interesting effect, making some sounds appear bigger and wider. Note you will only hear this if you are actually monitoring in stereo and the outputs from the sound source are directed to separate speakers.

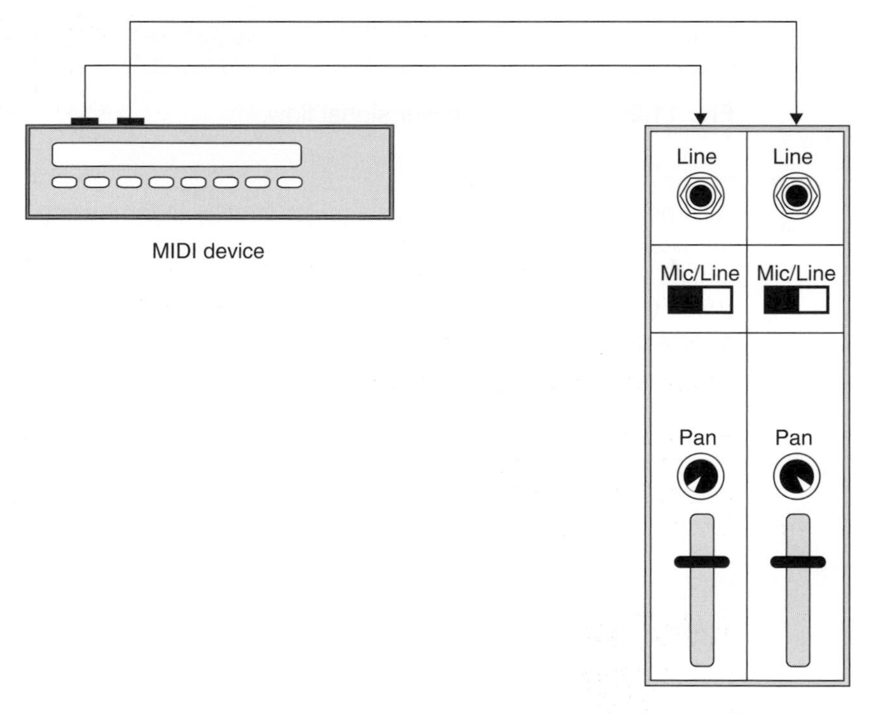

Fig. 11.2.6 – In this example both the left and right outputs of the sound source have been connected to the line inputs of two channels on the mixing desk. Note the pan settings for each channel.

note ▶ If you are monitoring a stereo signal using two channels of a mixing desk, make sure you pan each channel hard left and right.

Connecting an electric guitar via a DI box

Other types of instruments, such as electric bass and electric guitar, can also be connected directly to a mixing desk or soundcard. However, they will need to pass through a DI box (Direct Inject box). A DI box converts the signal produced by a guitar (without an amplifier) into a signal that is suitable to connect to an

XLR microphone input on a mixing desk. This eliminates the need to use a guitar amplifier and often provides a more convenient and simpler way to make a recording. Note: some audio interfaces may offer a special instrument or guitar input, so you won't need to use a DI box.

Electric guitars are normally connected to a guitar amplifier so sound can be produced from the guitar. By placing a microphone near the amplifier it's possible to record the sound being produced by the guitar. This is a perfectly normal way to record an electric guitar and can produce excellent results (see Exercise 11.1).

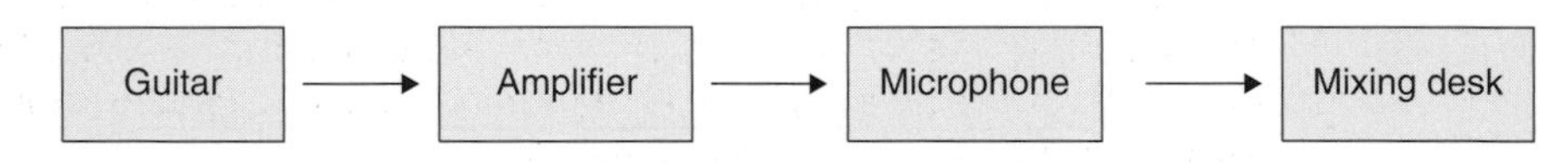

Fig. 11.2.7 – Guitar amplifier signal flow.

A DI box allows you to connect a guitar directly to the mixing desk, eliminating the need to use a guitar amplifier and microphone. This may be more convenient as you may not have a microphone or guitar amplifier available, or you may prefer the sound being produced this way, especially if the guitarist is using an effects pedal. This is a popular way to record a bass guitar.

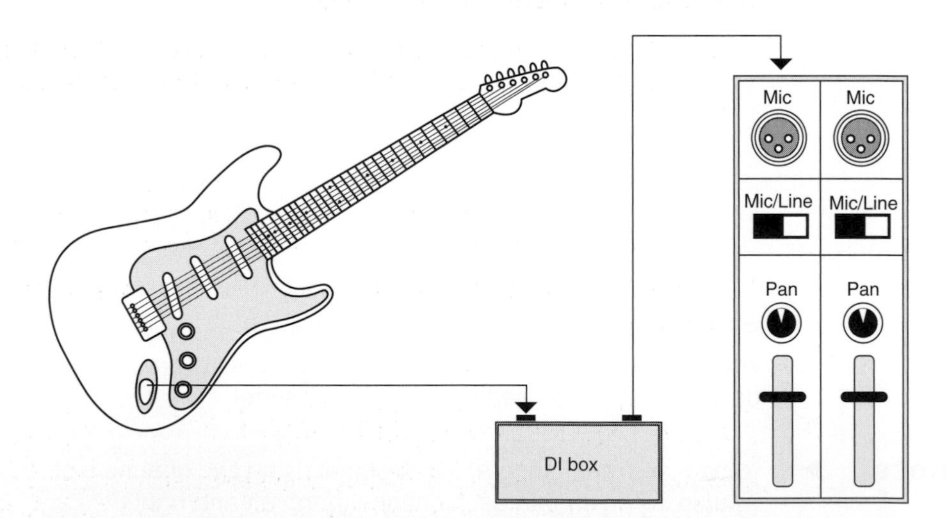

Fig. 11.2.8 – Guitar connected to a mixing desk using a DI box.

Connecting a DI box

The guitar's output will be connected to the line or instrument input on the DI box. The input on the DI box will be a jack input. The XLR output from the DI box will be connected to the mic or XLR input on the mixer or soundcard.

Cables

You will need two cables: a jack-to-jack cable to connect the guitar to the DI box, and a male XLR-to-female XLR cable to connect the DI box to the XLR input on the mixer.

Powering a DI box

A DI box will require power in order to work. This is normally achieved by placing a battery inside the DI box or by enabling the +48 V phantom power switch on the mixing desk or audio interface the DI box is connected to (see 'Phantom power' section in Chapter 9).

Fig. 11.2.9 – Mic/line Switch set to mic.

Switching between mic and line

As the DI box is connected to the mixing desk using an XLR, you will need to select the mic input. This will allow the signal from the DI box to enter the channel.

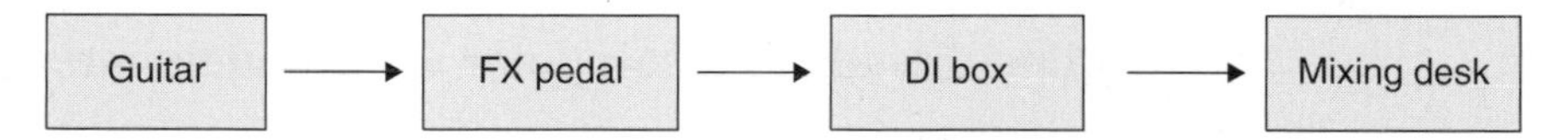

Fig. 11.2.10 – When using a DI box it is also still possible to use guitar effects pedals.

tip ►

Connecting a keyboard via DI box

It's also possible to use a DI box to connect an electronic keyboard or a sound module to a mixing desk or soundcard. This can be useful if only XLR inputs are available.

Connecting several instruments at the same time

When making a recording you will often find you will need to connect several instruments at the same time. An example is shown in Fig. 11.2.11.

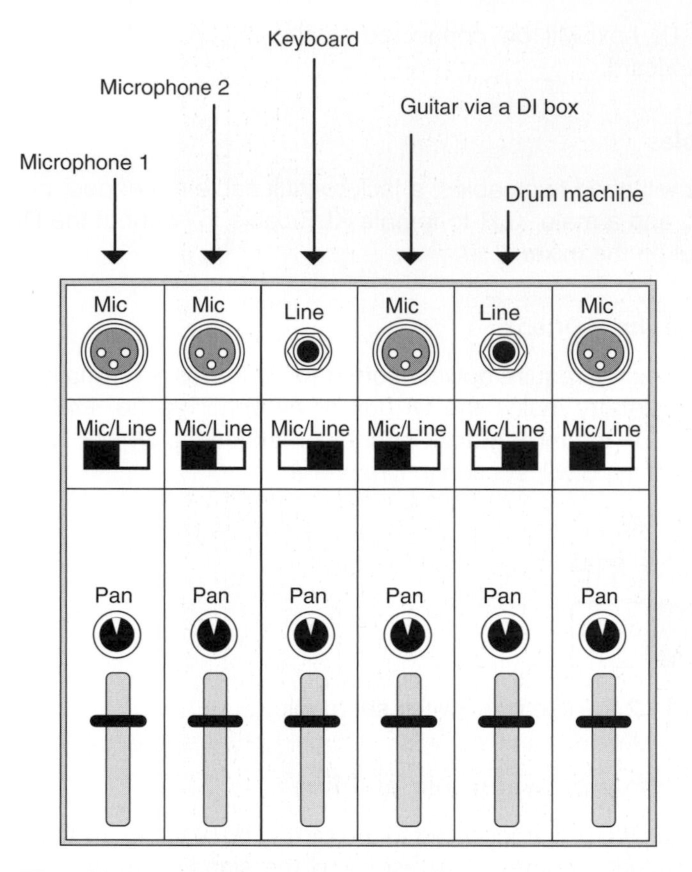

Fig. 11.2.11 – Diagram showing how several different devices can be connected to a mixing desk.

Connecting directly to a computer via an audio interface

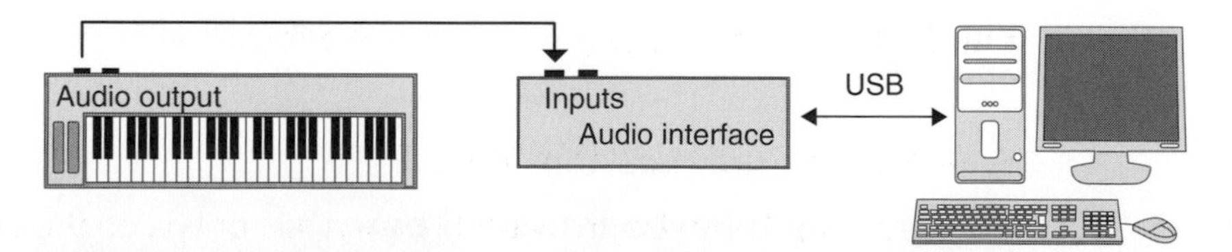

Fig. 11.2.12 – Sound source connected to an audio interface.

If you have connected a sound source directly to a computer via an audio interface then you will need to select the input you are using from within the software. This is usually found in the audio mixer page of the software.

Selecting an input will tell the software which input the signal is coming in on. Note: you will need to record enable the mixer channel in order to see the input signal.

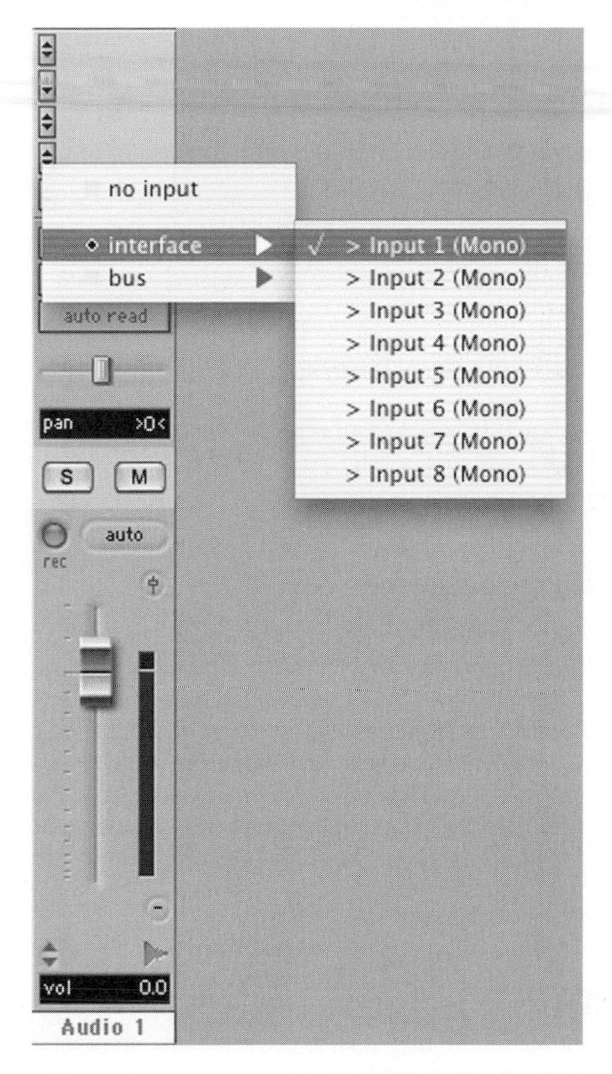

Fig. 11.2.13 – Input 1 being selected in Pro Tools.

Switching between mic and line

You may have the option to select between mic or line on the audio interface or this may have to be done from within the software itself.

Audio latency

You also need to consider whether latency will be a problem if you are planning to monitor a signal passing through a computer. Latency is a delay between the computer's input and output. This will make the sound you are listening to appear later than it really is, so it sounds out of time (see 'Audio latency' section on p. 312).

Now move on to signal routing.

Exercise 11.3 Routing and monitoring

Before attempting this exercise we recommend you read Chapter 10.

This exercise will cover the following:

- Routing a signal to the stereo bus
- Adjusting the channel fader
- Adjusting the master fader
- Adjusting the gain
- Setting up your input signal using solo
- Monitoring the sound
- Adjusting the control room monitor level
- Optimizing the input signal
- Monitoring a stereo sound source.

Listening to a sound source

Once you have connected a sound source the next stage is to set up how you're going to listen to it. This is normally done by using a pair of studio monitor speakers or by using a pair of headphones. However, before you can actually hear any sound you will need to make sure that the signal is routed, the gain control is set correctly, and any relevant faders and control room monitor levels are turned up. We therefore recommend that you have a basic understanding of a mixing desk and its signal flow before attempting this exercise (see Chapter 10).

The exercise ...

Choosing an output destination

In this exercise we are going to concentrate on routing a signal to the main left and right outputs or stereo bus, as this provides the simplest way to hear a signal. This will send the signal out from a channel directly to the master fader and then on to the studio monitor speakers or headphones. This is the ideal choice for listening to most sound sources, such as electronic keyboards, sound modules and effects units and should also be used to route microphones when making a stereo live recording. It can also be used to monitor back signals coming from a multitrack during a final mixdown. Any of the other output destinations you may have available, such as buses 3 and 4 or outputs 5 and 6, simply provide alternative routes for the signal and should only be used if you are planning to send the signal to an external multitrack or subgroup different channels together. Note that these are all covered in Exercise 11.4.

note ► Routing a signal to the stereo bus provides the simplest way to hear a signal.

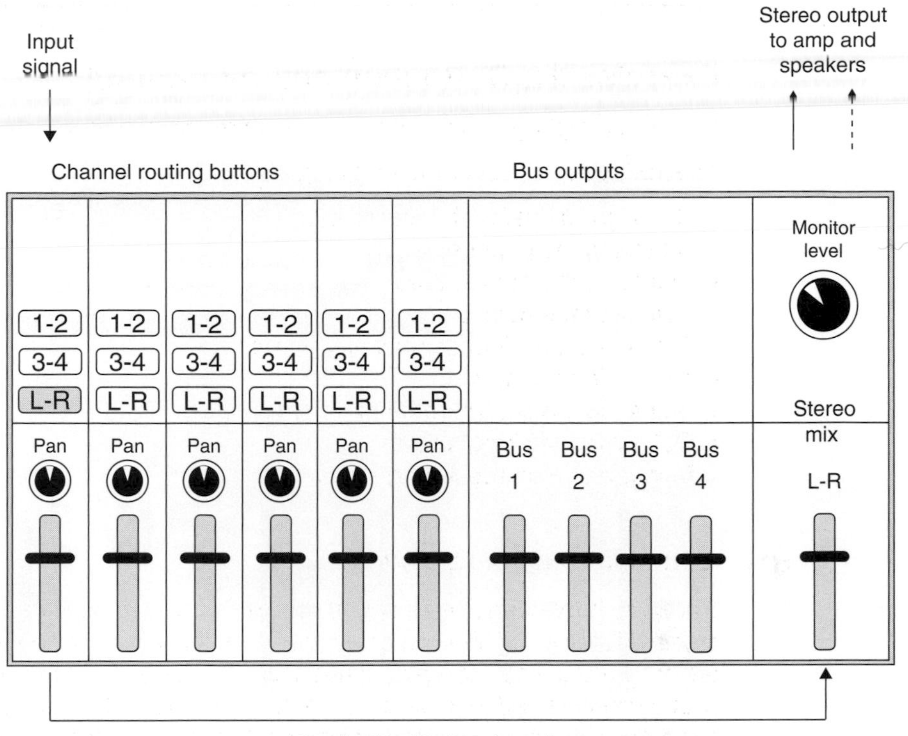

Fig. 11.3.1 – Diagram showing how an input signal can be routed to the stereo bus by pressing the L and R routing buttons.

Software routing

If you are using a computer system then the process of routing a signal to the stereo bus or master fader is basically the same.

Software mixing desks

Software mixing desks will allow you to route a signal to the main stereo bus by selecting output 1 and 2. They may also provide bus outputs to allow you to internally route signals within the software. If you have an audio interface connected to the computer that has several physical outputs, you will also be able to choose which output on the soundcard the signal gets sent to.

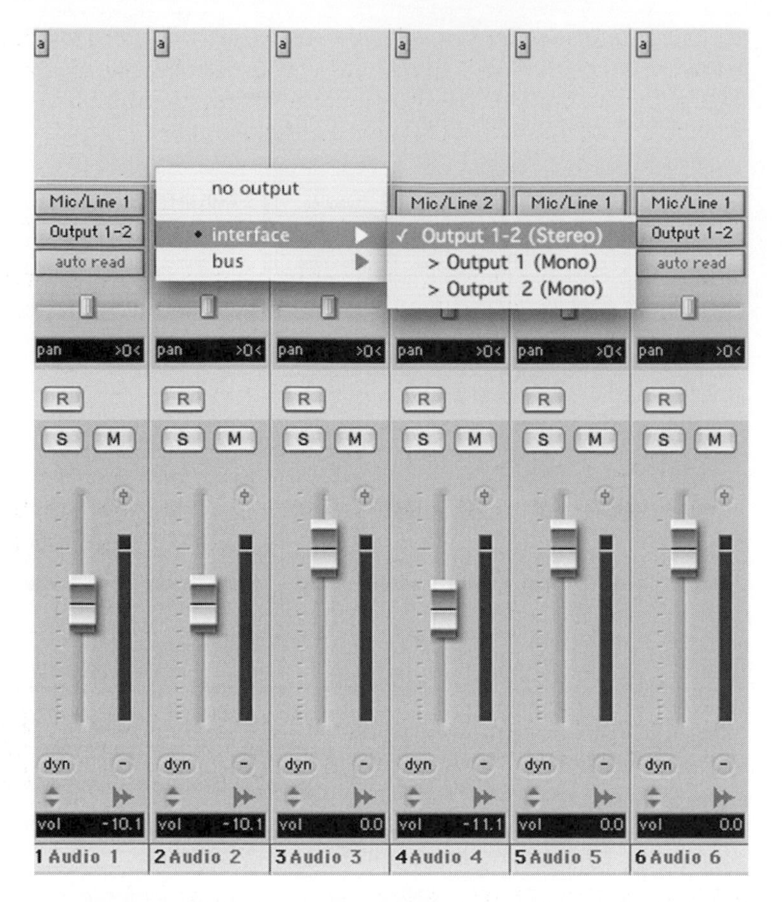

Fig. 11.3.2 – Selecting outputs 1 and 2 or Mix will route any signals produced by the channel to the main outputs. Note: when using a software mixer you may need to create a master fader.

Routing a signal to the stereo bus

In this exercise we are going to explain the different stages involved in actually listening to the sound source you have connected. Before you start we recommend that you turn your monitoring system down to help protect it from any sudden bursts of sound. You sould also ensure the gain controls on the channels you plan to use are turned down fully counter-clockwise (turn the gain fully left). This will result in the minimum amount of signal being sent into the channel.

Locating the routing buttons

Channels that have been used to connect microphones or instruments will need routing in order for you to hear them. Most mixing desks provide a set of routing buttons for each individual channel, allowing you to choose where you want each signal to go. The number of destinations available will vary depending on the flexibility of the equipment being used. If a channel is not routed or has no output you will not be able to hear it.

Routing the channels being used to the stereo bus

Press the routing buttons labelled Mix or L and R on the channels you want to hear. This will direct the signal contained in the channel to the master fader.

Adjusting the channel fader

Ensure that the channel fader is turned up. Try starting with the fader around 3/4 of the way up. This fader can be used later to make adjustments to the volume of the signal.

note ▶ The fader level of each channel will contribute to the overall volume level of the stereo bus.

Adjusting the master fader

Ensure that the master fader is turned up. This fader controls the overall volume level for each individual channel.

note ▶ You may have to create a master fader when using a software mixer.

Generating a sound

At this stage you will need to generate a sound in order to adjust the levels, so encourage the performers to rehearse or, if they are using a microphone, to simply count numbers. This will provide you with a signal so you can start setting levels and adjust your monitoring. It is also possible to connect a CD player or generate a simple tone from the mixing desks oscillator for the purpose of this exercise.

Look for a visual indication

Look for a visual indication to see if the input signal is registering on the meters. Some mixing desks may have an individual meter to show the signal strength for each channel, while others may simply have a stereo meter to show the overall signal strength arriving at the left and right stereo bus.

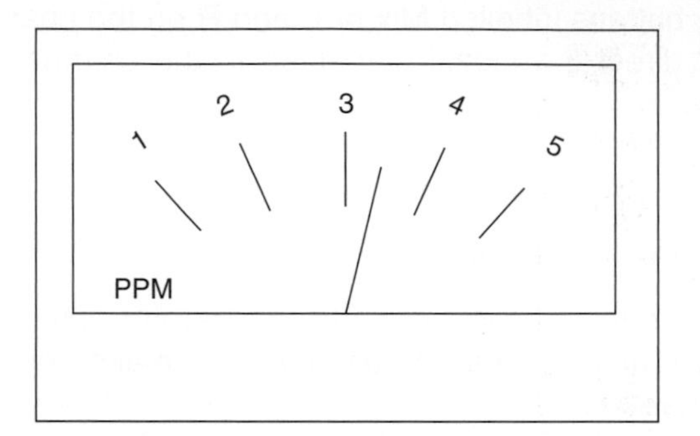

Fig. 11.3.3 – Signal meter.

note ▶ At this stage you should not be hearing any sound through the monitor speakers.

Adjusting the gain

Before starting this exercise you should have turned the gain control down fully counter-clockwise on the channels you were planning to use, so the minimum amount of signal is being sent into the input channel. The next stage is to adjust the gain in a way that allows the strongest signal possible to enter the channel without overloading it. This can be more difficult than it sounds, as most sound sources naturally vary in level.

note ▶ The gain control allows you to adjust the level of the input signal.

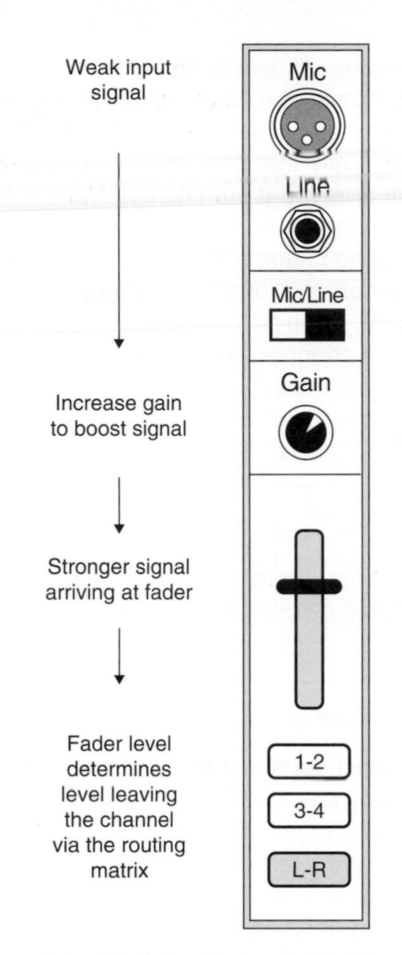

Fig. 11.3.4 – Diagram showing how the level of an input signal can be adjusted.

Try slowly increasing the gain control on the channel by moving it clockwise. If your channel routing is correct, you should see the signal start to increase on the master meters. You should aim to adjust the gain until the signal strength registers around halfway up on the meters. Avoid setting the signal level too high at this stage, as you will leave no room for any sudden rises in volume and may risk overloading the channel. Note that most sound sources will fluctuate, so look for an average level as you can always make any finer adjustments later.

note ▶ Setting the gain correctly is essential in order to achieve a good recording.

producer says ▶

Adjusting the gain

Setting up the gain correctly is essential once a sound source has been connected. You may find that you will need to adjust the gain on the input channel each time a new device is connected, whether it be via the mic or line input as most sound sources will vary in level. Your objective should be to use the gain control to optimize the input signal, so the signal entering the channel is as loud as possible without overloading it. We therefore recommend you always start with the gain control turned down fully counter-clockwise, as this will give you a starting point with the input signal and allow you to gradually turn up the gain if needed. The other thing to bear in mind is that moving the channel fader will also adjust the level of the sound. It's important to note that this will only adjust your listening level and not the gain. Therefore, if you set the gain too high and the channel starts distorting, moving the channel fader down will not get rid of the distortion – it will simply turn down the volume of the sound with the distortion. You would need to turn the gain control down or reduce the output level from the sound source to get rid of the distortion!

Warning lights

Most devices have a peak LED. This is a red light that flashes when the input signal overloads the channel. Avoid a constantly illuminated light, as this may result in distortion.

Adjusting the volume on the sound source

Most electronic sound sources such as keyboards and sound modules have a volume control that allows you to determine the strength of signal they output. Some microphones even have attenuation switches that will reduce their output level. These controls will have a direct effect on the strength of the signal that will arrive at the input channel and will therefore affect how you will adjust the gain control. As a general rule you should always try and get the strongest signal possible from the sound source, as this will help with the overall quality of the signal and will avoid having to overuse the gain control. Computer mixing desks usually don't have a gain control, so it's unlikely that you will be unable to adjust the level of the input signal from within the software. You will therefore have to adjust the volume on the sound source itself or you may find a gain control on the computer's audio interface.

note ▶ Setting up the gain correctly is an essential part of the recording and monitoring process.

Setting up your input signal using solo

Pressing the solo button on the channel where the sound source is connected provides an alternative way of adjusting the gain. It also gives you the advantage of monitoring the signal before it has reached the channel fader or routing buttons (see 'Different types of solo' section in Chapter 10). This is ideal for checking for distortion or overloading at the input stage and for making detailed adjustments to a channel's gain settings.

tip ▶ Pressing the solo button allows you to monitor the input signal before it has reached the channel fader.

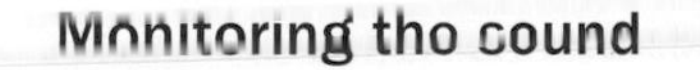

Monitoring the sound

The final stage is to actually listen to the sound. If you are using a stand-alone mixing console, make sure you are actually set up to monitor the stereo bus in the centre section of the console and that the amp that is powering the speakers is switched on. If you are using headphones, make sure you are listening to the stereo output and not the foldback mix.

Stereo output to amp & speakers

Master section

Ext

2Trk

Mix

Monitor level

Mix L-R

Fig. 11.3.5

Adjusting the control room monitoring level

Slowly turn up the control room monitoring level. This is the overall level from the mixing desk that goes to the amp and speakers. As you increase this level you should start to hear the sound sources that are routed to the stereo bus. Alternatively, if you are using headphones, slowly increase the headphone volume level.

If you're not hearing any sound, here's a list of things to check:

- Look for a visual indication to see if the input signal is registering on the meters
- The sound source is routed to the stereo bus
- The channel fader is turned up
- The master fader is turned up
- You are monitoring the stereo bus in the centre section
- The control room monitor level is turned up
- The amp or powered speakers are switched on.

Signals arriving at the stereo bus

Each channel that is routed to the stereo bus will send its signal directly to the master fader. If several signals arrive at the same time, then the overall level will start to increase. It's therefore important to regularly monitor the meters for the master fader to avoid overloading it, as this may lead to distortion. Any problems here will be transferred on to the monitor speakers and headphones and will ruin your mix if you're making a stereo live recording.

tip ▶ You need to take care when increasing the fader level of individual channels to not overload and distort the stereo bus.

Optimizing the input signal

At this stage you should be hearing the input signal. You may want to go back and make finer adjustments to the gain and channel fader.

Monitoring a stereo sound source

If you have connected a stereo sound source or are planning to make a live stereo recording using two microphones, you will have made two connections to the mixing desk: one for the left and one for the right. To hear each connection you will need to route each channel to the stereo bus and adjust the gain accordingly. You will also need to use the pan control to move the left input signal all the way over to the left and the right input signal all the way over to the right. Note that some mixing desks may have stereo input channels, so the pan pot will need to be set in the centre.

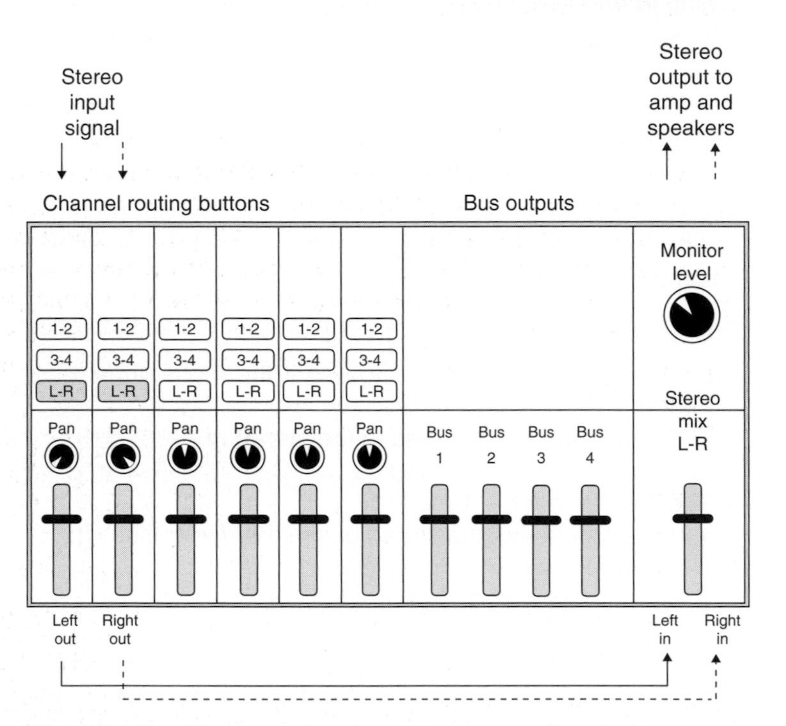

Fig. 11.3.6 – Diagram showing how a stereo input signal can be routed to the stereo bus.

tip ▶ If you are monitoring a stereo signal using two channels of a mixing desk, make sure you pan each channel hard left and right to hear it in stereo.

Stereo live recording

If you are planning on doing a live stereo recording using a mixing desk, then you will need to route the channels where the microphones are connected to the stereo bus. You will also need to set the pan control for each channel correctly.

Exercise 11.4 Routing a signal to a multitrack

Before attempting this exercise we recommend you complete Exercise 11.3 as this will help you set up the correct gain structure for an input signal.

This exercise will cover the following:

- Routing the input signal
- Setting the recording levels
- Monitoring the sound
- Recording a stereo sound source
- Direct monitoring.

Recording a sound source

In the previous exercise we covered routing and monitoring a sound source via stereo buses. The stereo bus is fine for simply monitoring sounds or making a stereo live recording. However, if you are planning to record individual sounds separately or build up a recording in stages using a multitrack, then you will need to set up how you are going to send any signals to a multitrack and how you are going to monitor them during the recording process. This can be a more complex process and may require the signal to pass through several different stages before you can hear it. The first example we are going to look at is how to route a signal through a mixing desk to a stand-alone multitrack while monitoring it.

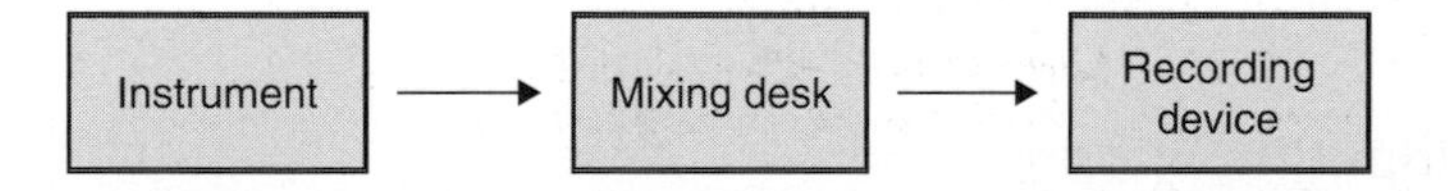

Fig. 11.4.1 – Instrument routed to a recording device using a mixing desk.

note ▶

Different types of multitracks

A multitrack is a device you actually record onto. This may be a stand-alone device such as an ADAT or Portastudio, or a digital audio workstation that consists of a computer with soundcard.

The exercise ...

In this exercise we are going to explain the different stages involved in routing a sound source to a multitrack and monitoring the sound.

Selecting a track to record onto

When using a multitrack you will have to decide which track you want to record onto. If this is your first recording then it doesn't really matter which track you choose. However, if this is an overdub or you are planning to record several sounds at the same time, you will need to plan ahead more carefully and make sure you are not going to record over anything you want to keep.

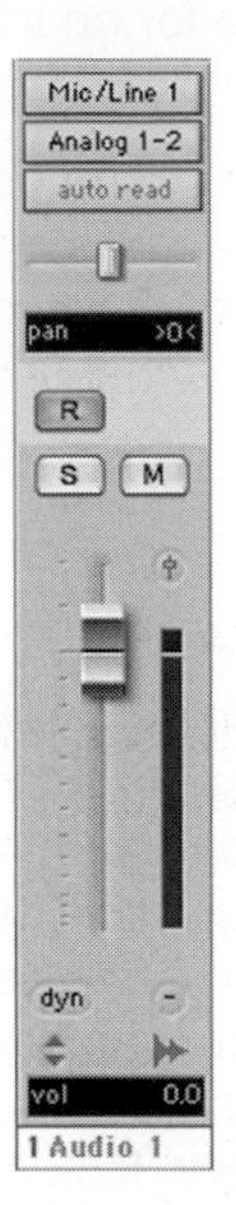

Arming a track

On the multitrack, arm the tracks you are planning to use. This is usually done by pressing one of the numbered buttons on the front panel of the multitrack or by pressing the red 'R' on the track you plan to use if recording into software. When using an audio sequencer you will need to arm the track where the input signal is connected. Once a track is armed you should see a flashing red light.

Fig. 11.4.2 – Track arm for recording.

tip ▶

Input monitoring

Arming the tracks you are planning to use is essential, as it allows signals to enter the recording device.

Choosing an output destination

The next stage is to use the channel routing buttons to send the signal to the track you have just armed on the multitrack. As a general rule you should choose the routing button that corresponds to the track you wish to record onto. For example, if you have armed track 1 on the multitrack, press routing button 1. Note that some routing buttons work in pairs, 1–2, 3–4, 5–6, 7–8, and will route a signal to two different destinations. If this is the case, use the pan control to select between each destination. For example, to send the signal to 1–3–5–7, pan the channel left. To send the signal to 2–4–6–8, pan the channel right.

tip ► Routing buttons allow you to send a signal to a multitrack.

Routing the input signal

In our example we have routed channel 1 where the sound source has been connected to buses 1 and 2. We have then moved the pan control all the way over to the left so that the signal only goes to bus 1. We have also made sure that the signal is not routed to the stereo bus. The channel fader is turned up and so is the fader on bus output 1.

Routing several different instruments at the same time

If you are planning to record several instruments at the same time, each individual instrument will need routing to a different audio track.

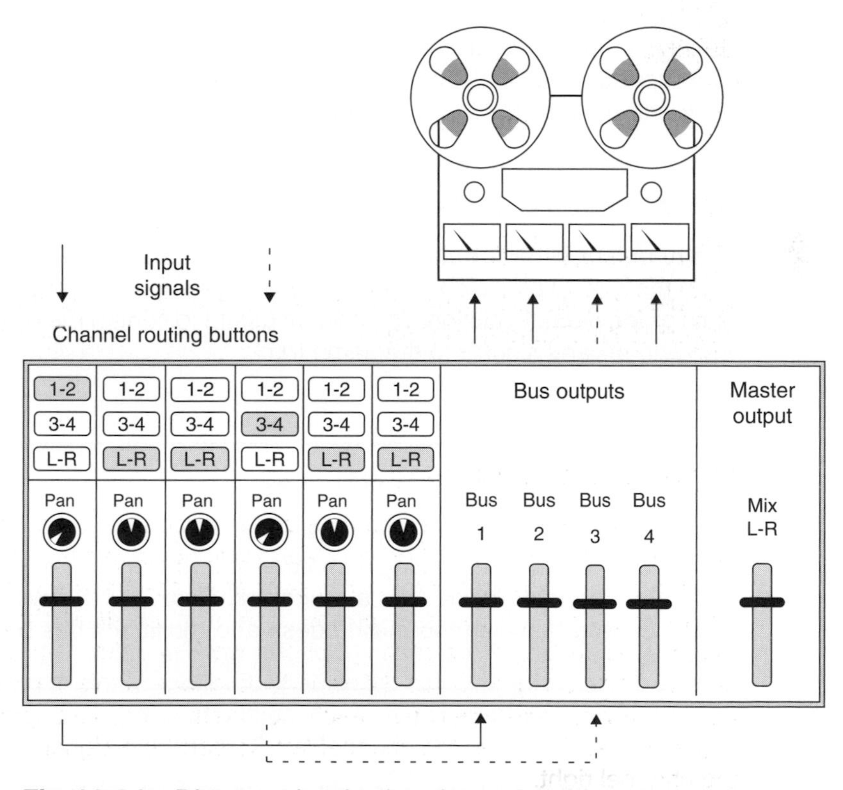

Fig. 11.4.3 – Diagram showing how input signals can be routed to a recording device.

Avoiding double monitoring

When using the routing buttons to send a signal to a multitrack, make sure the channel is not routed to the stereo bus and is only routed to one destination at a time. This will prevent hearing the same signal twice and allow you to monitor the signal correctly.

tip ►

Routing buttons beware

Before you start a recording session it's always a good idea to ensure that none of the channel routing buttons have accidentally been pressed down.

Adjusting the channel fader

Ensure that the channel fader where the sound source is connected is turned up.

Adjusting the bus output

When routing a sound source to a multitrack using a mixing desk, the signal will normally pass via a bus output. This is yet another stage the signal passes through before it reaches the multitrack. You therefore need to ensure that the corresponding bus fader or dial is also turned up in order for the signal to pass on to the multitrack.

What are bus outputs?

A bus output fader (or dial) will determine the level of the signal that is sent from the mixing console to the multitrack. This should normally be set to unity gain or the default position. This can be useful to control the overall level when sending several signals to the same track.

producer says ►

Corresponding tracks and buses

Routing buttons are usually configured to correspond to a multitrack's inputs. For example, routing a channel to bus 1 on a mixing desk should route a signal to track 1 on a multitrack. Note that this will only work if the cabling between the mixing desk and multitrack has been correctly configured.

note ►

At this stage you should not be hearing any sound through the monitor speakers.

Connections between mixing desk and multitrack

If you are using a mixing desk and a stand-alone multitrack recorder, you will be sending the signal out from the mixer into the multitrack. This will require a physical connection between the two devices (see Fig. 11.4.3). Each connection should be correctly configured so the bus outputs of the mixing desk correspond to a multitrack's inputs.

Software-based multitracks

If you are using a computer system with an audio interface, you may be able to connect the sound source directly without having to use a mixing desk.

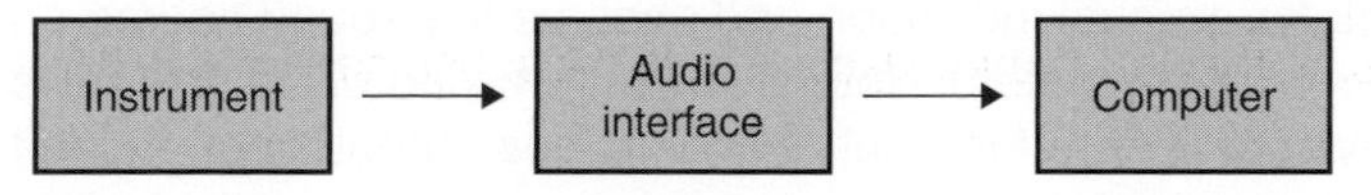

Fig. 11.4.4 – Instrument connected directly to computer.

Setting the recording levels

The next stage is to set the level going into the multitrack recorder. Before you attempt this, you need to ensure you have used the gain control to get the strongest signal possible from the sound source into the input channel (see Exercise 11.3).

Creating a sound

You will also need a sound source, so encourage the performers to rehearse or, if they are using a microphone, to simply count numbers. This will provide you with a signal so you can start setting up the recording levels.

Adjusting the recording level

If your signal routing is correct, you should see the sound source registering on the multitrack's meters. You can now use the appropriate channel or bus output fader to adjust the signal level.

Setting a good level is really important. If your levels are too high you may distort the signal (distortion adds undesirable harmonics to the input signal, often making it unusable). If your levels are too low you may struggle to hear the signal and it may start to lose quality due to noise. Your objective should be to get the levels as high as possible without overloading the multitrack's input. This is not as easy as it sounds, because most sounds naturally move up and down in level, so aim for somewhere in between.

tip ►

Leave some headroom

If you set the recording level high to start with, you will leave no space for the level to get any louder. You ideally need to find the loudest or highest level in the sound and work back from there. This isn't always practical, so always make sure your recording levels aren't at maximum just in case the signal suddenly gets any louder. A device called a compressor may help you automatically control any varying level, as this will automatically turn down the loudest parts of the sound (see 'Compressors' section in Chapter 7).

Visual indications

Look for a visual indication and observe the sound source on the multitrack meters. Note that some mixing desks will have individual meters for input channels, bus outputs, and the left and the right stereo bus.

Monitoring the sound

The next stage is to set up how your going to actually listen to the sound that has been routed to the multitrack. This is usually done by monitoring the signal coming back from the multitrack. When using a stand-alone multitrack this will probably be a different channel or dedicated tape monitor input from where the sound source is connected. This may seem quite confusing at first, as you will not use the channel where the sound source is connected to control its volume. However, this method allows you to adjust the volume level of the sound you are recording without affecting the recording level. Note: some multitrack returns may need routing to the stereo bus in order to hear them, but you should take extreme care not to route any of the returns to any of the other output destinations, as you may accidentally send the signal back to the multitrack or combine it with another signal.

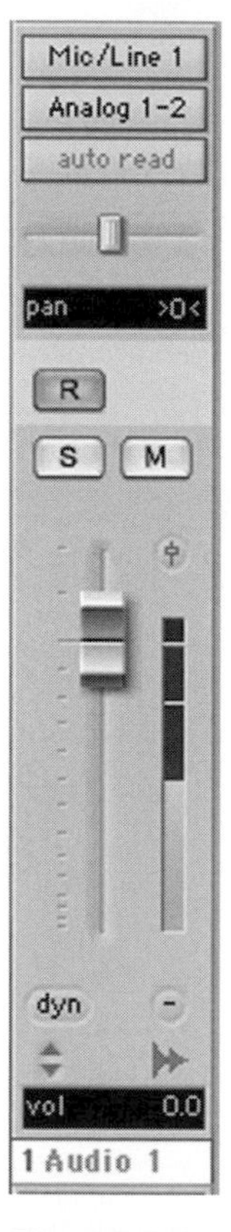

Fig. 11.4.5

Multitrack returns

Multitrack returns are the outputs from a recording device that allow you to hear each individual track. They should normally be connected to dedicated monitor returns or spare input channels on a mixing console, however this may vary depending on the specification of the mixing desk (see 'Tape monitors' section on p. 270). Once connected, they operate like line input signals.

producer says ▶

Monitoring a multitrack

Traditionally, you should always monitor the sound you are recording from the multitrack's outputs as this is the final stage of the signal and you can hear if there are any problems in the signal path as you record. Also, monitoring this way allows you to hear the signal as it will eventually play back. This makes it easier to switch between Play and Record, and do punch-ins, as any levels, EQ and pan settings will remain the same. If you're using a computer system for recording, see 'Audio latency' section, p. 312.

Adjusting the multitrack returns

Turn up the fader or dial where the multitrack returns are connected. This will determine the volume of the sound being returned from the multitrack.

Adjusting the master fader

Turn up the master fader – this is the overall volume level for each individual channel.

Adjusting the control room monitoring level

Slowly turn up the control room monitoring level. This is the overall level from the mixing desk that goes to the amp and speakers. As you increase this level you should start to hear the sound sources that have been routed to the multitrack.

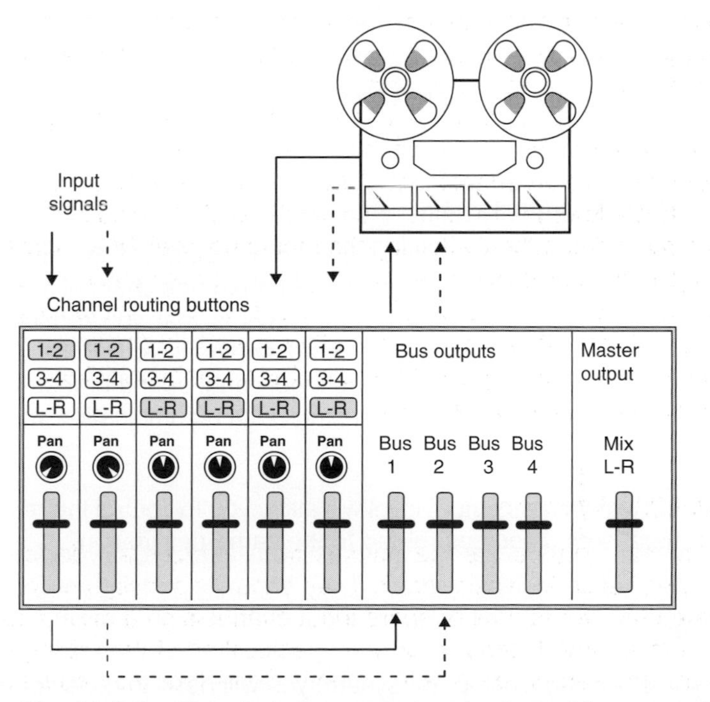

Fig. 11.4.6 – Signals being routed and returned from a multitrack.

In the example in Fig. 11.4.6 the two input signals have been routed to a different bus. The bus outputs from the mixing console send each signal to a different input on the multitrack. The signal then travels from the multitrack's outputs back into the mixing console, where the multitrack returns have been connected. In this example the multitrack returns have been connected to spare input channels. You may, however, be able to use a dedicated monitor return. These channels are routed to the stereo bus, where the signal travels to the master fader. The master fader has been turned up to allow the signals to pass out of the mixing desk to the studio monitor speakers or headphones.

note ► It's important to understand the difference between the input signal and the monitor return.

If you're not hearing any sound, check the following:

- The sound source is routed to bus 1
- The sound source is panned correctly
- The sound source is only routed to one destination
- The channel fader is turned up
- The bus output faders being used are turned up
- The master fader is turned up.

Recording a stereo sound source

If you want to record a stereo sound source you will need to use two separate tracks: one for the left and one for the right. Both outputs will need routing to separate tracks (see Fig. 11.4.6).

Combining audio signals

It is possible to combine and record more than one instrument or sound on the same track, simply by routing more than one channel to the same destination. The main advantage of doing this is that you reduce the amount of tracks or channels needed. The disadvantage is you will be unable to change the volume balance between each individual sound or instrument once they have been merged and may get stuck with a combination you don't like.

note ► Adjusting the bus output level will allow you to control the recording level of several sound sources routed to the same destination.

Audio latency

If you are using a computer system you will need to consider if latency will be a problem during recording. This is a delay between the computer's input and output that makes the sound you are recording appear later than it really is, so it sounds out of time. This problem only occurs when you record enable a track and try and monitor the signal from the computer's outputs. Sometimes the computer's audio buffer size can be changed to reduce this or a low latency mode may be available from within the software. However, this will only reduce the delay and it may not solve the problem. Some computer audio interfaces allow you to monitor the input signal directly, to avoid this latency problem while recording. They may have a dial so you can decide how much of the input signal you want to hear directly from the soundcard's outputs. This avoids having to monitor the input signal through the computer and therefore avoids the latency problem.

producer says ▶

Direct monitoring

Traditionally, you should always monitor the sound you are recording from the multitrack's outputs, as this is the final stage of the signal and you will be able to hear if there are any problems in the signal path as you record. Also, monitoring this way makes it easier to switch between Play and Record and do puch-ins. However, if latency is a problem you sometimes have to break these rules and monitor the input signal directly. If you are using a mixing console and recording into a computer system, you could route the input signal directly to the mix bus, so you would hear the sound source before it entered the computer and therefore avoid the latency. This is not an ideal situation but offers a solution. In this situation you would have carefully balanced the recording level with the monitoring level and avoided moving the fader where the sound source is connected while recording. You would also need to mute the output of the track being used in the computer.

Exercise 11.5 Setting up a foldback mix

This exercise will cover the following:

- Concept of creating a foldback mix
- Creating an independent foldback mix
- Using auxiliary sends
- Pre or post fader selections
- Mono or stereo foldback
- Creating a foldback mix
- Monitoring the foldback mix in the control room
- Talkback and communication.

The exercise ...

This exercise will explain how to connect and set up a foldback mix.

Concept of creating a foldback mix

If you are planning to make a recording while the performers/musicians are wearing headphones, then you will need to set up a foldback mix. This can be either a duplicate of the stereo mix being sent to the studio monitor speakers, or a completely separate mix set up exclusively for the performers/musicians to hear in their headphones while they are recording.

note ► If the performers are planning to wear headphones then you will need to set up a foldback mix.

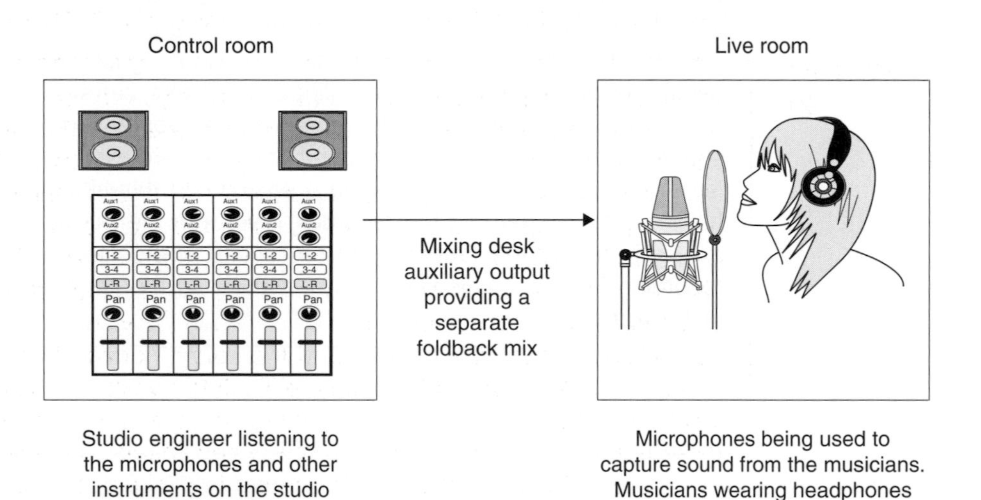

Control room
Live room
Aux1
Aux2
1-2
3-4
L-R
Pan
Mixing desk auxiliary output providing a separate foldback mix
Studio engineer listening to the microphones and other instruments on the studio monitor speakers
Microphones being used to capture sound from the musicians. Musicians wearing headphones listening to the foldback mix

Fig. 11.5.1 – A separate foldback mix allows you to decide exactly what the performers/musicians will hear in their headphones independently of what the engineer is hearing through the control room monitor speakers.

note ► The headphone mix for performers could simply be the same mix that you are hearing in the control room or a completely different mix created by using a cue or aux sends.

Accurately monitoring the sound

Most recording studios are based around two separate rooms: a live room and a control room. The main purpose of using separate rooms is to isolate the different sounds coming from each room. This is an essential part of the monitoring process, especially when using microphones, as it allows the recording engineer sitting in the control room to only hear the sound that is being picked up by the microphone, without being influenced by the sound coming from the live room or the actual instruments themselves. It also prevents the sound which is being produced by the studio monitor speakers being picked up by any microphones being used in the live room. You should always try to monitor this way if possible, as it's the only way to make any accurate judgements when using a microphone.

Recording in the control room

Sometimes it will be possible to make a recording without using foldback. This is because powered electronic instruments, such as keyboards, sound modules and even guitars connected via a DI box, don't produce any actual sound until they are connected to a mixing desk. This gives you the option of allowing the musicians/performers to be located in the control room, so they can monitor their performance directly through the control room monitor speakers.

producer says ▶

When using a microphone to record a vocalist your objective should be to use the microphone to only capture the sound of the voice. If the vocalist and microphone are placed in the same room as the studio monitor speakers, then the microphone will also capture the sound coming from the monitor speakers as well (this is because the studio monitor speakers will be providing the recording engineer with the sound of i.e. the vocal and any other sounds being used). This is not a good idea and should be avoided if possible, as combining any sounds at this stage i.e. the sound from the vocalist and the sound coming from the studio monitor speakers will make it more difficult to control the volume of the voice independently of any other sounds, and give you less creative flexibility when processing the voice with EQ and effects. The solution would be to place the vocalist and microphone in a separate room from the monitor speakers, so that the microphone only records the voice. The vocalist will then need to wear headphones in order to hear the other instruments play back, so a foldback mix will need to created.

Connecting headphones

Most headphones connect via a stereo 1/4-inch jack plug and can usually be connected directly to a mixing desk. However, this isn't always possible and the performers may be located some distance away in a live room. You therefore may need to find an alternative way to supply a foldback mix. Some examples are outlined below.

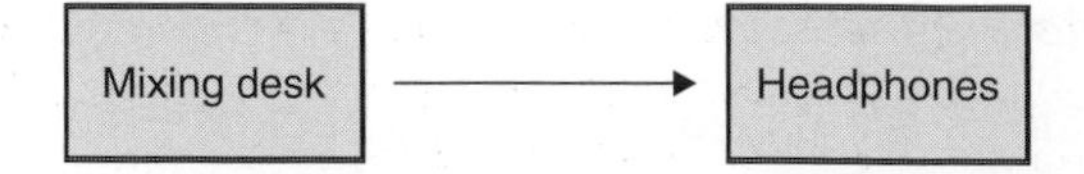

Fig. 11.5.2 – Headphones connected directly to the headphone output on a mixing desk.

Mixing consoles that provide dedicated headphone outputs will allow you to control the volume level being sent and may also allow you to choose which signal will be directed to the headphones. For example, you may be able to select the main stereo mix or an auxiliary.

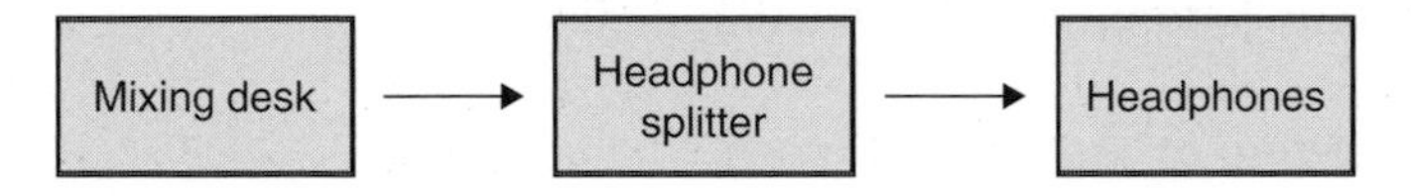

Fig. 11.5.3 – Headphones connected via a splitter box.

To provide foldback in a live room you will need a long cable to send the foldback signal from the control room to the live room. Most live rooms use a headphone distribution box or headphone splitter box, so several headphone inputs are available.

Fig. 11.5.4 – Headphones connected using an auxiliary send, headphone amplifier and splitter box.

A foldback mix can also be set up by using auxiliary or cue sends on a mixing desk. Note that auxiliary sends can also be used to send signals to effects units, such as reverbs and delays (see Exercise 11.7). To make this possible the auxiliary or cue send from the mixing desk may need connecting to a headphone amplifier to provide the correct signal level for the headphones. Note that using an amplifier without a headphone output will require a specially made cable. Most auxiliary or cue send connections are located towards the back of a mixing desk and are labelled 'Aux out' or 'Cue send'. These are outputs from the mixer and should not be confused with auxiliary inputs. If you are using a software mixing desk, see below.

Software foldback

It is possible to set up a foldback mix when only using a computer system running audio recording software. However, this is only practical when using an audio interface or soundcard with the computer. If the soundcard provides a headphone output you will be able to directly monitor the sound being produced by the computer using headphones. Alternatively, if the audio interface has several separate outputs you could simply designate one of these to be used as an auxiliary send and connect it to the input of a headphone amplifier. You would then need to designate which auxiliary send within the software you wanted to use for foldback. Providing the send was routed to the designated output it would work in the same way as a conventional mixing desk. This method is fine for playing back and monitoring previously recorded tracks; however, you may encounter a latency problem when recording (latency will delay the input signal, making it sound out of time – see 'Latency' section in Chapter 3).

Creating an independent foldback mix

During the recording process you will use the channel faders on a mixing desk to create a balance in level between each sound or instrument so you can clearly hear what is being recorded. Routing this stereo mix to the foldback may be an option in some circumstances but more often than not the performers/musicians will require a different balance. To achieve this you will need to use the auxiliary or cue sends located on each channel of the mixing desk. This will allow you to set up two independent mixes, one using the channel faders and the other using the auxiliary sends.

note ▶ Sometimes you will have to create a separate headphone mix for the musicians. This means they hear a different balance in volume of the instruments to you.

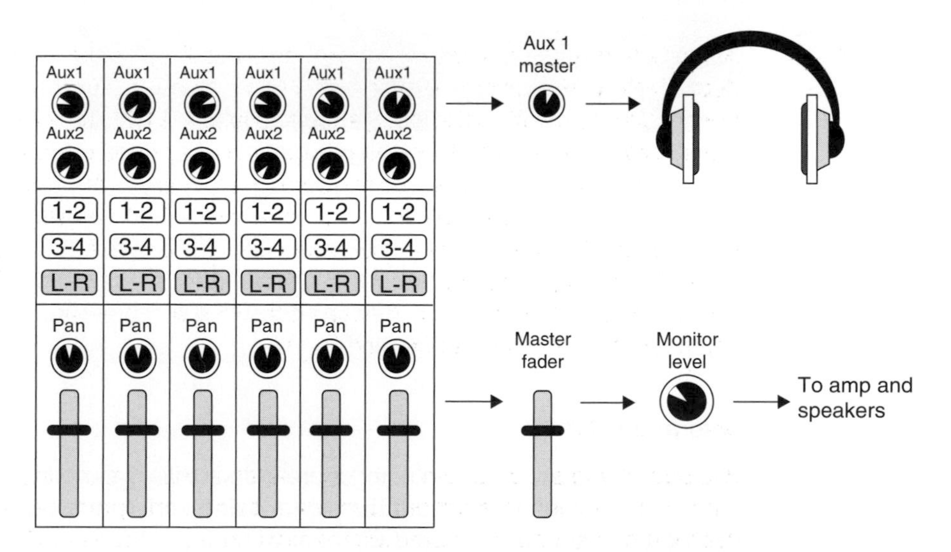

Fig. 11.5.5 – Diagram showing how an auxiliary send can be used to send signals to a pair of headphones independently of the channel faders.

Using auxiliary sends

Auxiliary sends are located on each channel of a mixing desk and you will find most mixing consoles will have two or more. A small dial or fader allows you to choose how much signal is sent from each channel to the headphones. You should take care to only use the aux sends on the monitor path of a signal while recording. Note that when using a software mixer you may have to create each auxiliary send yourself.

note ▶ An auxiliary or cue send will only work as a foldback control if it is physically connected or routed to the headphones in some way (see 'Connecting headphones', p. 315).

Pre or post

In order to make the aux send level totally independent to its channel fader, you will need to set each aux send to Pre.

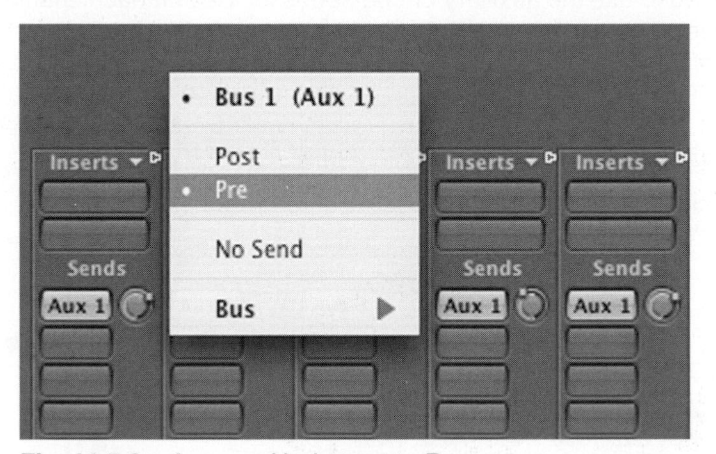

Fig. 11.5.6 – Aux send being set to Pre.

note ▶ Setting an aux send to Pre allows you to adjust the level of the channel faders without affecting the level being sent from the auxiliary sends to the headphones. Basically, the headphone mix will remain the same regardless of the channel fader positions.

Solo and cut

You need to be aware when using a pre fade auxiliary send that the channel's cut or mute button will not cut the signal being sent to the headphones. This can be confusing as a muted channel will not be heard in the control room through the studio monitor speakers but, if the auxiliary send is turned up and it is set to Pre, it will be heard in the foldback.

tip ►

When setting up a foldback mix only use the aux sends on the monitor path of the signal, not on the source channels.

Master auxiliary sends

Some mixing desks may have a master auxiliary control that will allow you to adjust the overall foldback level that is being sent to the headphones.

note ►

Aux sends are extra outputs from each audio channel that allow you to decide how much of the signal is sent to the headphones or an effects unit.

Mono or stereo foldback

Depending on how the foldback system has been configured, you may simply turn one dial and sound will appear in both sides of the headphones. If this is the case, it will probably be a mono auxiliary send where the same signal is being sent to both headphones. On other systems you may be able to use one aux send to send a signal to the left side of the headphones and another for the right side. Alternatively, you may have a stereo aux send with a pan pot incorporated. This will allow a single dial to send a signal to both ears and another dial to control the pan position.

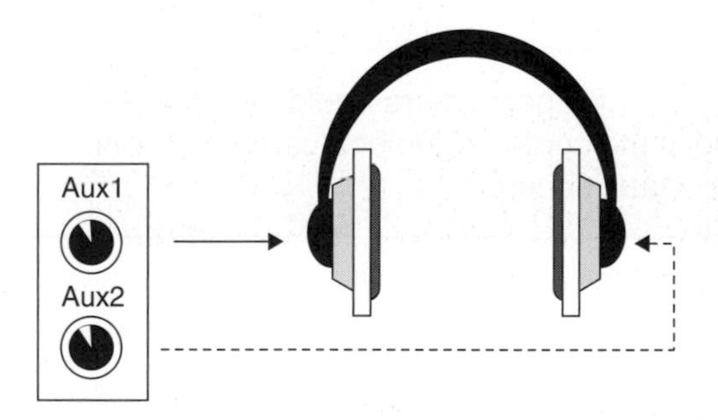

Fig. 11.5.7 – Example of a stereo foldback system using two aux sends.

producer says ►

Stereo foldback

A stereo foldback mix requires two separate signals: one for the left-hand side of the headphones and one for the right. This is the best option to choose if you can, as it allows you to place a sound anywhere in the headphones and gives you the same fallibility as using a fader and pan to create a stereo mix using the studio monitor speakers.

Creating a foldback mix

When creating a foldback mix you will need to consider what the performers will actually need to hear. This will vary depending on many factors and you will find that one foldback mix will generally not suit all. A good starting point, however, is to replicate the same mix that you are hearing through the studio monitor speakers. Here are some tips when creating a foldback mix:

- Start by duplicating the main stereo mix in the headphones
- Don't make the headphones too loud
- Make sure the performers can hear themselves clearly
- Ask the performers if they are comfortable with what they are hearing
- Don't add too much reverb to the foldback mix
- Keep the main elements in the centre when using stereo foldback
- Turn up the rhythm section if the performers are having trouble keeping in time with the other instruments
- You may find vocalists prefer to wear only one headphone
- Don't make any radical adjustments while the performers are wearing headphones
- Ask the performers what they would like to hear.

note ▶

Always make sure the musicians are comfortable with what they are hearing, as a poor headphone balance may affect their performance.

producer says ▶

The headphone mix for performers could simply be the same mix that you are hearing in the control room or a completely different mix created by using a cue or aux sends.

Monitoring the foldback mix in the control room

One essential stage that is often overlooked is how the recording engineer is actually going to hear the foldback mix. Some mixing consoles allow you to select the aux sends as the main listening source. If this is the case, the engineer will be able to temporarily listen to the foldback mix using the control room monitor speakers. However, the most effective way for the engineer to hear the foldback mix is to wear a pair of headphones so they can hear exactly what the performers are hearing.

note ▶

It is essential that the engineer can hear the foldback mix in order to set the levels correctly.

Talkback and communication

If the performers/musicians are located in a separate room and wearing headphones, it will become difficult for the engineer to communicate with them. This is why most stand-alone mixing desks provide a talkback microphone. This is a tiny microphone that is built into the console to provide a way of transferring any dialogue from the control room into the foldback. The talkback microphone is usually switched on by pressing and holding down a switch that is usually located in the centre section of the console.

producer says ▶

Avoiding headphone spill

You may find that some vocalists have difficulty hearing themselves and may prefer to wear only one headphone so they can hear the foldback and their voice acoustically. Care must be taken to ensure that any sound generated from any unused headphones is not loud enough to be picked up by the microphone.

Exercise 11.6 Recording and overdubbing

Before attempting this exercise we recommend you complete Exercises 11.2–11.5.

This exercise will cover the following:

- Transport controls
- Creating markers and memory location points
- Using a click track
- Recording for the first time
- Punching in and out of Record
- Overdubbing.

The exercise ...

This exercise is designed to guide you through the stage of the recording process when you actually press Record and introduce you to some of the options you will have when using a multitrack recorder.

Transport controls

It's always a good idea to familiarize yourself with a recorder's transport controls and practise using them before attempting an actual recording session. It is essential that you are able to stop and start a recording and navigate to different locations along a timeline. You may also find there are several preferences that can be adjusted that will help you to use the transport controls more efficiently.

Keyboard commands

If you are using a computer-based recording device, you will find most of the commonly used parameters will be assigned to a letter or number on the computer keyboard. Always try and familiarize yourself with the main transport controls, such as Play, Record and Stop, as this will allow you to use a device more efficiently.

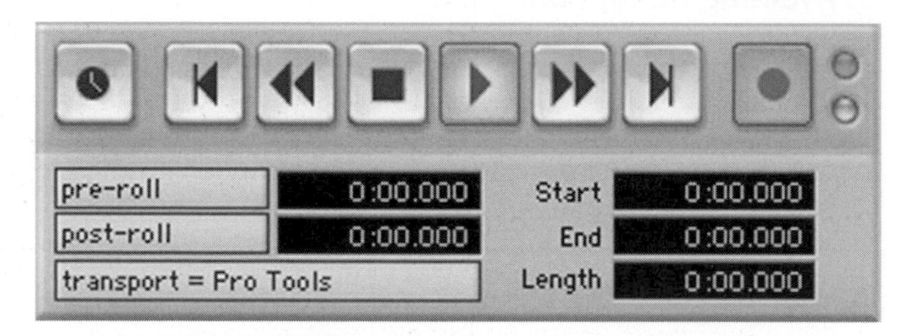

Fig. 11.6.1 – Pro Tools transport controls.

tip ► Always make sure you are familiar with a recorder's transport controls before attempting an actual recording session. It is essential that you are able to navigate to different locations efficiently.

Creating markers and memory location points

During the recording process you will regularly find that you need to navigate back to the same location within a recording. A digital recording device will usually allow you to place a marker or store a memory locate point anywhere on the timeline to help you do this. Alternatively, you may be able to use the left and right locators to select areas of time within your recording device.

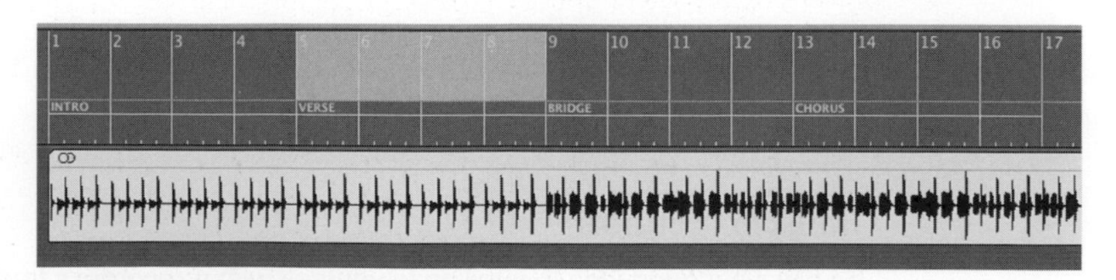

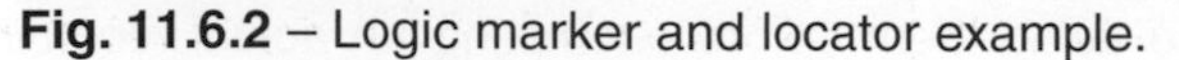

Fig. 11.6.2 – Logic marker and locator example.

Markers can be created to help you identify each different section of a song (for example, intro – verse – chorus) or used to mark any important locations throughout a recording. When starting a recording it is usually a good idea to mark the start point. Once a marker or locator has been created, you will be able to locate back to the same location easily. This will allow you to concentrate on the recording process rather than having to search for a location.

Using a click track

Before starting a recording you may want to consider using a click track. This can provide a regular timing reference to help musicians/performers keep in time during a recording. It will also help if you are planning to edit or copy the audio once it has been recorded, as audio recordings made while referencing to a click track or an audio sequencer's metronome will be a lot easier to edit as everything will be aligned to the audio sequencer's timeline. Note: recordings made without referencing to a click track or a recorder's internal metronome will render the beats and bars display on a recording device meaningless and will make moving and copying audio more difficult.

The decision whether to use a click track will depend on several factors, such as:

- Will it help improve the timing of a performance?
- Can the performers play in time with a click?
- Are the performers comfortable with using a click track and, if not, will it hinder and compromise their performance?

producer says ▶

Musicians who are not accustomed to using a click track may initially find it off putting and not perform well. In this situation you will need to give performers time to rehearse with it before you start recording. If you decide to use a click track then you will need to provide a suitable 'click' sound. A click can usually be produced internally from an audio sequencer or externally by a MIDI device. Typical click track sounds are rimshots, hi hats or cowbells.

tip ▶

Audio recordings made while referencing to an audio sequencer's metronome will be a lot easier to edit, as all the audio will line up with the beats and bars displayed on the timeline.

Calculating the tempo

If you decide to use a click track you will first have to determine the tempo of the performance or song. If the tempo varies considerably throughout a performance you may have to consider creating a tempo map. This is a list of tempo changes that will automatically change the speed of the click track throughout a recording (see Exercise 5.14 'Changing the tempo').

Setting the time signature

Depending on the complexity of a click track you may need to adjust a recording device's time signature. It is also possible to enter a time signature map if you need to make changes throughout the song.

Recording for the first time

Navigate to a suitable location on your recording device (you may want to set up a locator to mark the start point). You should now give the performers a visual indication that you are ready for them to start. You may need to wait for a count-in or any pre-roll that has been set up, so make sure you are actually in Record before you ask the performers to start.

Your responsibilities whilst recording

During the recording observe the recording levels for each individual track. Remember the recording level is independent of the monitoring level. For example, you can record a high level without listening too loud. The recording level is determined by the strength of the signal travelling from the bus output to the multitrack's input. The monitoring level is simply the level you choose to listen at and can be adjusted at any time without affecting the recording levels, either using the monitor faders or the control room speaker level. Also listen out for mistakes and note their location.

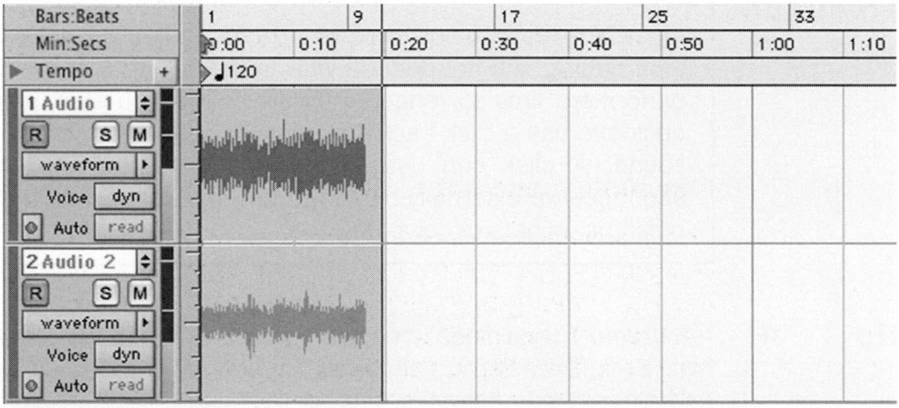

Fig. 11.6.3 – Audio being recorded into Pro Tools.

Listening back to a recording

Once the performers have finished *press Stop*. Don't be too anxious to press Stop, as any additional time at the end of a song can be edited later. It's always better to wait until the sound has completely faded away before pressing Stop, as you may cut off the natural decay of the instruments. Locate back to the beginning of the recording and press Play. It's always a good idea to disarm any tracks that have been used for recording to prevent accidentally recording over them while listening.

Analysing the recording

You now need to analyse the recording and listen out for any mistakes and see if there is anything you can do to improve the quality of the recording. If you hear any mistakes or sections you want to replace, you will need to note their exact location, so observe the multitrack's transport counter display and either write them down or create a marker.

- Are the performers happy with the recording?
- Are you happy with the recording?
- Are all the instruments playing in time?
- Are there any clicks or buzzes on the recording?

If everyone is happy with the recording you can now proceed to the next stage. However, if the musicians are not happy with their performance, or the timing is bad, you may consider recording the performance again, providing the musicians are prepared to perform again and think they can achieve a better result. You could also consider keeping this recording if it's OK and recording an alternative version (providing you have enough disk space or tape available). This would allow you to mix and match and pick out the best sections from each recording (see Exercise 11.8 'Editing audio'). However, if there is only a small mistake or the performer feels they can do a better job just in one particular section of the recording, then you could consider *punching in and out.*

Punching in and out of Record

This is a method of recording where you only drop into Record over the section you want to replace. If it's done correctly, you will be able to repair an existing recording by making a new recording over the mistake without having to re-record the whole performance again. To do this successfully you will have to find a suitable punch-in point (where the recording will start) and a suitable punch-out point (where the recording will end). Note that a bad join will sound unnatural and may potentially ruin the whole recording.

producer says ►

> Sometimes when listening back to a recording you may hear a small mistake or the performer may feel they can do a better job in one particular section of the recording. Instead of having to re-record the whole performance again you could consider punching in and out.

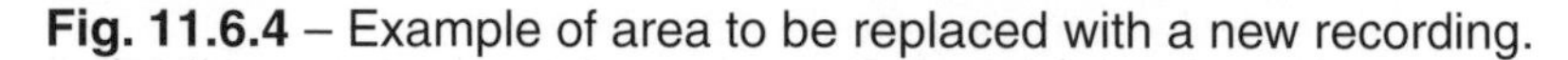

Fig. 11.6.4 – Example of area to be replaced with a new recording.

Finding a suitable punch-in and -out point

The location where you will actually start and end this recording will need to be identified. This should normally be at the start of a phase or new section, even if it means punching in several seconds before the actual mistake. It is important to choose a location in the recording that can be easily identified, so the performer will instantly know where they are once playback commences and will have time to prepare for the recording. It is a good idea to set up a locate point or marker, as you may have to return back to this location if the punch-in is unsuccessful. Most digital multitracks will allow you to punch in and out of Record automatically. This is normally achieved by setting up a drop zone using the locators and enabling the auto drop recording function. Once this has been set, it may also be possible to loop round this section and record multiple takes.

Listening back

Once a section has been replaced you will need to review the new recording and decide if it is an improvement. The new section should seamlessly replace the mistake, so check to see if you can hear the join. If you are unhappy with the new recording you could try another punch-in or press Undo to revert back to the original.

producer says ▶

Creating a smooth transition

Sometimes it may be difficult to achieve a smooth transition when punching in and out and a join may sound unnatural and stand out from the rest of the recording. It is essential that you match the settings and recording levels that were previously used and always encourage the performer to start before the punch-in point, so they are already performing when you actually punch in to record. If you're still having problems, try erasing or muting a mistake before punching in and out, or try punching in on a new track (see 'Keeping alternative takes' section below). Another option may be to repair a bad join by editing. As most digital audio is based around regions, it is usually possible to adjust the length of a region to find a better edit point or cross-fade two regions together (see Exercise 11.8).

Overdubbing

If you are using a multitrack recorder you will be able to add additional instruments to an existing recording by overdubbing. This process allows you to record new (or replace existing) instruments while listening back to existing tracks. This makes it possible to build up a recording in small sections by gradually layering different instruments on top of each other which gives flexibility to try out new ideas over an existing recording, as you can always erase them without affecting any of the other tracks. The creative possibilities are endless – you may even come up with an idea or part that only works on the recording and would be impossible to play live with other musicians.

producer says ▶

> **Overdubbing**
>
> You will generally find that most professional records are made this way, as overdubbing gives you a lot of flexibility. Sometimes it isn't possible to record all the different instruments at the same time, so the solution is to record each sound separately and overdub. Sometimes the music will be recorded in one studio and then the vocals overdubbed in another studio. Then it is usually mixed in a third studio. On some records, the musicians/vocalists never actually meet each other.

Performing an overdub

Select the next available audio track and record enable it. Route the signal you plan to record and set your recording levels. Locate back to the beginning of the song and record the next instrument.

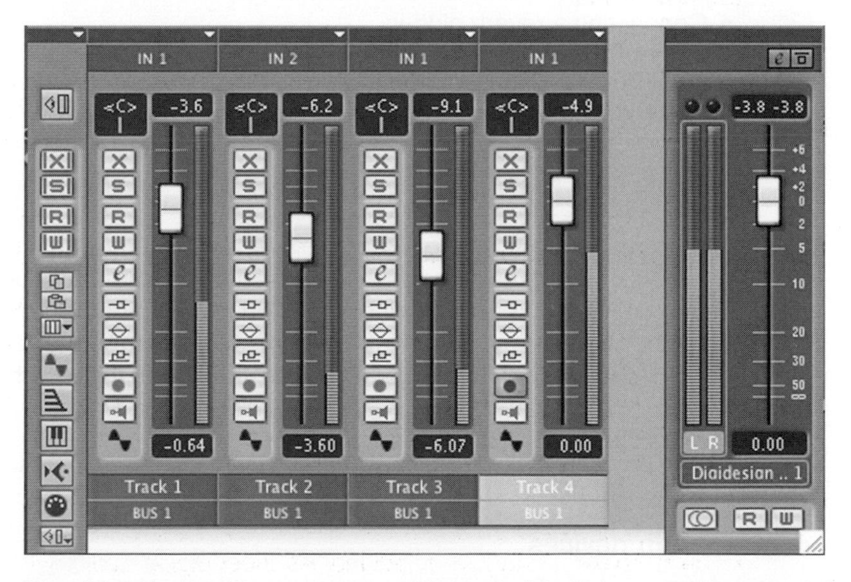

Fig. 11.6.5 – In this example audio track 4 is being used to overdub a new performance while listening back to tracks 1, 2 and 3.

Double tracking

If you are using a multitrack and have enough available tracks, you could try experimenting with double tracking. This is a process where you record the same instrument or performance twice, doing exactly the same thing on two separate tracks (so it's like recording two separate people). This can create an interesting effect and usually adds a thicker texture to the sound when it is doubled.

Keeping alternative takes

During the recording process it may be difficult to make decisions on what to keep and what to erase. If you have enough spare tracks it is possible to record and keep several recordings of the same part/instrument. Each recording is referred to as a take and must be muted, so you only hear one version playing at a time. The advantage of doing this is you may be able to mix and match different takes to produce a perfect recording (see Exercise 11.8 'Editing audio'). This is a common process used by most professional studios.

Exercise 11.7 Adding reverb

This exercise will cover the following:

- Connecting a reverb device
- Connecting an external reverb device
- Connecting a reverb plug-in
- Using auxiliary sends
- Reverb parameters.

What is reverb?

Natural reverb exists all around us in everyday places such as rooms, corridors, halls, etc. and is created when a sound reflects off different surfaces. As most environments are unique, the sound of the reverb will also vary depending on the size of the room and the time it takes for a sound to be reflected and absorbed. Generally, the larger the room, the longer the reverb will be. This is why sounds created in a large hall or cathedral take longer to fade away than sounds created in a small room.

Reverb devices

When dealing with audio you will have more control if you can determine the amount of reverb that is associated with a particular sound. Using a stand-alone reverb device or plug-in will allow you to generate reverb artificially, therefore making it possible to change the ambience around a sound at any time, even after a sound has been recorded. For example, if you have recorded a vocalist in a small room or vocal booth using a microphone, the overall sound will probably be quite dry and intermittent. Adding reverb will allow you to make the vocalist sound like they were singing in a large concert hall or cathedral. Reverb devices give you a huge amount of creative flexibility and allow you to experiment by placing sounds in different spaces. Most reverb devices have presets, such as Hall, Room, Plate, etc. and allow you to store and recall different set-ups in their memory.

note ▶ Adding reverb to a sound can make it appear bigger or more distant.

The exercise ...

The most efficient way of communicating with a reverb device is to set up an auxiliary send and return. This initially may seem rather complex to set up, but will pay dividends by allowing each separate channel on a mixing desk to have access to the same reverb device by allowing you to choose how much reverb you wish to add to each channel. Before you start this exercise you will need a sound source available to add the reverb to.

Connecting a reverb device

Before you can use a reverb device you will need to physically connect it. This is basically a two-stage process. This first stage is to get a signal into the reverb device. The second stage is to monitor the signal coming from the reverb device. How you actually make these connections will vary slightly depending on whether you are using an external hardware reverb device or software plug-in. Some examples are outlined below.

note ▶ Connecting a reverb device is a two-stage process:

1. Sending a signal to the reverb device.
2. Monitor the signal coming from the reverb device.

Connecting an external reverb device

If you are planning to use a stand-alone reverb device with a mixing desk, you will need to make two connections: one to send a signal out from the mixing desk into the reverb device and one to return the signal back from the reverb device into the mixing console.

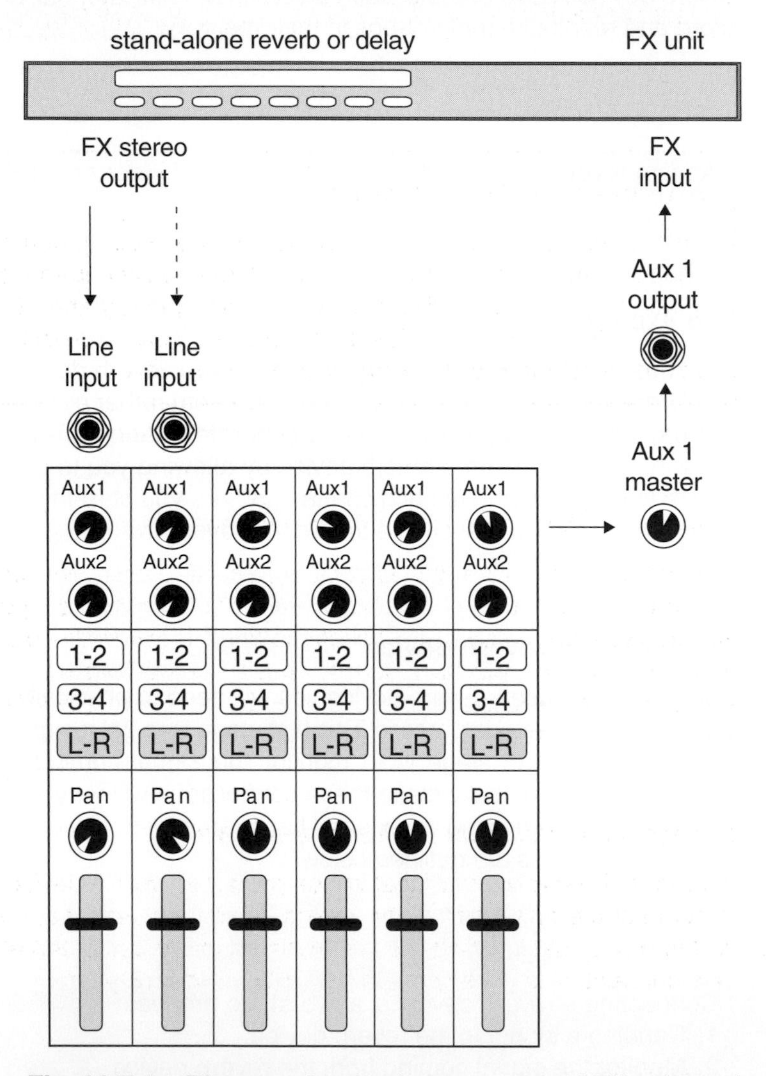

Fig. 11.7.1 – Diagram showing how an external FX unit can be connected to a mixing desk.

In the example in Fig. 11.7.1, auxiliary output 1 on the mixing desk has been connected to the input of the reverb device. The outputs from the reverb device are returned in stereo to two spare input channels.

tip ▶

When returning a reverb's output into spare input channels, care must be taken not to use the auxiliary sends on these channels, as this would create a feedback loop.

producer says ▶

External hardware effects units

An external reverb device is a separate piece of equipment that will produce reverb. These stand-alone boxes are usually capable of producing a variety of different effects, such as reverb, delay and chorus, sometimes at the same time. Most external effects units are digital and allow you to store and recall settings from their memory.

Connecting the inputs to the reverb device

To send a signal from a mixing desk to a reverb device, you will need to connect one of the auxiliary outputs on the mixing desk to the input of the reverb device. Auxiliary outputs are usually located towards the back of a mixing desk and labelled auxiliary sends. These are outputs from the mixer and should not be confused with auxiliary inputs. This connection is usually made with a 1/4-inch jack plug.

Connecting the outputs from the reverb device

In order to hear any reverb that will be produced from a device, you will need to connect the outputs from the reverb device into any spare input channels on the mixing desk (using the line input) or into a dedicated auxiliary return. This connection is usually made with a 1/4-inch jack plug.

Auxiliary returns

Some mixing consoles may provide you with auxiliary returns. These are additional inputs designed only for returning effects (they save you having to use input channels to hear the outputs from the effects). They are basically like scaled-down input channels, but only give you basic functions such as level and pan, and usually they don't include any auxiliary sends to prevent you from sending the effects back to its own input. When connecting a reverb device, don't confuse aux sends with aux returns.

Returning reverb in stereo

Most reverb devices will be able to output their sound in stereo. This will make any reverb they produce appear fuller and have a richer texture. To achieve this you will need to use two separate channels on the mixing desk to connect the left and right outputs from the reverb device to the mixing desk. Note that when two channels are used for returning reverb, each channel should be panned in opposite directions to make use of the stereo field (see Fig. 11.7.1).

Connecting a reverb plug-in

If you are using a computer system with a software mixing desk it is likely that you will have access to a reverb plug-in. This is a piece of software that will allow a computer to produce reverb. To make a reverb plug-in available to all channels, it will need inserting on a bus or auxiliary channel. These are special types of audio channels that don't play back audio files directly that can be used for inserting and returning plug-ins, such as reverbs and delays.

tip ► In order to hear the reverb in stereo, make sure you create a stereo auxiliary track

In the example in Fig. 11.7.2 a reverb plug-in has been inserted on a stereo auxiliary track (note that some sequencers, such as Logic, allow you to achieve the same results by inserting a plug-in on a bus). The input to the reverb plug-in on the auxiliary track has been set to bus 1 and 2. This will determine the path each channel will need to use when sending a signal into the reverb plug-in.

tip ► In order to hear a reverb plug-in, make sure an output is selected on the auxiliary or bus channel and the fader is turned up. The fader on the auxiliary channel will control the level of the reverb's output.

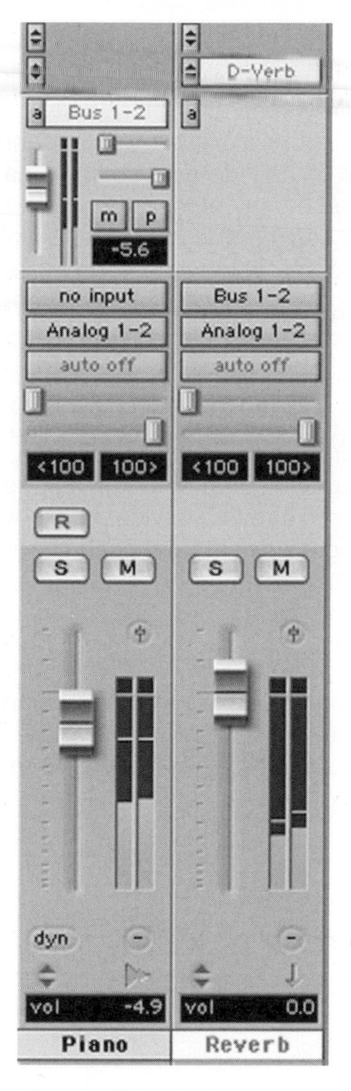

Fig. 11.7.2 – In this example the small fader on the piano track is being used to add reverb via bus 1 and 2. The level of the small fader determines the amount of reverb that will be produced.

Adjusting the wet and dry balance

When a reverb plug-in has been connected to an auxiliary or bus track you will need to ensure that the reverb balance setting on the plug-in is set to 100% reverb. This will ensure that only reverb is produced from the auxiliary or bus track (see 'Reverb parameters' section below).

producer says ▶

Using a reverb plug-in

Always try and set up an auxiliary send and return when using a reverb plug-in, as this will make the reverb available to all channels. Try and avoid inserting a reverb plug-in on an individual channel, as this will prevent the reverb from being used by any other channel and force you to open a new plug-in each time you want to add reverb. Note that using several reverb plug-ins at the same time will use up more of the computer's processing power and put more strain on the computer.

Using auxiliary sends

Once you have connected a reverb device you are now ready to use the auxiliary sends. These are like extra outputs from each channel of the mixer and provide a pathway to send signals to a reverb device. A dial or small fader will allow you to choose how much signal is sent from each channel to the corresponding effects unit. The more the auxiliary send level is increased, the more reverb or delay will be added to the signal.

note ▶

When you turn up the auxiliary send you will hear the original sound plus the reverb. Most mixing consoles usually have two or more aux sends, so you can connect and send to several different effects units at the same time.

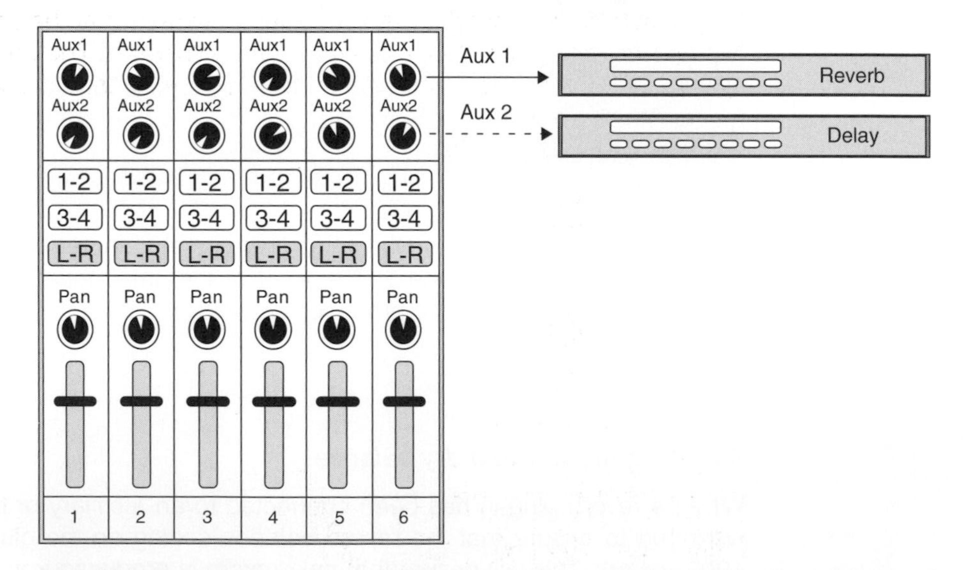

Fig. 11.7.3 – Moving the auxiliary send dials will determine the amount of reverb or delay that will be added to the sound on each particular channel. Notice how each channel has access to both reverb and delay.

producer says ▶

Auxiliary sends and returns

Auxiliary sends provide the most efficient way of connecting a reverb device, as they provide each individual channel access to the same reverb device. This is particularly useful during mixdown, as adding the same reverb to several different tracks at the same time can help blend a recording together.

Using several auxiliary sends at the same time

Once you have become familiar with how to set up a reverb device using an auxiliary send and return, you will be able to set up and use several different reverb devices at the same time or use other effects such as delay and chorus. For example, you may want to set up a small room reverb to add to drum and percussion sounds and a larger reverb and delay to add to a vocal.

tip ▶

Computer auxiliary sends

When using a software mixing desk you may need to enable and route the auxiliary sends on each individual channel in order to send signals to a reverb plug-in.

Master auxiliary sends

As you start to add reverb to several different channels at the same time, the overall level being sent to the reverb's input increases. The auxiliary master allows you to increase or decrease the overall level that is being sent to a reverb device.

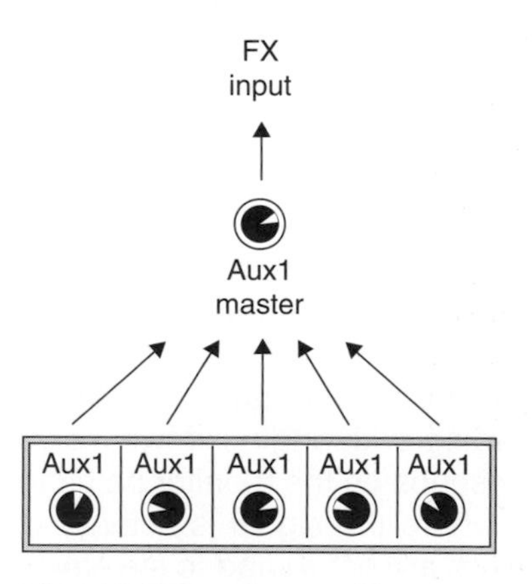

Fig. 11.7.4 – The auxiliary master allows you to control the overall level being sent to an FX unit.

Pre or Post auxiliary sends

The level of each auxiliary send can usually be configured to work in two different ways in relation to the channel fader. Some mixing desks provide a switch for this function.

- Pre – When Pre is selected the auxiliary send level becomes independent and is not linked to the channel fader level in any way. This is more commonly used when creating a 'foldback mix' and should not be used when sending a signal to a reverb device.
- Post – When Post is selected the auxiliary send level is linked to the level of the channel fader. For example, as you move the channel fader down, the auxiliary send level will also move down. This mode should be selected when using an auxiliary to send a signal to an effects unit such as reverb or delay and is also commonly used during mixdown.

Fig. 11.7.5 – Selecting between Pre and Post when using an auxiliary send.

producer says ▶

> ### Monitor reverb
>
> If you are at the recording stage you may want to add some reverb to temporarily enhance what you are hearing. For example, if you are recording a solo instrument or vocal it may help the performer if they can hear themselves with some reverb. You may, however, not wish to use this reverb on the final recording or during mixdown, so you will need to ensure that you use the aux sends on the monitor path of the signal and not on the source channel where the signal is actually connected and that the returns from the reverb are not routed to the stereo bus or to the multitrack.

tip ► Try and avoid rooording oounde with largo amounto of rovorb ao it will bo impossiblc to rcmovc it latcr.

Reverb parameters

Once you have successfully added reverb to a sound you may want to experiment and adjust some of the reverb parameters. Here is a list of some of the most commonly used:

- *Reverb Type*. This determines the type of environment the sound of the reverb is based on – for example, Room, Hall, Plate, etc.
- *Reverb Time*. This parameter allows you to control the time it takes the reverb to fade away. It is usually a time value, so if you set it to 500 ms the reverb will take half a second to decay ('ms' stands for milliseconds and is a 1000th of a second).
- *Pre Delay*. This parameter allows you to determine the amount of time it takes before the reverb starts. In a large space it will take longer for a sound to travel and be reflected before reverb is created. Increasing this parameter will generally make the sound appear more distant.
- *Diffusion*. This determines how the reverb is reflected and builds up within an environment. A highly diffused reverb will have more reflections and produce a thicker texture to the reverb.
- *High Frequency Damping (EQ + Filters)*. Reverb devices often include an equalizer section that allows you to change the tone of the reverb. This can be useful for making the reverb sound brighter or duller, which can often help separate it from the original signal.
- *Wet and Dry Balance*. This control allows you to adjust the balance between the input signal and the amount of reverb that is produced from the reverb's outputs. It is important you set this control to 100% reverb if you are sending to the reverb device using an auxiliary send and return, as you don't want to hear the dry input signal coming from the reverb's outputs. Note that a signal with reverb is usually referred to as the wet signal.

Fig. 11.7.6 – Reverb parameters.

Exercise 11.8 Editing audio

Before attempting this exercise we recommend you complete Exercises 5.3 and 5.4.

This exercise will cover the following:

- Non-destructive audio editing
- Destructive audio editing
- Basic audio editing
- Moving audio
- Cutting and removing audio
- Copying and repeating audio
- Normalizing an audio file
- Reversing audio
- Time stretching.

The exercise ...

Audio editing is normally carried out after the recording stage in preparation for mixing. The complexity of the editing can vary from simply removing a noise or mistake from a recording, to completely rearranging all the instruments so they play back in a different order. Before undertaking any digital editing, it's a good idea to understand the difference between destructive and non-destructive editing and how audio files are recognized and structured within a digital system, as this will help prevent any unexpected surprises or permanent loss of data. We also recommend completing MIDI Exercises 5.3 and 5.4, as many of the techniques used to edit MIDI can also be used to edit audio.

tip ▶ Once you have made a recording you may feel it can be improved by editing.

Non-destructive audio editing

Each time you press Record on a digital multitrack, an audio file is created and stored on disk (unless you're using a tape-based system). When you access these files the multitrack or sequencer is simply instructing the hard drive which audio files to play back and in what order. This process is called random access and is the main principle behind non-destructive editing.

A non-destructive edit function will allow you to make changes to the way a piece of audio appears or is used, without affecting the actual audio file stored on the hard drive. This is an important factor to bear in mind when you copy or repeat an audio file, as you are not actually copying the original audio file stored on the hard drive – you are simply instructing the device to play that piece of audio again. This saves space on the hard drive and generally gives you a lot of flexibility for manipulating and changing the way the audio is played back. For example, it becomes possible to:

- Play back the audio files in a different order
- Move the audio to a new location
- Copy and repeat the same section over and over
- Remove sections of audio
- Change the arrangement of a piece of music so the sections play back in a different order
- Try out different arrangements of the same song
- Join different songs together
- Create a composite by picking out the best bits to create a perfect recording.

All non-destructive editing functions allow you to use the magic function of Undo, so if you make a mistake or change your mind you can easily revert back to the original audio file.

note ► A non-destructive audio edit function does not change the original audio file in any way and can be undone at any time.

Destructive audio editing

Some of the more complex audio editing functions, such as time stretching, normalizing and reversing, do actually change and modify the original audio file that is stored on disk, so you have to be careful when using these functions, as they cannot be undone if you change your mind! Before trying out a destructive edit function, it is advisable to make a back-up or duplicate of the audio file so you can revert back to using the copy if it all goes wrong!

note ► Destructive audio editing means the original audio file that is stored on disk is changed or modified in some way.

Basic audio editing

The audio waveform

Once an audio file has been recorded or imported into a digital multitrack you should be able to see the audio visually as a waveform. Note: you may have to zoom in to see this. The height of the waveform represents the amplitude or volume of the audio file. The higher the waveform, the louder the audio file. Seeing the waveform will also allow you to see where each audio track starts and stops, and help you visually when trying to find a suitable edit point.

Identifying the beat

By looking closely at the waveform it's usually possible to identify where the main beats of the music fall. In the example in Fig. 11.8.1 the regular peaks represent each beat of the music.

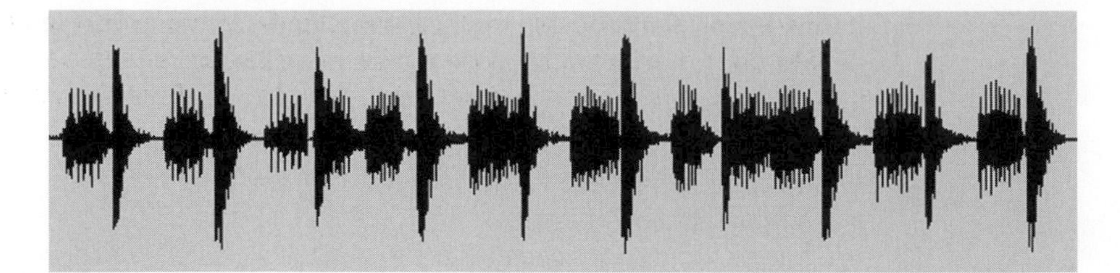

Fig. 11.8.1 – Audio waveform.

tip ► Zoom in on the audio waveform to help you identify a suitable edit point.

Audio editing tools

Most audio editing programs allow you to select different tools so you can manipulate the audio in different ways.

Fig. 11.8.2 – Editing tools.

Moving audio

With the pointer (or Pro Tools hand tool) you will be able to move the audio from left to right, anywhere along the timeline. Try moving the audio left or right to a different position on the timeline. You can also move audio up or down onto different audio tracks.

1 bar
1/2 note
✓ 1/4 note
1/8 note
1/16 note
1/32 note
1/64 note

Fig. 11.8.3 – Snap values.

Snap and Grid

When moving audio regions you need to be aware of the Snap or Grid value. This will limit how freely you can cut or move the audio against the timeline. For example, when set to 'bar' this would only allow you to move the audio to the beginning of each bar. However, this value can be changed in order to move regions by smaller amount.

Editing audio

Often when you record or import an audio file you may need to trim it down to size or remove a section you don't want to keep. The scissors will allow you to divide an audio file into separate sections, allowing you to remove a mistake or an unwanted noise. When you cut or separate a piece of audio in half, two new regions will be created – one either side of the cut. As these are just references to an audio file, each region can be resized or edited back to its original form at any time.

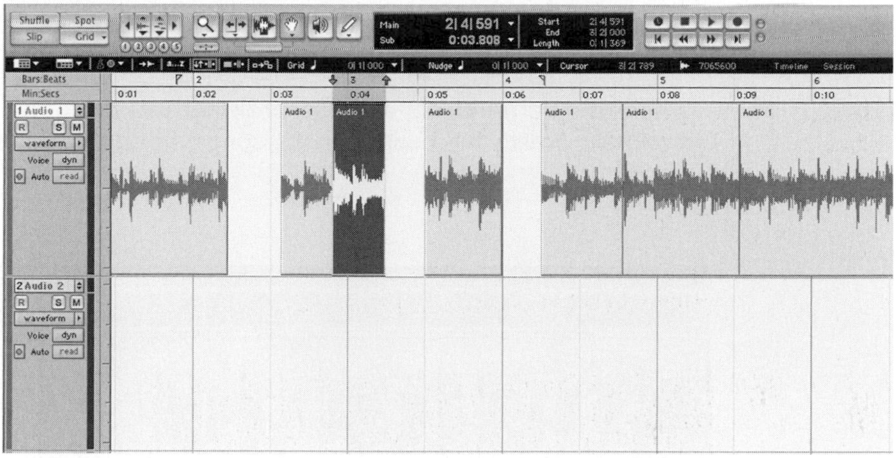

Fig. 11.8.4 – Audio regions being selected and deleted.

Use the scissors or Separate Region command to separate the sections of audio you want to keep. Once this is done, select the unwanted sections and press Backspace or use the erase tool to remove what you don't want to keep. The audio does not get deleted, it is just removed from the screen. At any point you will be able to get it back, as this is a non-destructive command that can be undone.

producer says ▶

Audio files and regions

Audio files or parts of audio files are usually referred to as regions. A region can be any size from a single drum sound to a complete track from a CD. You can move, cut, copy, repeat and paste regions, and resize them at any time to reveal more or less of the original audio file they are referenced to.

Copying and repeating audio

The Copy and Paste commands allow you to repeat sections of audio. This allows you to build up a song arrangement and try lots of different ideas quickly. For example, highlight the audio file you want to copy. From the Edit menu select Copy (this command makes a copy of the selected audio region to an invisible clipboard, where it is temporarily stored). It leaves the original audio region unaffected. Move the Song position marker to the new location (i.e. where you want the file copied to) and select Paste. A copy of the audio file should now appear on the selected track. As this is a non-destructive edit command, duplicates made of the same audio file will not require any more disk space and can be easily undone if you change your mind.

tip ▶

When pasting an object it is essential that you move the Song position marker to exactly the position in the song where you want the copy to be placed.

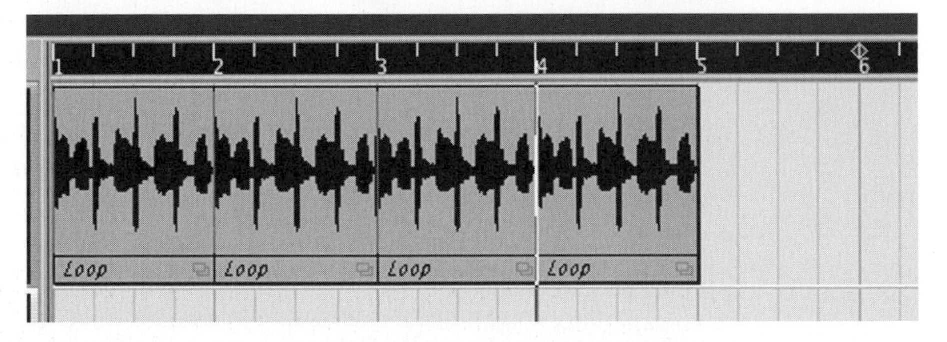

Fig. 11.8.5 – A one-bar audio region being duplicated using Copy and Paste.

Once an audio file is copied to the clipboard it will stay there until you press Copy again. This allows you to repeat the same piece of audio again and again. So if you have a one-bar section of audio or a drum loop you want to use for eight bars, or throughout the whole song, simply copy and paste it.

producer says ▶

Once you become familiar with some of the basic audio editing techniques you should start to see how powerful editing can be. For example, during the recording process it may be difficult to make decisions on what to keep and what to erase. It's therefore common practice to record an alternative version, so you end up keeping several recordings of the same part/instrument (each recording is referred to as a take and must be muted so you only hear one version playing at a time). Once the recording is complete you can use various editing techniques to mix and match and pick out the best sections from each recording to create the illusion of a perfect recording. This is a common process used by most professional studios.

Cross-fading

During the editing process you may find it difficult to get a smooth transition between audio regions that have been rearranged and then pushed together. Creating a cross-fade over the join (edit point) may help combine the different sections together more smoothly.

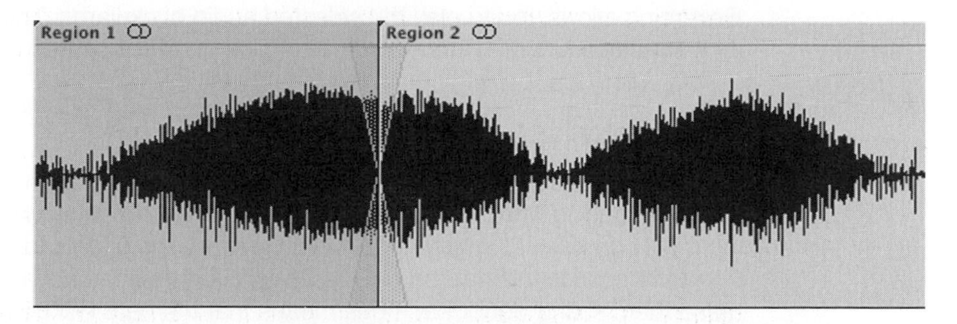

Fig. 11.8.6 – Two audio regions joined together using a cross-fade.

When two audio regions are pushed together without a gap you may get a click or the transition may sound unnatural. Creating a cross-fade over the join (edit point) may help.

tip ▶ Sometimes it's difficult to get a smooth transition between two different sections of audio. Using a cross-fade may help.

Destructive edit functions

You may find some of the more complex audio editing functions do actually change and modify the original audio file in some way, so you have to be careful when using these functions, as sometimes they cannot be undone if you change your mind! Before trying out a destructive edit function it is advisable to make a back-up or duplicate of the audio file so you can revert back to using the copy if it all goes wrong!

note ▶ Destructive audio editing means the original audio file that is stored on disk is changed or modified in some way.

Normalizing an audio file

This function will increase the overall level of an audio file by making it as loud as possible without overloading the capabilities of the recording device. The normalization process scans the selected piece of audio and finds its highest peak. It then raises the overall level of audio proportionally, so the peak of the waveform is at the highest possible level (you will see the height of the waveform increase). The downside of using this function is that it also increases the level of any noise the file may contain.

Reversing audio

Reversing allows you to play the selected audio backwards. This can be used as a special effect or to create something unusual sounding.

Time stretching

This is a very useful function as it allows you to change the timing of an audio file without changing its pitch. Normally, when you speed up or slow down a recording it will play back at a faster or slower speed, so the pitch and tempo change. For drum sounds this can be OK, but for pitched instruments and vocals it will make them sound out of key. (This means that the beat will be in time but the music will be out of tune). The solution is to time stretch the audio so the tempo changes but the pitch stays the same. In order to time stretch an audio file successfully to another tempo you will need to first calculate the tempo of the audio you want to change. Sometimes this information may be listed in the filename; alternatively, you may have to calculate the tempo yourself.

It is also possible to adjust the pitch of an audio file without adjusting its tempo. This opens up all sorts of editing possibilities as it allows you to change the key of an existing audio file and correct the tuning of a vocal or solo instrument.

Time stretching allows you to:

- Change the tempo without changing the pitch
- Change the pitch without changing the tempo.

Time Machine

	Original	Destination
Tempo Change (%)		- 12.5000
Tempo	120.0000	105.0000
Length (Samples)	88200	100800
Length (SMPTE)	00:00:02:00	00:00:02:07
Length (Bars)	1 0 0	1 0 549
Transposition (Cent)	Free	100

Fig. 11.8.7 – Time stretching parameter in Logic. In this example the tempo will be reduced from 120 to 105 BPM and the pitch will be increased +1 by a semitone.

Exercise 11.9 Mixing

This exercise will cover the following:

- Concepts of mixing
- Starting a mix
- Adjusting the channel faders
- Muting tracks
- Using the pan control
- Equalization
- Adding reverb and effects
- Compression
- Automation
- Using the master fader
- Storing and recalling a mix
- Creating an alternative mix.

Concepts of mixing

This is a very creative stage where time can be spent experimenting how to blend individual instruments together and exploring different effects and processors. The idea is to create the best possible blend of all the individual sounds or audio tracks. This is not as easy as it sounds, as you have to listen to the overall sound and not each individual instrument or component. The complexity of the mix will really depend on how many individual tracks or instruments you have recorded. Ideally, each instrument should be recorded to its own separate track and therefore have its own separate channel on the mixing desk. The advantage of keeping instruments separate is that you can set individual volume and pan levels and process each channel independently, whereas with instruments combined or recorded together you can't. Generally, the more options you have, the longer the mixing process will take. Also, there are no rules on what sounds good and each person will have their own idea of how the mix should sound. Some basic tips to get you started are outlined below.

producer says ▶

Time management

You should try and approach this stage after all the recording, overdubbing and editing has been completed. This way you can focus on blending the overall sound together without being distracted by any mistakes or missing instruments. A common mistake is to rush the mixing process and therefore jeopardize a potentially good recording by presenting a poor blend of the overall sound.

Starting a mix

During the recording and overdubbing stage you will have created a rough mix in order to hear each sound and instrument. This rough mix (the monitor mix) can often be used as a starting point for your final mix. Alternatively, you may prefer to start with all the faders down and create a completely new mix. Another technique is to start by only listening to the drums or rhythm section and then add each instrument one by one.

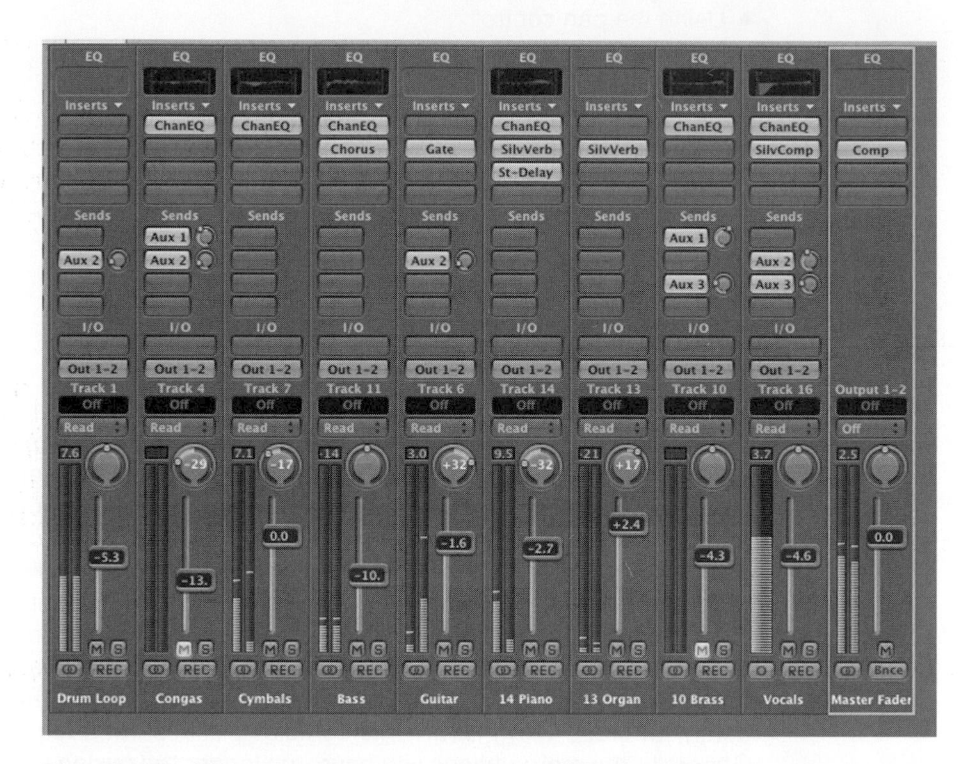

Fig. 11.9.1 – This example shows several different audio tracks and a master fader during mixdown. Notice how each channel is routed to output 1–2 so all the signals arrive at the master fader, and how the plug-in and auxiliary send settings vary for each channel.

Adjusting the channel faders

Experiment by moving the fader levels on each individual channel, and try and consider what's best for the overall sound, not just your favourite instrument. Try and create a balance so you can hear each instrument clearly. You may need to adjust the volume of any solo instruments, for example, so they cut through over the top of the rhythm track or turn up the bass line so it drives the music.

producer says ▶

You can adjust the level of each audio track by using the *channel fader*. A channel fader allows you to control the volume of the sound playing back on that channel. Always try to start with the fader turned up about halfway as this will leave you with some 'headroom' to adjust the level up or down. If you start with the fader at the top you will be unable to increase the level of that channel any further.

tip ▶

Using a reference CD

Before you start mixing try listening to a commercial CD of other popular music to get an idea of how they have blended different instruments together. This may help you decide what instruments or sections of the music you want to feature.

Muting tracks

Sometimes when mixing you may not want to listen to all the channels at the same time or you may want to add each instrument separately when building up a mix. The mute switch allows you to experiment, bringing instruments in and out throughout the song. Sometimes you may find it sounds better if some instruments don't play all the way through a song or in certain sections. As always, experiment.

tip ▶

Pressing solo provides a quick way to listen to a sound on its own.

Using the pan control

Make use of the stereo field using the pan control. Experiment by panning different elements of a mix from side to side and think about where you want to position each instrument. Experiment by panning instruments that occupy a similar frequency range in opposite directions, such as piano and guitar, as this will help with separation and create perspective. The main elements of a mix, such as drums, bass and vocals, should be placed in the centre to avoid the mix becoming one-sided. A typical example of this would be to position the kick drum, snare drum, bass and lead vocals in the centre, the hi hats left and the tambourine right. Also make sure that any stereo sounds are panned extreme left and right. This should make the sound rich and full sounding, and make the stereo image appear to widen.

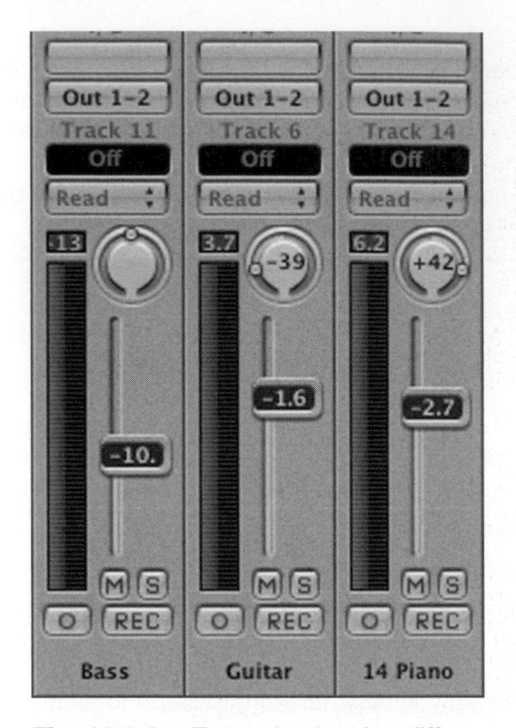

Think about where you want to position each instrument. Try panning instruments in opposite directions to create a wider stereo image, as this may help with separation.

Fig. 11.9.2 – Example showing different pan positions during a mix.

note ▶ The sound will be heard equally in both the left and right speakers when the pan pot is set to the central 12 o'clock position.

Equalization

Each channel on a mixing desk will normally have access to an equalization (EQ) section that will allow you to cut or boost certain frequencies of a sound or instrument. This may help if you are finding it difficult to blend certain instruments together – for example, a sound may need to become brighter in order to cut through and be heard more clearly. As a general rule, when mixing don't use EQ unless you have to.

Using EQ

In its simplest form EQ allows you to cut or boost the bass or treble of a sound, so if you want a brighter sound boost the high frequencies; for a baser sound, boost the low frequencies. However, EQ can also be used to remove unwanted frequencies. One way to approach this is by boosting the gain then sweeping through the different frequencies until you find the offending frequency you wish to remove. Once located, you can reduce the gain to cut the unwanted frequencies.

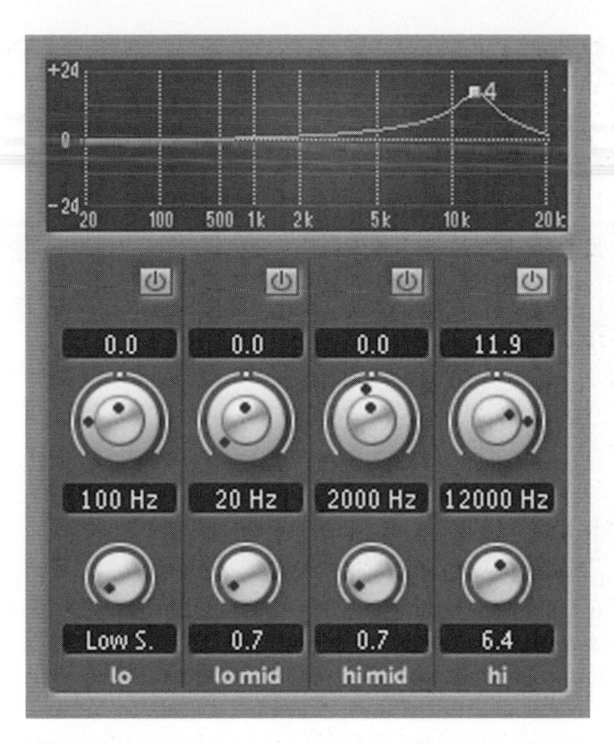

Fig. 11.9.3 – A four-band plug-in EQ boosting the high frequencies.

tip ▶

It is often easier to find an offending frequency by boosting the gain and then sweeping through the different frequencies until you find the frequency you wish to remove. Once located, you can reduce the gain to remove the unwanted frequencies.

producer says ▶

EQ is a very complicated area of sound and it can take time to understand how to use it, so before reaching for the EQ ask yourself how you want to change the sound. Is it too bright or too dull, or just lacking some sparkle? Sometimes it's easier to choose another sound or adjust the volume than trying to fix it using EQ.

Adding reverb and effects

Adding an effect to an audio track is a way to change or add final gloss to a sound during a mix. Adding reverb, for example, will make a sound appear bigger or more distant and will help create perspective. Also consider adding some reverb to any solo instruments to create more space and ambience around them (see Exercise 11.7).

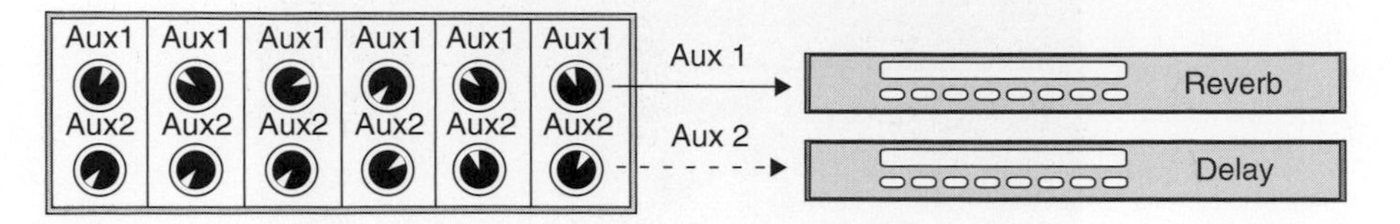

Fig. 11.9.4 – Diagram showing how separate effects can be set up using Aux 1 and 2. This is ideal in a mix situation as it allows you to experiment, as each channel has access to both reverb and delay.

Compression

You may want to consider using a compressor. This can be used as a level control or artistically as an effect to make things sound bigger or tougher. They are commonly used on vocals and solo instruments and can also be used to control the overall volume of a stereo mix.

Automation

Sometimes you may find the level of certain instruments or sounds may need adjusting at different locations throughout a recording to make them cut through. Mix automation is available on most digital audio workstations and allows you to record any parameter movements, such as faders, pans and mutes, into the sequencer. Once recorded, these movements will play back automatically just as if there was someone sitting next to you helping you move the parameters. This gives you a tremendous amount of control and allows a mix to be tweaked to perfection. Automation can usually be enabled for each track separately and will offer different modes of operation, such as record, touch or read. Care must be taken when selecting different modes not to record parameter movements accidentally.

Examples of using automation are:

- Creating gradual volume changes throughout the song
- Automatically muting tracks
- Automating the pan pot so it will automatically move the sound from left to right, creating an auto pan effect
- Automating the channel mute switch – this allows you to experiment, bringing instruments in and out throughout the song
- Automating the FX on/off button or the effects send level in certain places to create a sudden burst of reverb
- Creating a volume fade at the end of a song.

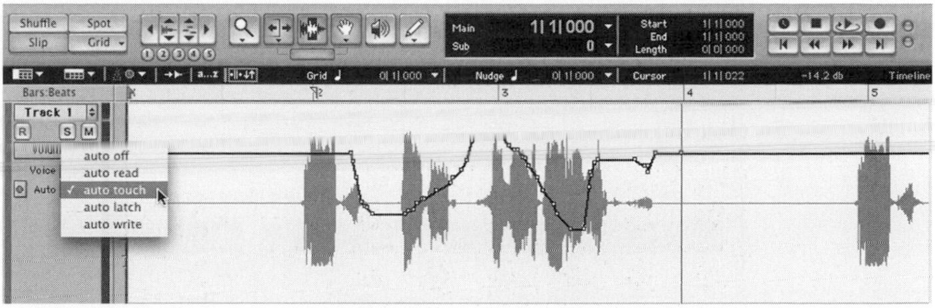

Fig. 11.9.5 – When using a Digital Audio Workstation you will be able to view and edit automation data graphically.

producer says ▶

Sometimes you can spend a lot of time automating and adjusting small details. You may need to listen back to the whole song to get a perspective and see how the changes fit with the overall song.

Snapshot automation

Many digital mixing desks allow you to store a scene or snapshot of all the mixing desk settings at a given time. This would allow you to store and recall different settings during mixdown or create a template so all your favourite settings could be recalled at the touch of a button.

tip ▶

Grouping tracks together

During mixdown it may be easier to group certain faders together, such as all the drums or all the guitar tracks. This will make it easier to adjust the level or process several tracks at the same time and simplify a complex mix.

The master fader

As your mix progresses the level of each channel fader will contribute to the overall level at the stereo bus. Therefore, it is a good idea to regularly observe the master fader LED when mixing. Be aware that setting individual channel levels too high will overload the input to the master fader and may cause 'distortion' or 'clipping'. You will see a red overload LED on the master if this level is too high! As this is the final output stage, any problems here will appear on the final master and when you make a CD it will distort! To ensure your music is in stereo (like on a CD), the master fader or master output has to be in stereo. On a software mixing desk it may be possible to make the master fader mono. This should be avoided as it will make the sound exactly the same in both speakers and you would lose any stereo imaging created by using the pan control or from any stereo sounds, such as stereo audio files, reverbs and strings.

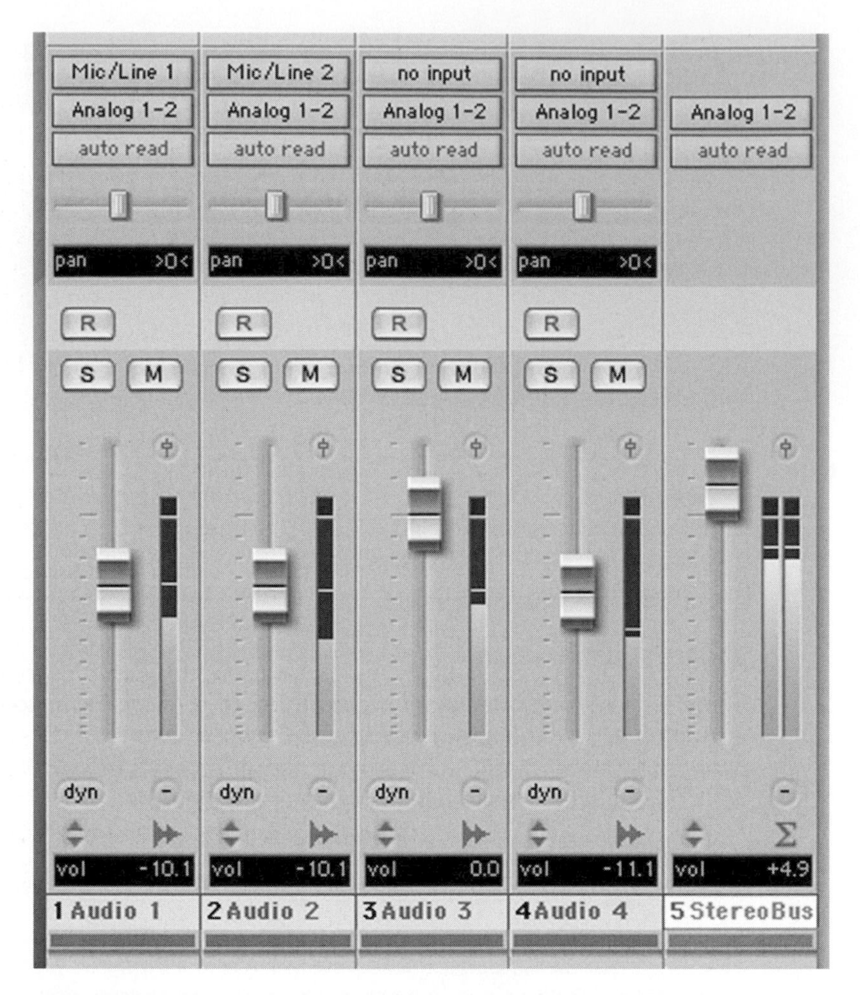

Fig. 11.9.6 – Four audio channels routed to the stereo bus in Pro Tools.

producer says ▶

> The master levels leaving the console should be set as high as possible without peaking! Sometimes this is very difficult to achieve, as each individual channel will add to the overall level differently throughout the song. One technique is to insert a compressor or limiter across the stereo bus so any sudden peaks can be controlled automatically.

End fade out

The master fader can also be used to create a fade out at the end of the mix. This will ensure a smooth transition to silence at the end of the mix.

Storing and recalling a mix

If you are using a digital audio workstation then all your mix settings will get saved with the song file. This is extremely useful as it allows you to go away and listen to the mix and reflect on any improvements or changes you may want to make knowing that when you return to the project it will sound exactly the same. If you are using external equipment you will need to manually write down the settings for each device used.

Creating an alternative mix

Sometimes it's difficult to make decisions during the mix. If you are really unsure about something then you could create two different mixes and choose the best one later.

Exercise 11.10 Creating a CD

This exercise will cover the following:

- Creating a stereo master
- Using an external stereo recorder
- Monitoring a stereo mix
- Bouncing to disk
- Transferring MIDI sounds into audio
- Mastering
- Creating an audio CD.

The exercise ...

In this final audio exercise we explain how to create a stereo master recording of a mix and explore some of the main issues when transferring music onto a CD.

note ▶ A stereo master recording combines all the individual audio tracks arriving at the stereo bus into a new stereo audio file that can be transferred onto CD or converted into an MP3.

Once you are happy with how your mix sounds you will need to make a recording of it so it can be transferred onto CD or converted into an MP3. This is achieved by making a stereo recording of all the individual elements arriving via the master fader at the stereo bus.

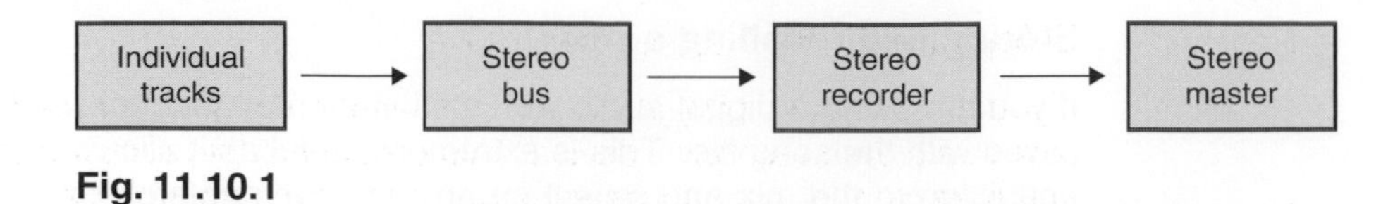

Fig. 11.10.1

Once a stereo mix file has been created it can be transferred onto CD.

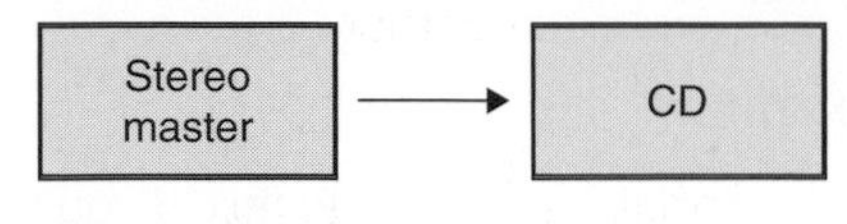

Fig. 11.10.2

A stereo master recording can be transferred directly onto CD. However, most professional recordings pass through another stage called mastering before being transferred onto CD (see 'Mastering' below). This stage allows additional processing such as EQ and compression to be applied to the stereo master to further enhance it before being transferred onto CD. This is usually carried out in a dedicated mastering studio and will be optional depending on the time and budget of each particular project. However, there are a number of things you can try yourself (see 'Mastering' below).

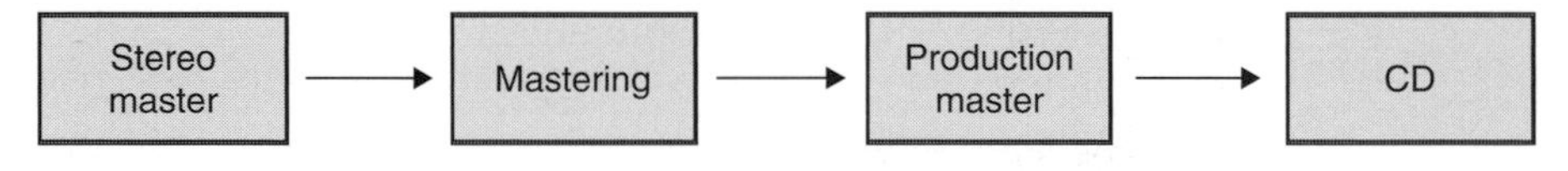

Fig. 11.10.3

Creating a stereo master

This is the first stage of transferring a completed mix onto a CD. How you achieve this will vary depending on the type of recording equipment that is being used. Popular formats used for recording a final stereo mix to are stand-alone DAT, mini-disk and CD recorders. It is also possible to record a stereo mix into a computer via a soundcard, and if you are using a computer system where all the mixing has been carried out internally you may be able to create a stereo mix file by bouncing down internally (see 'Bouncing to disk' below).

note ► A stereo master can be created and stored in a variety of different formats, such as DAT, CD, cassette or as a computer data file. Alternatively, if you are using a digital audio workstation you may be able to create a stereo master by internally bouncing to disk.

Using an external stereo recorder

If you are using a mixing desk you will need to connect the main stereo outputs to a stereo recording device. This should not be confused with the control room monitor output that is connected to the amp and speakers.

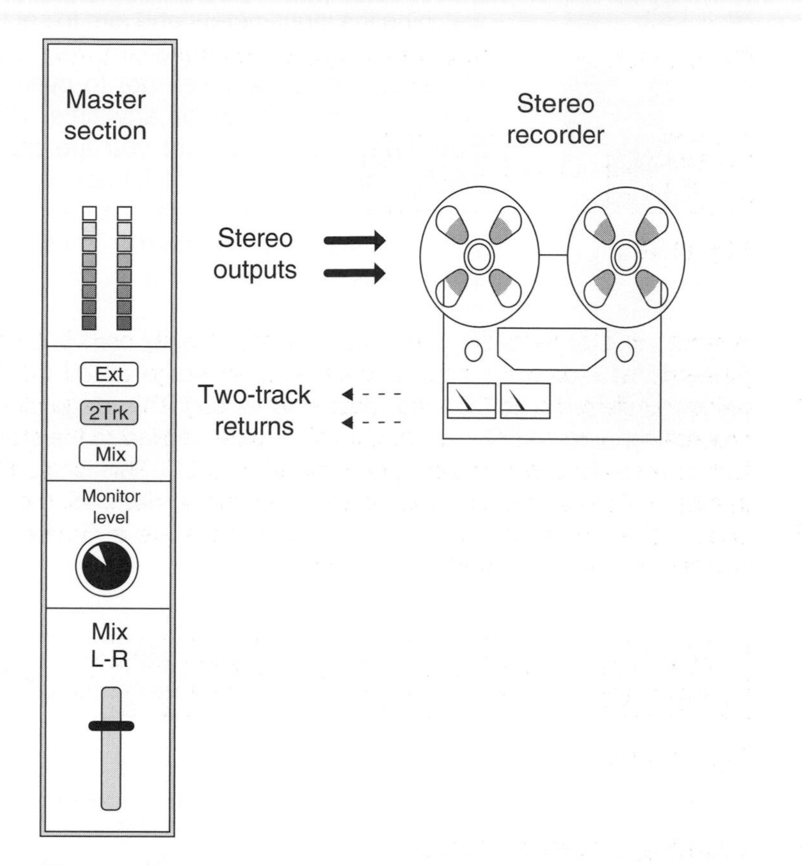

Fig. 11.10.4 – Diagram showing how a stereo recording device can be connected to a mixing desk.

Monitoring a stereo mix

The outputs from an external stereo recording device such as a DAT or external CD recorder should be connected to a dedicated two-track or external input on a mixing desk. This will allow you to listen to the outputs from the stereo recorder on the control room monitor speakers by enabling the external or two-track return switch located in the master section.

Setting the recording levels

The stereo recorder will need to be armed and recording levels set. You may also need to select the correct type of input, i.e. analogue or digital, and select the sample rate.

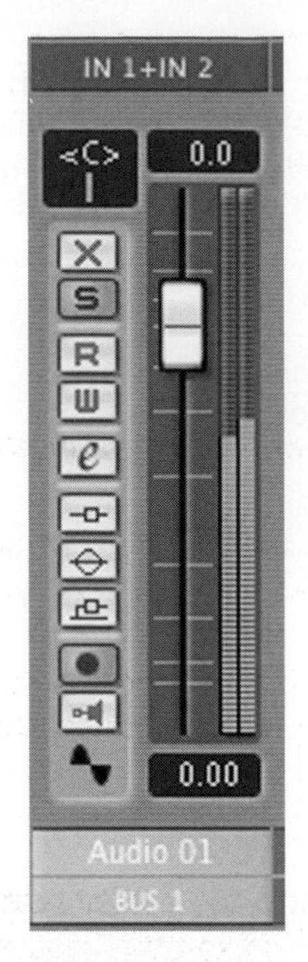

Fig. 11.10.5 – Stereo metering.

Finding an optimum recording level can be difficult as the overall signal level will fluctuate, so you could consider placing a stereo compressor over the mix bus (see 'Compression and limiting' below). As a general rule you should play the mix through while watching the input meters and aim to set the recording levels just under the maximum. As this is the final stage, care must be taken not to overload the input of the stereo recorder at any time. When suitable input levels have been set you are ready to record the mix, so cue up your multitrack, press Record on the stereo recorder and Play on the multitrack. Once your mix has been recorded it is essential that you listen back to it to ensure it has been recorded correctly.

note ► Care must be taken not to overload the input of the stereo recorder at any stage. If this level is too high it may distort when you make a CD!

tip ► Always play back a stereo mix to ensure it has been recorded correctly.

Bouncing to disk

If you are using a computer system and all the mixing has been carried out internally, you may be able to create a stereo mix file by bouncing to disk. This function combines all the individual channels that are routed to the master fader into a new stereo audio file leaving all the other audio files unaffected. Once you have bounced your mix to disk, a stereo mix file will be created that can be transferred onto CD. Note: all muted tracks will not be included or any MIDI tracks unless they are outputting to software instruments (see 'Transferring MIDI sounds into audio' below).

note ► The Bounce function can be used to create a stereo mix file for putting onto a CD.

producer says ▶

> If you are using a combination of audio and MIDI tracks at the same time, you may have to record any external MIDI sound sources as audio if you want them to be included when you bounce to disk (see 'Transferring MIDI sounds into audio' below).

Bounce parameters

The bounce to disk option can usually be located on the master fader on the audio mixer or from a drop-down menu.

1. Before you bounce to disk make sure you are happy with the balance between all the individual audio tracks and that you are not overloading the master fader.
2. Specify the duration of the song. This is normally done by setting the locators. You need to take into account how the song ends and allow for any reverbs or effects to end.

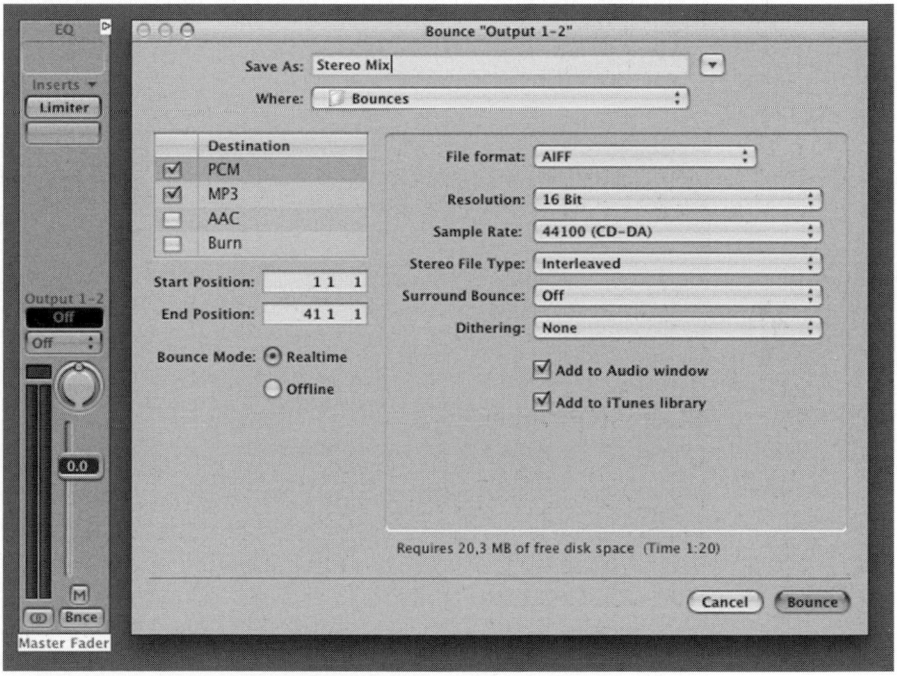

Fig. 11.10.6 – Bounce to disk parameters in Logic.

3. Choose the type of audio file that will be created, i.e. AIF or Wav, and select the bit depth as 16.

4. When you press Bounce you will need to choose a location where this new mix file will be created on the hard drive and give it a name. You may also get a choice to bounce in real time (which will take the same time as listening to the song) or offline, which will allow you to create the mix file quicker.

note ▶ The bounce function combines all the channels that are routed to the master fader into a new stereo audio file.

Transferring MIDI sounds into audio

Most digital audio workstations allow you to use a combination of audio and MIDI tracks at the same time. One potential problem you may encounter when creating a stereo master is how to include sounds generated from a MIDI sound source. This will depend on where your sounds are coming from. Are the sounds coming from inside the computer or from an external sound source, such as a MIDI keyboard or sound module? If the sounds are virtual instruments, i.e. sounds generated from a software program running on a computer, and are routed to the stereo bus, they will be included when bouncing to disk along with all the other audio files (see 'Virtual instruments' section in Chapter 3).

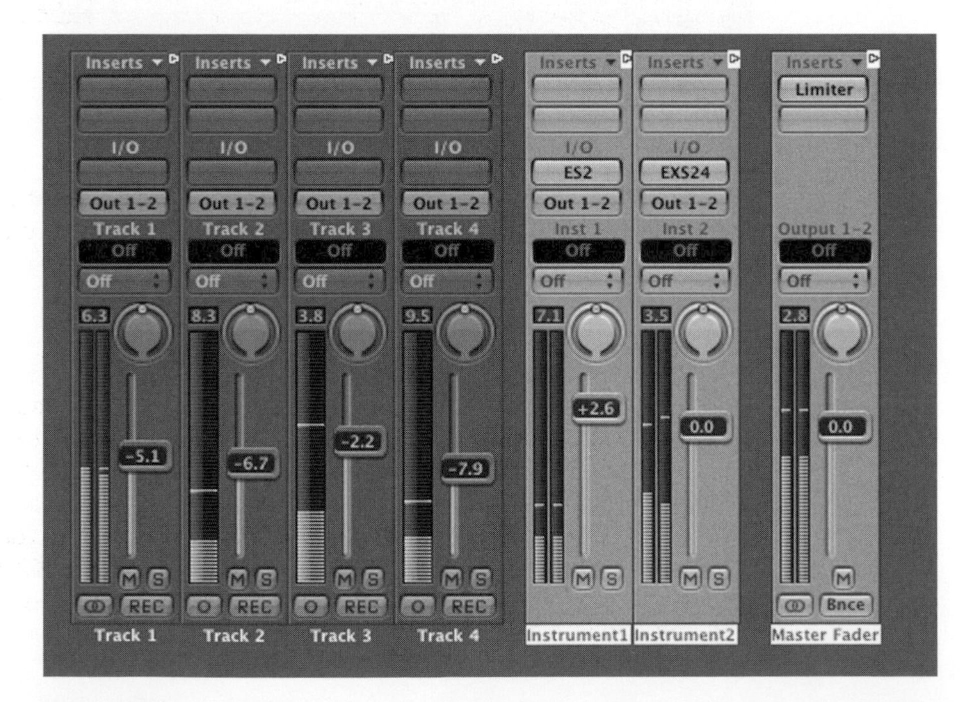

Fig. 11.10.7 – All virtual instruments that are routed to the stereo bus will be included when bouncing to disk.

External MIDI sounds

If the sounds are external, i.e. coming from a MIDI keyboard or sound module, then you could simply connect and record them directly into a standalone stereo recording device. However, if you are using a combination of different sound sources, i.e. sounds generated from the computer and sounds being generated from external MIDI sound modules, you will either need to have everything connected to a mixing desk and mix down via the stereo bus or record the MIDI sounds into the computer as audio so they can be bounced to disk.

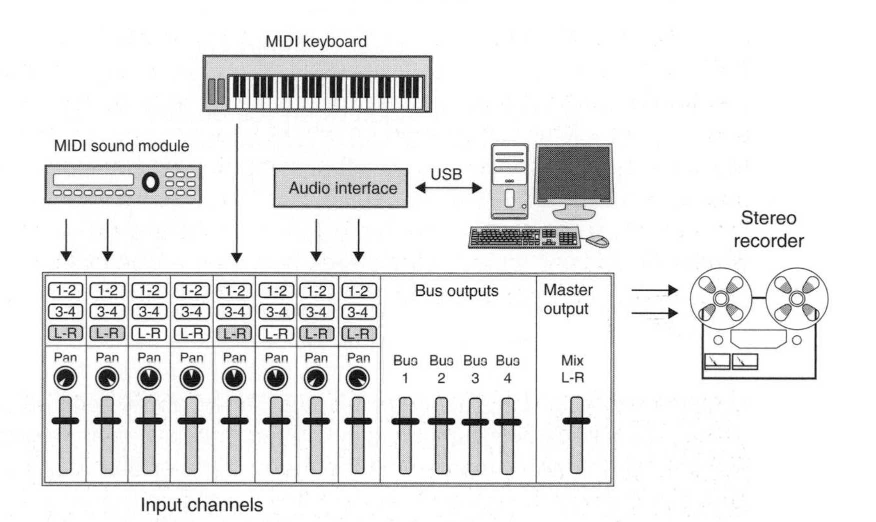

Fig. 11.10.8 – A combination of different sounds can be connected to a mixing disk and then routed to a stereo recording device to create a stereo master recording.

External MIDI sound sources can be recorded as audio, so they can be included when bouncing to disk (see Fig. 11.10.9). If you are using a computer with a PC or GM soundcard, you may be able to internally route the sounds from the card so they appear as input signals within your audio sequencer so they can be recorded as audio. However, if this isn't possible you will need to temporarily connect the soundcard's audio outputs into the audio inputs on the soundcard. Note that care must be taken when recording this way not to create a feedback loop.

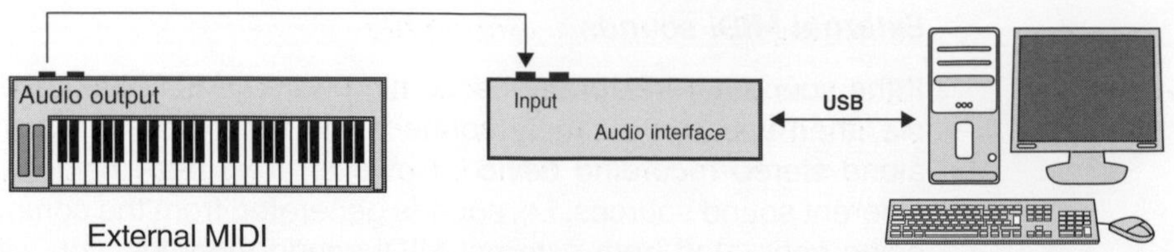

Fig. 11.10.9 – Diagram showing how an external MIDI sound source can be recorded as audio.

producer says ▶

> If you are using a combination of audio and MIDI tracks at the same time, you may have to record any external MIDI sound sources as audio if you want them to be included when you bounce to disk.

Mastering

Once you have created a stereo master recording of a mix, you may want to consider trying to enhance it further before it gets transferred onto CD. If you have already recorded or bounced the audio file to your hard drive, then you simply need to load this file into your DAW. Note: always ensure that you use a stereo audio track. Alternatively, if the stereo master has been recorded to DAT, cassette or mini-disk then you will need to transfer this into your DAW via a soundcard. This can be done by recording the stereo mix file back into a digital audio workstation in real time (i.e. the duration it takes to play the recording from the start to the end). If you have already recorded your mix directly onto a stand-alone CD recorder, you may still want to consider tweaking it further using the mastering process.

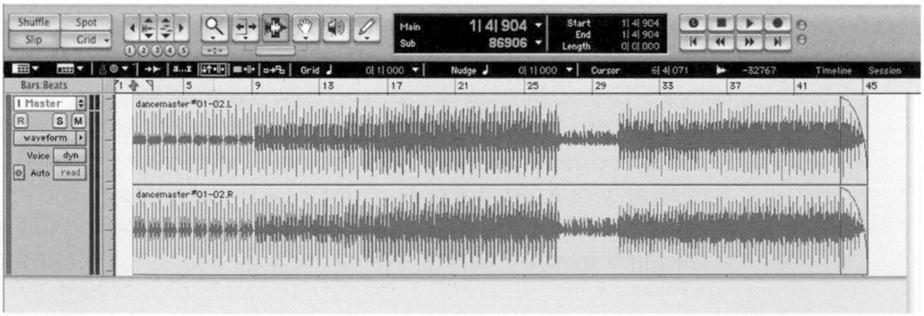

Fig. 11.10.10 – Example of a stereo mix being edited and faded at the end of the file.

Adjusting the start and end points

One of the first things to look at is how the audio file starts and ends, and ensure that there isn't a gap or any noise at the beginning or end of the audio file. If trimming the end of the file sounds unnatural, try using a fade.

Final equalization

At this stage you can also assess whether the overall recording is too bright or too dull, or if there is too much or too little bass by comparing it to a commercially available CD. If you feel the mix can be improved, experiment using a stereo equalizer. Care must be taken, however, not to make any drastic adjustments that could potentially ruin the recording.

Normalizing

This function can be used to increase the overall level of the audio file without causing distortion.

Compressing and limiting

Compression and limiting are commonly used at this stage to control the overall signal level and prevent distortion and clipping. When used correctly, gentle compression can help a mix gel together and make the overall sound louder. A special type of compressor called a multi-band compressor is often used. This divides the sound into areas of frequency (low bass, mid bass, mid, treble, upper treble), which can then be compressed independently of each other.

Production master

If you choose to make any adjustments using the mastering process, then you will need to make a new recording of the stereo master. This can be achieved by bouncing a new audio file that should be referred to as the production master.

producer says ▶

Transferring audio digitally

If the two devices being used to transfer a stereo mix both have a compatible digital connection, then this should be used rather than the analogue connection. A digital connection will keep the signal in the digital domain, which will help avoid any signal loss during the transfer and help retain the overall quality of the mix. As you are transferring rather than re-recording, you will not need to adjust the recording levels as these will be predetermined from the source. Popular digital formats are Optical and SP/DIF, which require specific digital cables.

Creating an audio CD

The final stage is to transfer the production master onto CD. If you have bypassed the mastering stage then you will need to use the stereo master instead.

Using a computer to create an audio CD

If your mix is already stored on a computer you can simply drag this file onto a CD burning program, such as iTunes, Toast or Nero.

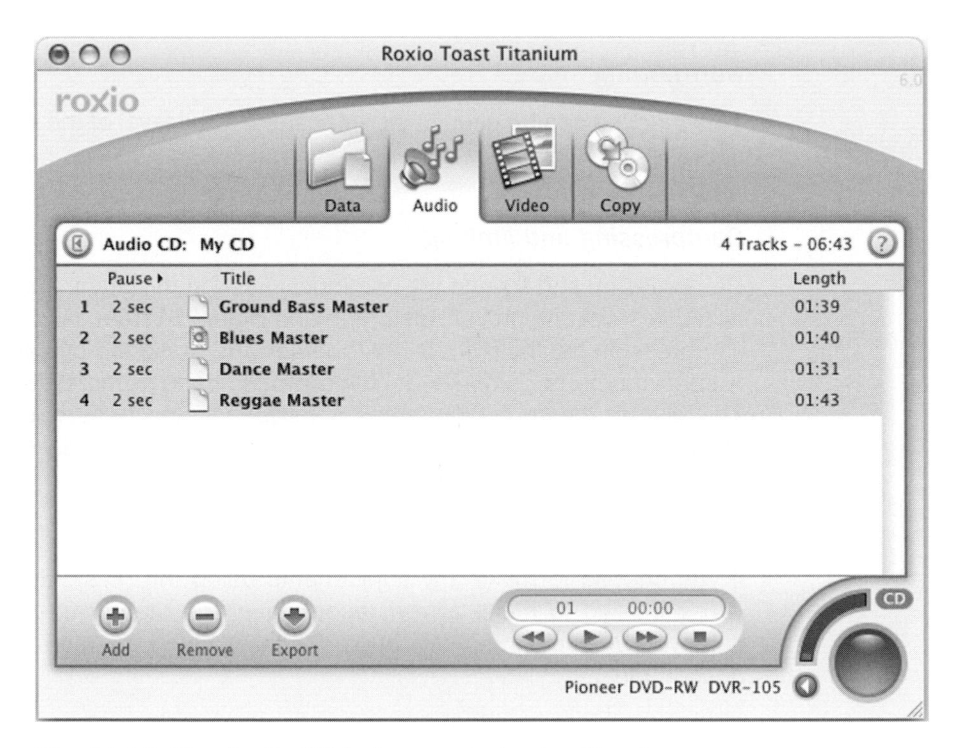

Fig. 11.10.11 – Audio CD being created using Toast.

When making an audio CD using a computer it is essential that you choose the audio CD format. CDs made in other formats, such as files and folders for data back-ups and MP3 and video CDs, will not play back on a standard CD player. If you are planning to place several different tracks on a CD you will need to determine their playback order and leave a suitable gap between each track. A standard CD will be able to store up to 80 minutes worth of stereo audio. You also need to bear in mind that once the CD has been created you will not be able to add any more to it.

Burn speed

When burning an audio CD we recommend you use 4× speed. However, many CD burners can now burn at 52× speed. In these cases experiment and listen back to the result. If you are burning at the wrong speed you will hear 'clicks' and 'pops' on your music that will indicate the settings you are using are incompatible with the type of CD being used.

12
Audio Projects

This chapter guides you through two different audio projects – live stereo recording and multitrack recording. Before attempting these two projects we recommend you complete all the audio exercises.

Project 12.1 Live stereo recording

This project is designed to guide you through the different stages of making a stereo live recording using two microphones. It can also be used as a check-list for setting up a pair of microphones when recording an instrument in stereo.

A live stereo recording is normally made when you need to capture a musical performance in real time. This may be a concert performance or even a rehearsal or a situation where it is impractical to use a recording studio.

This recording method is based around two microphones that are positioned so they capture a performance in stereo. This is one of the simplest and most economical ways to make a recording as no overdubbing or mixing stages are required. However, as the recording is made in real time and all the different sounds and instruments are recorded and mixed at the same time, you don't have the luxury or editing or overdubbing, so the overall quality of the recording usually relies on careful microphone placement.

As you are only using two microphones, you need to find the best location in the room to place the microphones and make use of the stereo field. A good stereo recording should allow you to hear the position of each instrument. (See, 'Stereo Microphone techniques' in Chapter 9). The musicians also need to provide a good balance between themselves before you start recording as you will be unable to change the balance between the different instruments after recording. If one instrument is too loud or if it is overpowering another there will be very little you can do to fix this later. So always make sure you can hear everything clearly before you start recording.

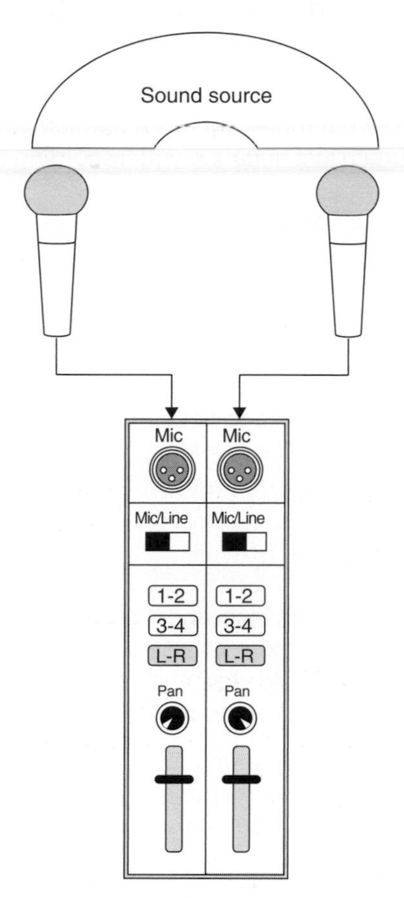

Fig. 12.1.1 – Signal flow diagram of a live stereo recording.

producer says ▶

A good live stereo recording should capture the ambience and energy of a live performance, so when you listen back it sounds as though you were actually there listening in the audience. This type of recording can often produce a very natural sound that is unlike a studio recording where you can overdub and process each individual sound separately. It may also be more natural for the musicians to record this way, as they don't have to wear headphones and can set up in the way they normally rehearse or perform.

Objectives

- Capture the performance using only two microphones
- Produce a high-quality recording without hiss or distortion
- Capture the full frequency range of the sound onto the recording medium
- Be able to hear each instrument clearly
- Balance the left and right channels equally.

Planning your session

You may only get one chance to record a live performance, so all the technical aspects need to be perfect. If there was an audience you could hardly ask the performers to play it again just because your recording levels were too high or you ran out of hard drive space. You therefore need to plan carefully and take the following points into consideration.

The venue

Consider the location of the venue. Where is it situated? How long will it take you to get there? What time is the performance? How much time will you need before the performance begins to set up the recording equipment? Always try to set up early so you can use any rehearsal time to set levels and test the equipment, and avoid getting in the way of the performers.

Room acoustics

The physical space where you place an instrument can have a dramatic effect on the way it sounds. This gets even more complicated when you start using a microphone as basically you have to deal with two sounds: the original sound of the instrument and the sound of the room.

For example, a large room will echo the sound and generate a natural reverb, making everything sound bigger and more ambient. This can make the sound appear more distant and further away from you. A smaller room on the other hand will make the sound more intimate and appear closer to you.

Different types of surfaces such as floors and ceilings will also change the characteristics of a sound as they will reflect the sound in different ways. Rooms with soft furnishings or curtains will dampen the sound making it less bright, whereas a live audience in a room will absorb more of the sound making it less ambient. All this needs taking into consideration when using a microphone.

producer says ▶

> A microphone will pick up the sound of an instrument and the sound of the room. Moving a mic further away from an instrument will make it pick up more of the room ambience. Moving a mic closer to an instrument will capture more of the direct sound and therefore less ambience.

Equipment

In order to make a live stereo recording you only require two audio tracks, so try and use equipment that is portable and easy to set up. There are a variety of different formats available that you can use. Here are some examples.

Two-track recorders

- *DAT* (digital audio cassette) – a good choice as it offers good sound quality, adjustable input level and peak hold metering
- *CDR* (stand-alone recorder) – recording direct to CD offers good sound quality, adjustable input level and metering
- *Mini-disk* – a good choice as it's a portable device but remember this is a compressed digital format
- *Analogue cassette* – should be avoided if possible, as tape hiss can cause a problem.

Multitrack recorders

- *Alesis ADAT* – digital eight-track
- *Tascam DA88* – digital eight-track
- *Portastudio* – combined multitrack and mixer (can be analogue or digital)
- *Stand-alone hard disk recorders* (Mackie/Fostex/Alesis)
- *Computer system* with an audio interface
- *Analogue multitrack*, which can be four- or eight-track – should be avoided if possible, as tape hiss can cause a problem.

note ► If you are using a multitrack for a live stereo recording, you will only need to use two tracks!

Analogue or digital

Audio can be recorded analogue or digital depending on the equipment available. A digital format such as CDR, DAT and mini-disk is probably the best choice for a stereo live recording, as it will offer longer recording times and be cheaper than reels of analogue tape. Alternatively you could use a DAW.

tip ► Make sure you check all the equipment you are using works before you start, as you don't want it to let you down during the session.

Recording time

How much recording time will you need to capture the entire performance? This will vary, so be prepared. Always check the available hard drive space if you are using a computer or any hard disk system tape and make sure you have enough available if you are using a tape-based system.

It is always a good idea to try and roughly calculate the duration of the performance before you start as you don't want to run out of tape or hard disk space halfway through the recording or be changing tapes or drives in between songs.

Example recording times:

- CD – 74/80 min
- DAT – 60, 90, 120 min
- Mini-disk – 70 min
- ADAT tape – 40/60 min
- Cassette – 60/90 min
- 1/4 reel-to-reel 30 min
- Hard disk – 50 megabyte at 16 bit/44.1 kHz = 5 min.

producer says ▶

> Remember, tapes can malfunction and computers can crash, so always consider a back-up plan, as some live performances only happen once. You could actually have two systems working in parallel (e.g DAT and mini-disk). This sometimes happens in professional studios when the session is really important, like recording an 80-piece orchestra, for example.

Do you need to use a mixing desk?

It's common practice to connect microphones directly to a mixing desk. This will allow you to set up the correct *gain structure* before sending the signal to a recorder. It will also allow you to control the recording level and balance the left and right channels equally.

Alternatively, you may be able to connect the microphones directly to your recording device or connect them using a channel strip or mic amp. Most recording devices such as stand-alone multitracks and hard disk recorders don't usually have microphone inputs, but you may find portastudios and computer audio interfaces allow microphones to be connected directly.

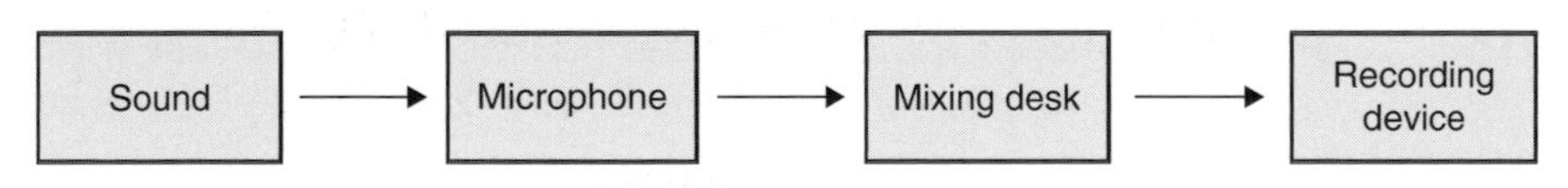

Fig. 12.1.2 – Signal flow using a mixing desk.

A 'stand-alone' microphone amp or channel strip offers an alternative to using a mixing desk.

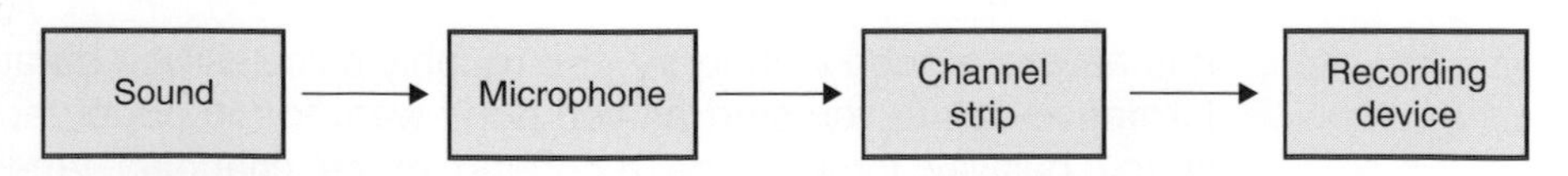

Fig. 12.1.3 – Signal flow using a mic amp or channel strip.

tip ▶

If you are monitoring a stereo signal using two channels of a mixing desk, make sure you pan each channel hard left and right to hear it in stereo! (See 'Stereo sounds' section in Chapter 9).

If you're using a computer with an audio interface you may be able to connect a microphone directly without using a mixer (see Figure 12.1.4).

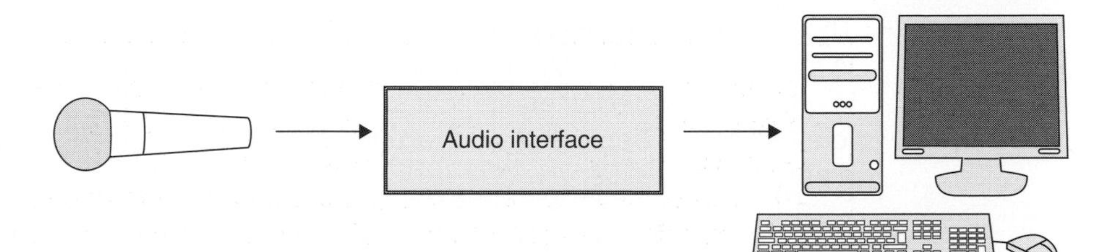

Fig. 12.1.4 – Microphone connected directly to an audio interface.

producer says ▶

It is also possible to make a live stereo recording without using microphones. If each instrument is separately miked or connected to the mixer, you could connect the main stereo output from the mixer directly to input of a stereo recording device. This method relies on everything being connected to the mixer and the fader levels being suitably balanced.

Connections

What type of inputs will you be connecting your microphones to? A microphone is normally connected using male to female XLR cable, so always ensure you have the correct cables available.

Fig. 12.1.5 – Male to female XLR cable.

Always try and use good quality microphone cables and make sure they work properly, as poorly maintained cables can pick up interference and cause unwanted noise.

tip ▶ Mark each cable or connection with a piece of tape or use different coloured cables so you can easily identify each microphone. When two cables look the same, things can get confusing!

The microphone cables need to be long enough to reach from the mixing desk to the microphones. With a live recording it is common to use long cable runs in order to position the equipment some distance away from the performers or to avoid cables becoming stuck around any obstacles such as a door or even a performer!

tip ▶ You can join two XLR microphone cables together to extend the length.

Microphone stands

Microphones usually require microphone stands, so make sure the stands offer good support and allow you to position the microphones high enough. You don't want microphones falling off stands or slowly changing position during a recording.

note ▶ Microphone stands are essential to hold the microphones in place.

Mic clips will also be needed to attach the mics to the stands. You should always ensure that you have the correct type of mic clips available as they are not all the same and some mics require their own special type of clip. A microphone cable clip may also be used to secure the mic cable to the stand.

tip ▶ Always ensure that you tighten the mic stand firmly to prevent the microphones changing position during a performance.

Phantom power

Will you need to use phantom power (+48 V)? Most condenser microphones require this to work, so make sure your mixing desk or recording device can supply this. Alternatively, you can get stand-alone boxes that supply phantom power or sometimes the microphone will be able to use internal batteries (see 'Phantom power' section in Chapter 9).

Mains power

You will need to locate a power source for your recording equipment (unless you are using batteries). When this is located make sure the mains cables and power supplies can be safely connected without trailing leads. You don't want to lose power in the middle of the recording.

tip ►

It's always a good idea to have some gaffer tape available so that any trailing cables can be taped down.

Where to set up and how to monitor

Consider where you are going to set up your equipment – on the floor, a table, etc. Avoid setting up anywhere people need to pass. Additionally, don't block access to doors and fire exits. Generally, keep out of the way.

Good access to the transport controls and metering will be essential during the recording, so position the equipment carefully. You also need to have visual contact with the performers to see exactly what's happening.

Headphones are the best choice for monitoring a live recording, as it's impractical to use studio monitors in the same room as the performers/microphones (see 'Studio monitor speakers and headphones', p. 245).

producer says ►

If you are in the same room as the performers it will be difficult to monitor the sound you are recording accurately, as you will be hearing both the live and recorded sound together. This makes it difficult to judge the recorded sound by itself. One solution would be to record a test section and then play back to check the sound quality and recording levels. Alternatively you could wear headphones.

Choosing a microphone

Depending on the equipment you have available your choice of microphone will generally be between a 'condenser'or 'dynamic'. As you are making a stereo recording, try and use a matching pair of microphones, i.e. two mics that are the same make and model, as this will help balance the left and right sides of the recording.

Condenser microphones generally pick up a wider range of the sound and offer a better frequency response than dynamic microphones. They also allow you to switch polar patterns (polar patterns allow you to select the recording area around the microphone). This will give you more flexibility when stereo miking.

Dynamic microphones are generally less suitable for ambient miking than condensers, as they only have a limited high-frequency response and start to lack sensitivity when moved beyond the range of one foot or more away from the sound source. Microphones such as the Shure SM58 are more robust than condensers but are better suited for close miking. Some microphones offer other options, such as attenuation switches or pads to allow you to reduce the output level from the mic and filters such as HPF (high pass filter) to remove very low frequencies.

note ▶ Even if a device has a built-in microphone, it's usually better to connect external microphones.

Microphone placement

Once you have selected a pair of microphones you need to find the best location in the room to position them. One of the main considerations when positioning a pair of microphones in a room is choosing the distance between the microphone and the sound source, as this will determine how much ambience or room sound is recorded in relation to the original sound of an instrument. The other main factor is how to make best use of the stereo field.

Positioning a pair of microphones in a room can be a very difficult job, as you have to consider the following:

- The sound of the actual room itself
- The balance between the original sound and the sound of the room
- How ambient you want the recording to sound
- How to make best use of the stereo field.

You also need to think about the overall sound and not just one instrument. A good stereo recording should allow you to hear each instrument clearly and make good use of the stereo field. We recommend that you should always experiment when using microphones and try moving them to see if you can improve the overall sound as you will only take away what the microphones pick up. Here are a number of things you may want to try:

- Vary the distance between the microphone and the sound source
- Try pointing the microphone in a different direction
- Adjust the height of the microphone in relation to the sound source
- Vary the distance between each microphone.

producer says ▶ Think about the type of recording you want to make. Do you want the instruments to appear close or distant? When the finished recording is played back do you want it to sound like you're standing on stage or in the audience?

Stereo microphone techniques

How the performers are arranged on the stage may limit where you're able to place the mics. Are the performers spaced widely apart or closely bunched together? You don't want the mics to get in the way of the performers.

There are several different stereo miking techniques you could use (see Chapter 9). One of the simplest examples is a spaced pair (see Fig. 12.1.6). This is where the microphones are placed at a set distance from each other, facing the sound source. Alternatively, if you are recording an orchestra you could place two mics above the conductor to record what the conductor is hearing (think of two ears!).

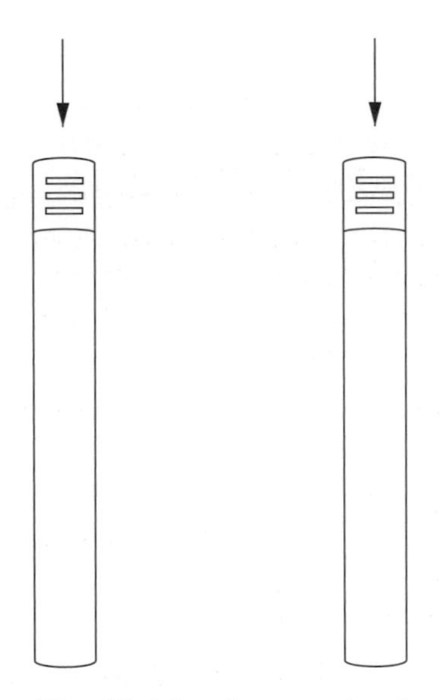

Fig. 12.1.6 – A spaced pair.

Preparing to record

Once you have decided which stereo miking technique you want to use, place the microphones on some stands and connect them to your mixer. Enable input monitoring on your recording device and set a basic recording level. You should now be able to monitor the microphones via headphones.

Sounds to avoid

Before you make a recording think about any other sounds that the microphones may also pick up, such as air-conditioning, telephones, opening and closing doors, aircraft noise, etc. Be aware that any external sound will potentially ruin a recording as it will be impossible to remove. You may be able to minimize this by carefully positioning the microphones away from any background noise. You should also encourage the musicians to avoid foot tapping and ensure that their music stands are secure to avoid them falling over.

tip ▶ Always try and minimize the amount of background noise if possible, as it will be impossible to remove it from a recording.

Setting the recording levels

Make sure you can see the input level meters clearly then adjust the gain control to get a good strong signal from the microphones into the mixing desk (see 'Gain' section in Chapter 10). Both inputs should be set to around the same level. Try and avoid overloading the input, as it will distort.

Balancing both channels equally

The input signals coming from each microphone should be around the same – if not, you will need to re-evaluate the mic placement. Avoid using the fader or gain to compensate, as it may unbalance the stereo image.

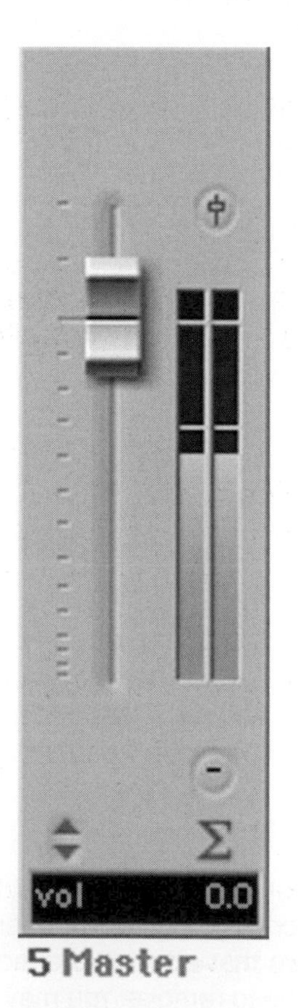

Fig. 12.1.7 – Stereo meter.

Keeping the recording levels constant throughout a recording can be tricky. You really need to know where the loudest part or section of the music is and work back from there. This is usually impossible to predict, so try and set a good average level with some headroom. If you set your recording levels high to begin with you will have no headroom so, if the signal suddenly increases in volume, you will overload the input. If you set your recording levels too low to begin with, you may hear more noise and reduce the sound quality.

tip ▶ Ask the performers to rehearse so you can test the equipment and set the recording levels. Why not do a trial recording at the same time, so you can check how it sounds?

If you are finding it difficult to keep the recording levels constant throughout the recording, a *compressor* could be used to help control the dynamic range of the sound (see 'Compressors' section in Chapter 7). Alternatively, if the signal is too loud:

- Turn down the recording level
- Move mics further away from the sound source.

If the signal is too soft:

- Ask the musicians to play louder (this may not be an option)
- Move the mics closer to the sound source
- Boost the gain on the mixer or mic amp
- Turn up the recording level on the recording device.

tip ▶ During the recording it is always a good idea to keep an eye on the recording levels and check that the left and right channels are equal.

producer says ▶ Often, performers do not play at maximum level when rehearsing, so when they actually play 'for real', the level is usually louder.

Preparing recording media

At this stage you should be ready to make your recording. However, here are a few points you should double-check:

- Make sure your recording medium has sufficient recording time
- Check you're not recording over anything you want to keep
- Make sure you are recording in stereo
- You are using the correct sampling rate (44.1 kHz or 48 kHz)
- You are using the correct Bit rate (16 or 24 bit)
- Turn off the pre-count, so you don't have to wait to drop into Record
- Cue the tape so it's ready to record
- Reset the tape counter to 00.00.

It is always a good idea to start recording just before the performance begins, as you can always edit this out later. It's better to have a long introduction than a clipped recording. This will ensure that you don't miss anything and allow you to capture some of the room ambience on the recording. Also, don't be too anxious to press Stop at the end of the performance as you never know what might happen (an encore maybe!). Any additional time at the end can be edited later so it's always better to wait until the sound has completely faded away before pressing Stop.

Once you have completed the recording you should save your project if you are using a computer system and log any settings you may need.

producer says ▶

> Listen back carefully to anything you record. If something is not sounding as good as it should, then try a different approach – move the microphones or point them in a different direction. No matter how much experience you have, each recording situation is unique.

Final format

Whatever format you have used to make a stereo live recording becomes your *original master*. This tape or disk should be duplicated before using it for general listening and playback, as it's the only copy. Also consider whether you need to do any of the following:

- Make a log of your settings
- Label the recording media
- Edit the recording so it starts cleanly
- Index the start of each track
- Make a copy and back-up the data
- Think about any improvements for next time.

tip ▶

> If you need to transfer your recording onto a different format such as CD or mini-disk then try and do this digitally, as this will avoid degrading the sound quality.

Project 12.2 Multitrack recording

Before attempting this project we recommend you complete Exercises 11.2–11.5.

This project is designed to guide you through the different stages of making a multitrack recording using a microphone and a DI box.

The idea of multitrack recording is that each individual sound or instrument is recorded onto a separate individual audio track. This is the most flexible way of recording, as it makes it easier to control the level and pan position of each individual sound or instrument, and generally gives you more creative flexibility. This is unlike a stereo live recording as you have the option of recording or erasing each track independently, so you don't have to record all the instruments at the same time.

A process called overdubbing can be used to add additional instruments to an existing recording at any stage, allowing a recording to be built up in small sections or by gradually layering different instruments sound by sound on top of each other to create complex textures of sound. This process would effectively allow one person to play more than one instrument on a recording (not at the same time, of course) and build textures that would be impossible to reproduce or create during a live performance.

note ▶ Overdubbing allows you to add new instruments to an existing recording.

The amount of tracks you will have available for recording each sound will vary depending on the type of multitrack you plan to use. In our example, we are going to use an eight-track; however, multitracks with 16, 24 or more tracks are commonly used. Note: when using a computer system the number of audio tracks is usually determined by the software and the power of the computer.

tip ▶ Always try and record each individual instrument onto a separate audio track, as this will give you more creative flexibility.

Things to do before you start the recording session

Before you start a recording session it's always a good idea to check that all the equipment you are planning to use is actually working. Faulty equipment can add extra stress to a recording session and waste valuable time. Pay particular attention to headphones and talkback, and always check the foldback is working. It's also a good idea to have a pen and paper and a chinagraph pencil available so you can make a note of all the connections you make and label the channels that are being used on the mixing desk.

Planning your session

Time management

Before starting a recording project it is always advisable to allocate enough time to each stage of the recording process:

- Recording
- Overdubbing
- Mixing.

A common mistake is to rush the mixing process and therefore jeopardize a potentially good recording by presenting a poor blend of the overall sound.

Arrange in what order you will record each instrument

You will need to choose which instruments will get recorded first and which instruments will be overdubbed. Traditionally, it's best to record the rhythm section first, as this will provide a solid base to add other instruments to. Examples are drums, bass, piano holding down chords, and rhythm guitars. Avoid recording solo instruments first, such as vocals and lead guitars, as these can be easily overdubbed later.

tip ► Always try and record the rhythm section first, as this will provide a solid base to add other instruments to.

How many tracks do you need?

In order to have complete control over a recording, you ideally need to record each sound or instrument on a separate audio track. However, this isn't always possible or practical, as you may be restricted by the number of tracks you have available or by the amount of inputs on your mixer or soundcard. You may therefore need to plan ahead and carefully consider if you will need to combine certain instruments or sounds together. You don't want to find yourself running out of audio tracks during the recording session.

Imagine you have eight tracks available to record a drum kit, bass guitar, rhythm guitar, keyboards, lead vocals and backing vocals. Each instrument can easily be allocated to its own track; however, if you allocate a separate audio track to each drum sound you will quickly run out of tracks and be unable to overdub any additional instruments, such as vocals and guitar solos. One solution would be to combine the individual drum sounds onto a single audio track, therefore leaving plenty of tracks available for overdubs. Alternatively, you could create a stereo drum track by using two audio tracks.

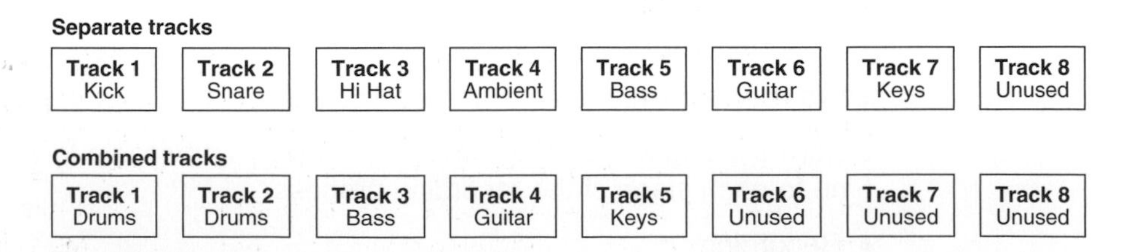

Fig. 12.2.1 – Example showing how individual drum sounds can be combined together.

Deciding which sounds or instruments to combine can often be a difficult decision. This is because you will be unable to change or process them independently once they have been merged. Therefore, you should only combine sounds or instruments that will blend together, such as all the drum sounds or lead and backing vocals, or all the guitars. Note that it's also possible to combine sounds together after they have been recorded by merging audio tracks together (see 'Merging audio tracks', p. 388).

producer says ▶

If you choose to combine several instruments or sounds together, you will be unable to process or change the volume, pan or EQ of the combined sounds individually. It is therefore essential that you get a suitable balance between each instrument before you combine them together.

Stereo sounds

You also need to consider if any of the sounds you plan to record will benefit from being recorded in stereo (see 'Stereo sound' section in Chapter 8, p. 260). Remember, stereo sounds require two audio tracks, so only do this if it is absolutely necessary and if it's actually improving the sound.

Arranging the musicians/performers

How you position musicians/performers for a studio recording may be quite different from how they are used to setting up for a rehearsal or live performance. It may be necessary to move the musicians further apart or isolate them in a separate rooms in order to achieve better separation between each instrument during the recording. This often makes the performers feel uncomfortable at first, so you may need to explain to them why this is necessary.

tip ▶

Visual contact is essential when several musicians are performing at the same time. If they can't see each other they will miss any visual cues, or they may not perform as well.

Room acoustics

The physical space where you place an instrument or microphone can have a dramatic effect on the sound that it will produce. For example, a large room will echo a sound and make it seem more distant, whereas a smaller room will make the sound seem more intimate and make the sound appear closer to you. Different types of surfaces will also change the characteristics of a sound. For example, rooms with soft furnishings or curtains will dampen the sound, making it less bright. All this needs taking into consideration when positioning a performer and using a microphone.

Microphone or DI box?

You will need to choose the most suitable way of recording an instrument and decide whether to use a microphone or DI box (see Exercise 11.2).

Instruments that will require a microphone:

- Acoustic guitar
- Strings
- Brass or woodwind
- Drums
- Percussion
- Piano
- Voice

Instruments that can use a DI box:

- Bass guitar
- Electric guitar.

Most other electronic sound-producing devices, such as electronic keyboards and drum machines, can be connected directly using a jack plug (refer to Exercise 11.2. 'Connecting a sound source').

tip ►

Tuning an instrument

Always make sure all the instruments you plan to use are in tune before you start recording. Also be aware that some instruments may drift out of tune during the recording session due to changes in room temperature.

Choice of microphone

Depending on the equipment you have available, your choice of microphone will generally be between a condenser or dynamic microphone. You will also need to decide if you are going to close mic an instrument or use a more distant ambient miking technique.

Condenser microphones are generally more suitable for recording solo instruments and vocals, as they offer a good frequency response. They can also be used for ambient miking. Some condenser microphones allow you to adjust their polar patterns, allowing you to determine the direction they will be most sensitive to (see Chapter 9). This will give you more flexibility and help reject unwanted sounds.

Dynamic microphones are more robust than condenser microphones and are better for close miking (close miking means placing the microphones close to the sound source, whereas ambient miking means further away from the sound source). Dynamic microphones are generally used for miking drums and guitar amplifiers.

Microphone placement

When using a microphone your objective should be to only capture the sound you want to use. This can be more difficult than it sounds, as there may be several sound sources close together or the microphone may have a wide pick-up area. When miking an instrument you should always try to point the mic directly at the sound source, so it rejects any other sounds around it. When using several microphones at the same time you should always try to get the best separation you can while recording as this will help later when mixing. If a microphone is picking up two sounds at the same time it will be impossible to separate them later. Professional studios often use screens to improve separation between instruments or provide separate isolation booths.

producer says ▶

Microphone placement is a real art, so don't expect great results straight away. If you're having problems achieving a good result, try varying the distance of the mic from the sound source or adjusting the height of the microphone over the sound. You can even try pointing the mic in a different direction. You will find that moving a microphone a small amount can considerably affect the way in which it picks up the sound. As always, experiment.

tip ▶

When using a microphone with a cardioid pick-up pattern, try pointing the back of the microphone at the sound you don't want to record.

Setting up a DI box (see Exercise 11.2)

A DI box allows an electric guitar or bass to be connected directly to a mixing desk or audio interface. Normally, an electric guitar is connected to a guitar

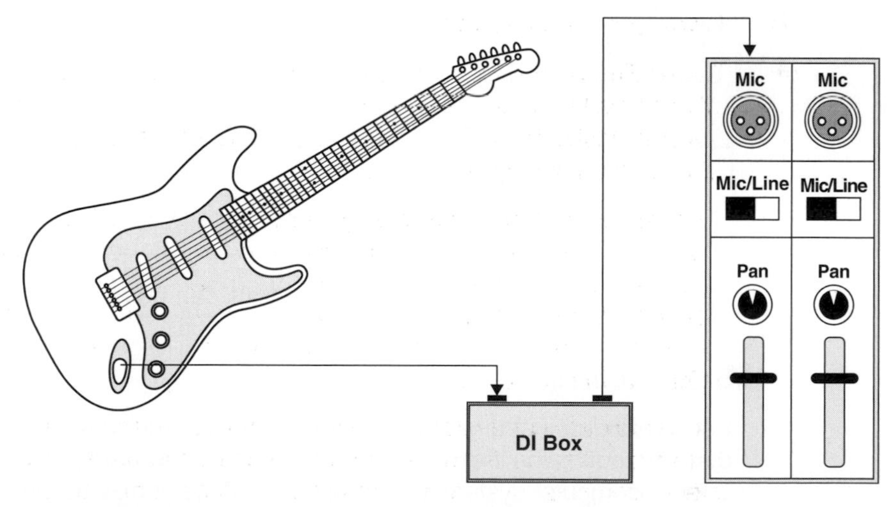

Fig. 12.2.2 – A DI box adjusts the signal from a guitar so it can be connected directly to a microphone input of a mixer.

amplifier or combo to amplify the signal produced from the guitar. If you wanted to record the guitar you could place a microphone in front of the amplifier. The disadvantage of using a DI box is you bypass the tone and processing effects on the amp, so the signal can sound thin and dry. The advantage is that it's more convenient to use, as you may not have a guitar amplifier available, or you just don't want all the noise a guitar amp creates. Note that a DI box may require 48 V phantom power.

Making a connection

Each instrument or sound source will need connecting to a mixing desk or soundcard. You will then need to select the correct type of input, Mic or Line, and adjust the gain control to get the optimum signal level into each channel (see Exercise 11.2).

- *Mic inputs* – for XLR connections from microphones and DI boxes
- *Line inputs* – for jack connections from powered signals such as electronic keyboards and sound modules.

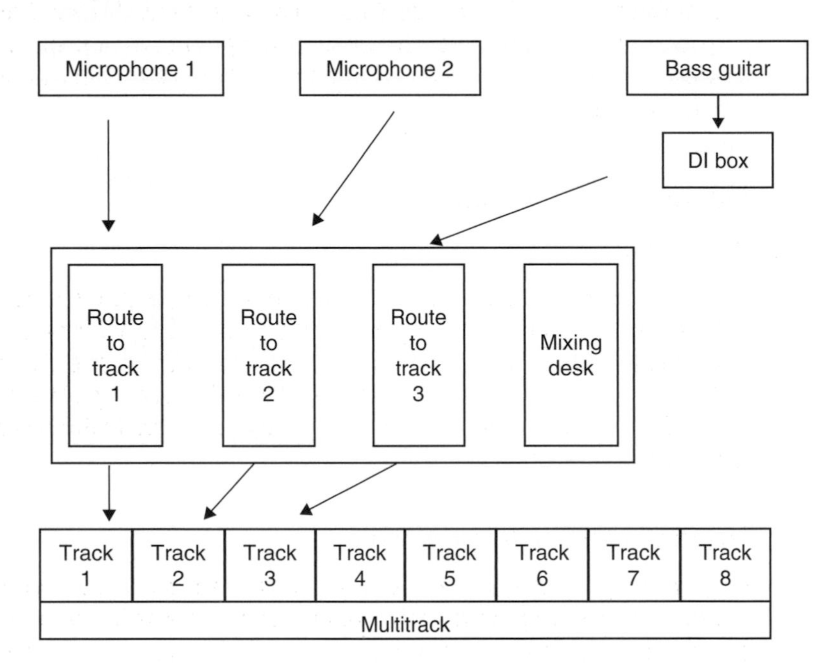

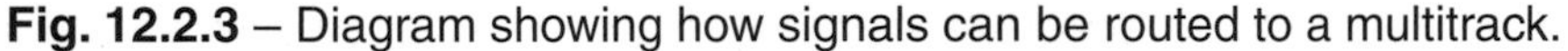

Fig. 12.2.3 – Diagram showing how signals can be routed to a multitrack.

Signal routing

Each individual instrument will need routing to the multitrack. Usually, this can be done by routing the instrument to a bus output on a mixing console or, if you are using a computer system with an audio interface, it may be possible to connect the sound source directly and select the input via the software (see Exercise 11.4). Remember to record enable the tracks you wish to use on the multitrack.

tip ▶

Avoid double monitoring

When routing a signal via a bus to a multitrack, always ensure that the signal is not routed to the stereo mix bus. This will ensure the signal is only routed to one destination at a time and will avoid any potential double monitoring.

Setting the recording levels

If you are recording several instruments or sound sources at the same time you will need to observe a meter for each individual audio track. This will allow you to gauge the signal level being sent to each individual track on the multitrack. If your gain has been set up correctly and your signal has been routed to the multitrack, you should now be able to see the signals registering on the multitrack's meters.

Levels being sent to the multitrack are normally adjusted by moving the bus output fader on the mixing console or adjusting the volume on the sound source itself. Note that if the signal gets recorded too high it may overload and distort, and make the recording unusable. A *compressor* may be used at this stage to help control the recording level (see 'Compressors' section in Chapter 7). However, too much compression at the recording stage can be undesirable and cannot be undone.

Monitoring the sound

Studio monitoring

At this stage you need to set up how are going to monitor the recording process. This is normally done by using the studio's control room monitor speakers as, traditionally, recording studios have separate rooms: one for recording live instruments and one for the engineer (see Chapter 7). Alternatively, headphones can be used if the performers/microphones are in the same room as the recording engineer.

Monitoring the multitrack returns

It's generally standard practice to monitor the signals you are recording via the multitrack's outputs (not the input signal) as this is the final stage of the signal and you can hear if there are any problems in the signal path as you record (see Exercise 11.4). If you are using a computer system you will need to consider if latency will be a problem during recording. Latency is a noticeable delay between the input signal and the output signal that can appear when monitoring audio via a computer system.

Monitor mix

During the recording process you will create a balance in level between each instrument so you can clearly hear what is being recorded. This is called a monitor mix, and is purely designed to help you and the performers with the recording process. It should not be confused with the final mix, as it usually doesn't represent how the final recording will sound.

tip ► Always listen out for any clicks, hums or hiss and try and eliminate them before you start recording.

Foldback

If the performers/musicians are planning to wear headphones during recording or they are located in a dedicated live room, then you will need to provide a foldback mix. This is a separate mix or blend of all the sound sources being used that is sent only to the headphones. It allows the performers/musicians to hear themselves and the playback of any other tracks/instruments, and helps with sound isolation when using microphones.

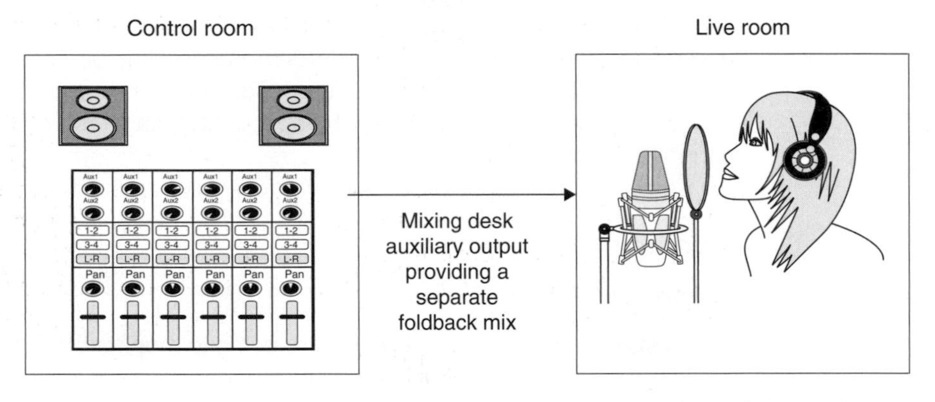
Control room
Live room
Aux1
Aux2
1-2
3-4
L-R
Pan
Mixing desk auxiliary output providing a separate foldback mix
Studio engineer listening to the microphones and other instruments on the studio monitor speakers
Microphones being used to capture sound from the musicians. Musicians wearing headphones listening to the foldback mix

Fig. 12.2.4 – A separate foldback mix allows you to decide exactly what the performers/musicians will hear in their headphones independently of what the engineer is hearing through the control room monitor speakers.

Some performers may find it difficult to record while wearing headphones, as it will seem unnatural and not represent their normal listening environment. Therefore, you should always make sure that the musicians are comfortable with what they are hearing, as a poor foldback mix may affect their performance (see Exercise 11.5).

note ► Musicians/performers should always try and use headphones for monitoring when using microphones, as this will help prevent any unwanted sounds being picked up by the microphone.

Monitoring a DI signal

If you are using a DI box to connect a bass or guitar directly to a mixing desk or soundcard, then you will provide a way for performers to hear themselves. For example, a bass or guitar is usually connected to an amplifier and speaker when used on stage; however, when used in a recording studio, it can be connected directly to the mixing desk via a DI box. This means it can only be heard through the control room studio monitors or via the foldback.

Using a click track

Do you need to set up a click track to provide a timing reference while recording? This depends on several factors such as: can the performers play in time with a click or will it put them off and affect their performance? Can you calculate the tempo of the song? Can you set up a click and provide a sound? Always try to record audio while referencing to a click track if possible, especially if you're using a computer or hard disk system as it will make editing the session a lot easier and allow you to line up the recording with the beats and bars within your digital audio workstation.

tip ►

Creating an ending

Before you start recording, you need to make sure the musicians are clear about how they are going to start and end their performance. If a song doesn't have a natural ending you may want the musicians to repeat the last section over and over several times to provide you with enough time to create a gradual fade out during the final mixdown.

Making a recording

Dropping into Record

Make sure you have located to a suitable position on your multitrack, press Record and then give the performers a visual indication that you are ready for them to start. You may need to wait for a count-in or any pre-roll that has been set up, so make sure you are actually in Record before you ask the performers to start.

During the recording

During the recording keep an eye on the recording levels for each individual track. Remember, the recording level is independent of the monitoring level. For example, you can record a high level without listening too loud. The recording level is determined by the strength of the signal travelling from the bus output to the multitrack's input. The monitoring level is simply the level you choose to listen at and can be adjusted at any time, without affecting the recording levels, either using the monitor faders or the control room speaker level.

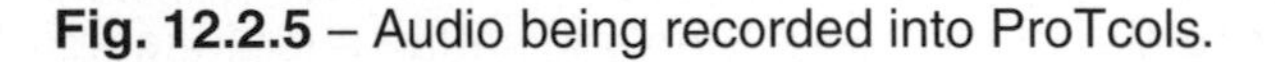

Fig. 12.2.5 – Audio being recorded into ProTcols.

After the recording

Once the performers have finished, press Stop. Don't be too anxious to press Stop, as any additional time at the end of a song can be edited later. It's always better to wait until the sound has completely faded away before pressing Stop, as you may cut off the natural decay of the instruments.

Play back what you have just recorded

Locate back to the beginning of the song and press Play. It's always a good idea to disarm any tracks that have been used for recording to prevent accidentally recording over them while listening. While listening back you may need to make some adjustments to the monitor mix in order to hear each instrument clearly.

Analysing the recording

You now need to analyse the recording and listen out for any mistakes:

- Are all the instruments playing in time?
- Are you happy with the recording?
- Are the performers happy with the recording?
- Are there any clicks or buzzes on the recording?

All the instruments in the rhythm track need to be in time in order to provide a solid base to overdub any other instruments on. If the musicians are not happy with their performance or the timing is bad, you can always consider recording the rhythm track again, providing the musicians are prepared to play again and think they can achieve a better result.

note ► You can always edit what has been played. However, it may save hours of endless fiddling and tweaking and make your music sound more natural if you can simply capture a great live performance.

Punching in and out

If there is only a small mistake or the performer may feel they can do a better job in one particular section of the recording, then you could consider punching in and out. This is a method of recording where you only drop into Record over the section you want to replace. This allows you to correct any mistakes without having to re-record the whole rhythm track again.

To do this successfully you will have to find a suitable *punch-in point* (where the recording will start) and a suitable *punch-out point* (where the recording will end). This is essential as a bad join will sound unnatural and may potentially ruin the recording.

tip ▶ A small mistake in a recording can usually be replaced by punching in and out of Record.

Once you have got a recording of the rhythm track you are satisfied with, you have now completed the first stage of your recording. If you're using a hard disk system press Save to update your session file.

Overdubbing

The next stage is to add some additional instruments to your recording. This process is referred to as overdubbing and allows you to record new instruments while listening back to the rhythm track.

Select the next available audio track and record enable it. Route the signal you plan to record and set your recording levels. Locate back to the beginning of the song and record the next instrument.

tip ▶ Always check the tuning of each instrument before you start overdubbing, as each new instrument will need to blend with the rhythm track.

producer says ▶ Overdubbing makes it easy to try out ideas over the rhythm track, as you can always erase them without affecting any of the other tracks. This can make it time-consuming since you now have the luxury to experiment as you don't have to play the song live with the other musicians. Overdubs can even be added to a recording in small sections or recorded out of sequence in a completely different order – the creative possibilities are endless. You may even come up with an idea or part that only works on the recording and would be impossible to play live with other musicians. Overdubbing even allows the same person to play more than one instrument on a recording. For example, if a musician played the piano or drums when recording the rhythm section, the same person could then overdub a solo instrument while listening to the previously recorded rhythm section. Using this method it is possible to create a one-man band!

Merging audio tracks

If you find that you are running short of tracks during the recording process you may have to consider combining certain instruments or sounds together. This can be achieved by routing and recording the sounds or instruments you want to merge onto another audio track. The advantage of doing this would be to free up more tracks. For example, if you only had four audio tracks available you could combine/merge the first three tracks onto just one audio track; so, instead of using three tracks for the rhythm track, you would only use one track. You could then use the original tracks to record something else.

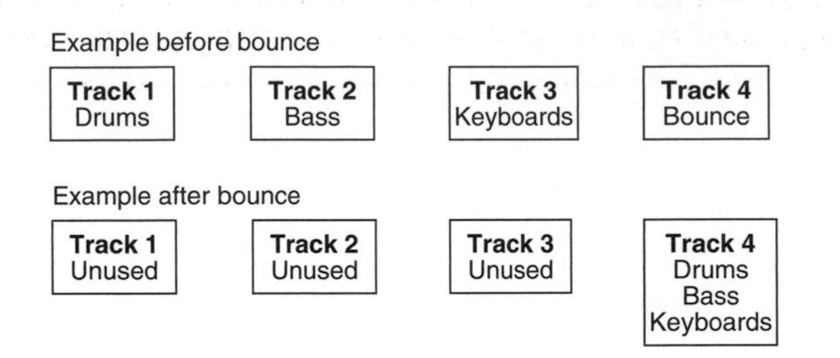

Fig. 12.2.6 – Example showing how the rhythm track could be merged on to a single track.

producer says ▶

> If you choose to combine several instruments or sounds together, you will be unable to process or change the volume, pan or EQ of the combined sounds individually. It is therefore essential that you get a suitable balance between each instrument before you combine them together.

Mixing

Once you have finished recording and overdubbing all the instruments and have done any editing, you are ready to mix. The complexity of the mix will vary depending on how many separate audio tracks you have recorded and what you are looking to achieve from the mix (see Exercise 11.9).

This mixing stage involves setting the individual levels and pan position of each individual audio track, and trying to achieve an overall blend of the sounds that have been recorded. You also have the option of adding EQ to tonally change or enhance each sound by cutting or boosting its frequency, or adding reverb to create distance or add a final gloss to a sound (see Exercise 11.7).

Before you start mixing, try and decide what instruments or sections of the music you want to feature. Also try listening to a commercial CD of other popular music to get an idea of how they have blended different instruments together. Mixing is a personal thing, so each person will have their own ideas how the mix should sound.

Final format

Once you are happy with how your mix sounds you will need to make a recording of it. How you achieve this will vary depending on the type of recording equipment that is being used. Popular formats used for recording a final mix to are DAT, mini-disk and CD. Note that if you are using a computer system and all the mixing has been carried out internally, you may be able to create a stereo mix file by simply bouncing down (see Exercise 11.10).

producer says ▶

> The last bit of advice on this project is something which has been reiterated throughout the book – that of ideas and creativity. You will find that these are equally as important as the recording process itself.

Glossary

ADAT A tape-based modular eight-track digital recorder made by Alesis.

AIFF (audio interchange file format) An uncompressed digital audio file format.

Ambient miking A recording technique where microphones are positioned away from a sound source so they capture the natural acoustics and reverberation of a recording environment.

Amplitude Refers to the level or strength of a signal; when displayed visually the louder the sound the higher the waveform.

Analogue Refers to a continuous unbroken signal as opposed to a digital signal that is broken up and represented by a stream of numbers.

Arrange window The main window you will see on MIDI sequencer or DAW whilst making music.

Attenuate To reduce the level of a signal.

Audio Sound or reproduction of sound.

Audio interface A device that connects to a computer to provide the necessary connections for a computer to record and playback audio.

Audio sequencer (See **DAW**)

Automated mixing A system of mixing that allows parameter movements to be recorded and played back using a computer. This allows complex parameter movements to be built up in several stages and tweaked to perfection.

Auxiliary return (Aux Input) A specially designed mixing desk input that can be used for returning effects such as reverb and delay.

Auxiliary send (Aux Send) A dial or small fader that can be used to control the amount of signal that gets sent to an effects device, such as a reverb or delay can also be used to create a foldback mix. (See **Foldback**)

Balanced connection An audio connection that uses three cables and a clever system to help reduce interference.

Bass notes Term used to describe notes with a low frequency.

Bi-directional (figure of eight) A type of microphone polar pattern that makes a microphone most sensitive to sounds arriving from front and rear while rejecting sounds approaching either side of the microphone. Sometimes called figure-eight because of the shape of its polar pattern.

Bit rate (bit depth) The number of bits (ones and zeros) used to calculate the amplitude of a digital signal each second. Commonly used bit rates are 16 and 24.

Blues Term used to describe a form of music often structured around a 12-bar sequence and regular rhythm. Often includes flattened notes that clash or are outside a more traditional major scale that emphasize the feeling of depression in the music.

BPM Beats per minute. (See **Tempo**)

Bus A facility for combining signals together. A stereo bus combines all the individual channel signals together into stereo, whereas bus outputs on a mixing desk are normally used to send signals to a multitrack recorder.

Cardioid microphone Heart-shaped microphone polar pattern that only picks up sound from one direction.

CD (Compact Disc) A read-only optical disc medium for storing digital audio. The compact disc stores and plays back audio at 44.1kHz.

Channel Can refer to the input signal path of a mixing desk (usually, each channel contains a different signal) or to one side of a stereo signal.

Channel fader A sliding control that will adjust the level of an audio signal for a particular channel.

Channel routing Refers to the output signal path of an individual channel on a mixing desk. This is usually determined by a set of routing buttons that allows you to choose where you want each signal to go.

Chords A combination of notes (usually three or more) played at the same time to create harmony.

Chorus A special effect that can be used to thicken a sound. It can also refer to the main section of a song.

Click track A regular pulse or beat that is used during a recording to help keep in time.

Close miking A recording technique where a microphone is placed close to the sound source.

Coincident pair A stereo microphone technique where two separate microphones are positioned closely together one directly above the other.

Compressor A device that reduces the dynamic range or gain of a signal by means of automatic volume control. When used correctly can keep signals at a consistent volume.

Condenser microphone Condenser microphones are commonly used in a recording studio for miking solo instruments and recording vocals as they have a good frequency response.

Control room A separate room where the engineer and the recording equipment are situated.

Controller Device such as a keyboard used to transmit MIDI data.

Controller data A type of MIDI message that allows you to control a variety of different parameters such as modulation, volume, pan.

Cross-fading An audio editing function that can help create a smooth transition when joining two different sections of audio together.

Crosstalk The unwanted transfer or spill of an audio signal that appears on an adjacent track or channel.

Cue send An output control that can be used to control the amount of signal that gets sent to a pair of headphones. (See **Foldback**)

DAW Abbreviation for digital audio workstation. A computer system that allows you to record, edit and mix audio entirely in a digital form.

Delay A device that can create time-based effects.

Destructive editing Digital editing that rewrites or modifies the original data stored on disk. A destructive edit cannot be undone unless a copy of the original data is saved before the edit is carried out.

DI box (direct injection box) A device used for connecting an instrument such as an electric or bass guitar directly to a microphone input. It also adjusts and matches the required input and output levels.

Digital Signals or information represented by a stream of numbers.

DIN plug Type of connector that can be used to transmit and receive MIDI data.

Double tracking A recording technique where two identical performances are recorded and played back to thicken the sound.

Drum machine Hardware device or software instrument that plays samples of drums. Also may include a sequencer to record and play back rhythm patterns.

Dynamic microphone Dynamic microphones are usually quite robust, which makes them suitable for live use and for miking loud sound sources such as drums and guitar amplifiers.

Dynamic range The range in volume level between the loudest and softest sounds.

Edit windows Additional windows on a MIDI sequencer or DAW that allow you to view and edit information in more detail. Typical examples are: Drum, List, Piano Roll, Score.

Editing Modify or change material in some way. Most MIDI sequencers and DAWs allow complex editing features to cut, delete unwanted material, insert spaces and rearrange recorded material into a desired order.

Effects Term used to describe sound processing devices such as reverb and delay.

EQ (See **Equalization**)

Equalization Control which can boost or cut specific frequency ranges.

Fader A sliding control usually used to adjust the level of an audio signal.

Firewire A versatile computer connection that can be used to connect audio interfaces and hard drives to a computer and transfer large amounts of data quickly.

Foldback A studio monitoring system that allows musicians to hear themselves and previously recorded tracks through headphones.

Frequency Term used to describe how many oscillations a sound makes. A sound that completes 100 cycles each second would have a low frequency (100 Hz) and a low pitch, whereas a sound that completes 10,000 cycles in a second (10,000 Hz) would have a high frequency and pitch. In order for our ears to hear a frequency it must fall somewhere between 20 and 20,000 Hertz.

FX (See **Effects**)

Gain A control that allows you to increase or decrease the level of an audio signal.

Ground bass A style of music where a short musical phrase is constantly repeated in the bass or lower register while the upper parts of the music vary.

Group The ability to bring two or more signal paths together. The collective volume of grouped signals can then be adjusted with a single fader and be sent to a multitrack.

Hard-disk recorder A device dedicated to recording digital audio using a hard disk drive.

Harmony A combination of notes played simultaneously to produce a pleasing musical sound.

Hertz (Hz) The unit of measurement for frequency.

Insert point Point where additional devices such as compressors and noise gates can be inserted into the signal path.

Latency This is the time delay between a computer's audio input and audio output. This can cause problems on some computer systems when recording, as signals monitored through a computer may appear later than they really are and therefore sound out of time.

Live room A room that provides the performers/musicians with their own space to set up and perform while making a recording. Live rooms are usually soundproofed to isolate any sound made by the musicians from the recording engineer in the control room.

Master fader A volume control that adjusts the level of all signals that are routed to the stereo bus. On a mixing desk a stereo master fader controls the overall signal level leaving the console. Moving the master fader will adjust the level being sent out of the mixer to the amp and speakers and the stereo recorder.

Master keyboard (See **Controller**)

Mastering A process where a finished mix is further enhanced using EQ and compression before getting transferred on to CD.

MIC An abbreviation for microphone.

Microphone A device that converts variations in air pressure caused by sound into electrical signals that can be recorded and amplified. (See **Condenser microphone, Dynamic microphone, Cardiod microphone**)

MIDI (Musical Instrument Digital Interface) An agreed standard that allows electronic devices such as synthesizers, drum machines and computers to connect and exchange information with each other.

MIDI cable A cable with a five-pin DIN plug at each end that can be used to physically connect different MIDI devices together.

MIDI channel A numbering system that allows MIDI devices to be identified independently. A standard MIDI system allows up to 16 separate streams of information to travel down one MIDI cable at the same time.

MIDI controller A musical performance device such as an electronic keyboard or set of drum pads that outputs a MIDI signal.

MIDI interface Device that can be connected to a computer via USB or Firewire to provide MIDI input and output connections to a computer.

MIDI ports Connections found on MIDI devices so they can be connected to other MIDI devices. There are three types of MIDI connection available: (1) MIDI IN is for receiving MIDI data from another MIDI device, (2) MIDI OUT is for transmitting MIDI data to another MIDI device and (3) MIDI THRU sends out a duplicate of the incoming MIDI data to allow the MIDI data to be passed on to another MIDI device.

MIDI sequencer Device that will record, play back and edit MIDI information.

Mixing The process of blending individual sounds or instruments together to shape the overall sound.

Mixing desk (mixer) A device that can combine audio signals together and control their volume. Usually has many additional functions such as equalization, pan pots, auxiliary sends and monitoring controls.

Monitoring system A loudspeaker system or pair of headphones that are used for listening and judging the overall sound quality during the recording process.

Mono Sound reproduction using a single channel. When an identical signal is fed to both speakers it is mono.

Monophonic Describing a synthesizer that plays only one note at a time (not chords).

MP3 (MPEG Layer-3) A data compression format commonly used on the Internet. Audio in an MP3 file is compressed so it downloads faster; however, the sound quality is reduced.

Multitambral A MIDI device such as a synthesiser or sound module that has the ability to produce two or more different patches or sounds at the same time.

Multitrack Audio recording device with more than two tracks that allows sounds to be recorded onto their own separate tracks.

Multitrack recording A recording technique where additional sounds or instruments can be added to an existing recording allowing a recording to be built up in sections.

Near-field monitoring Studio monitor speakers that are placed very near the listener in a control room (usually just behind the mixing console).

Noise gate A device that can be used to automatically control and eliminate unwanted sounds such as noise and tape hiss from a recording and ensure that no sound is heard when an instrument stops playing.

Non-destructive editing Digital editing that can be undone and returned to its original state. This type of editing does not rewrite or modify the original data stored on disk.

Octave A regular musical interval of eight notes where the upper frequency is twice the lower frequency.

Omni directional A microphone polar pattern that makes a microphone become equally sensitive to sounds arriving from all directions.

Overdubbing Recording technique where a new musical part is recorded while listening back to previously recorded tracks. Commonly used to add additional sounds and instruments to an existing recording.

Pad Switch commonly found on mixing desk input channels and microphones that prevents overloading by reducing the signal level.

Pan control Dial that moves a signal from left to right within a stereo field.

Phantom power A DC voltage (usually 12–48 volts) that can be sent down a balanced audio cable to power a condenser microphone or DI box.

Pitch The perceived lowness or highness of a sound. Each note on a keyboard corresponds to a set frequency. For example, a frequency of 440 Hertz would produce the note A.

Pitch bend A MIDI message that gives you the ability to continually vary the pitch of a note and create a similar action like bending a note on a guitar. It is usually transmitted by using a sprung wheel or joystick on a MIDI controller keyboard.

Plug-in A piece of software that runs within a DAW to provide additional functionality. These can vary from simple effects such as reverb or delay to complex processors such as EQ and compressors. There is a wide variety of different plug-ins available and most DAWs allow additional plug-ins to be added.

Polar pattern Determines the direction in which a microphone will be most sensitive.

Polyphonic Describing a synthesizer or sound module that can play more than one note at a time.

Portastudio A self-contained recording device that contains a mixing desk and a multitrack recorder. Usually portable which makes them ideal for location recording.

Program change A special type of MIDI message that allows you to select and recall different program numbers using MIDI.

Quantize An editing function that automatically moves notes to the nearest beat or sub-beat within your sequencer. When used correctly can make your music sound more in time and generally tighten up a performance.

Reggae Reggae music originated from the island of Jamaica and is a combination of traditional mento and American style R&B. The rhythm is often characterized by regular chops on the off-beat, which are usually played using a piano or guitar chord.

Reverberation Natural reverberation is caused when sound bounces off surrounding surfaces in an enclosed space. It can be generated artificially and added to a recording to simulate real spaces such as rooms, halls and auditoriums.

Scale A predetermined sequence of notes that are played in an ascending or descending order.

Sequencer (See **MIDI sequencer**)

Snap A function which will restrict movement of audio and MIDI information against a timeline in a DAW. Snap values can be adjusted to the nearest bar/half bar/beat or to an even smaller value.

Solo Pressing solo allows you to listen to the selected channel on its own.

Song position marker Normally a thin line that will move horizontally across the window when the DAW plays so you can visually see the current position. It can display beats and bars or actual time in hours, minutes and seconds. This makes lining up parts much easier as it provides a common reference point when copying and moving MIDI or audio data.

Sound module A synthesizer without a keyboard, containing several different timbres or voices. These sounds are triggered or played by MIDI signals from a sequencer program or by a MIDI controller.

Stereo A two-channel system that uses two speakers to simulate how we perceive sound coming from different directions.

Synthesizer A musical instrument (usually with a piano-style keyboard) that can create and modify a variety of different sounds electronically.

Synthetic Refers to a sound that is created electronically to imitate a real sound artificially.

Talkback An intercom system that is usually built in to a mixing console so the recording engineer and producer can talk to the musicians in the studio.

Tempo Fundamental speed or pulse of a piece of music.

Two-track A recording device that uses two independent tracks, usually for stereo.

Unbalanced An audio connection that uses two wires.

USB (universal serial bus) A computer serial interface for connecting external devices such as MIDI interfaces and audio interfaces to a computer.

Velocity Velocity is part of a MIDI 'note on' message and is sent each time you press a key on a MIDI keyboard. It measures how hard a key is pressed and determines the individual volume of each note.

Virtual instrument (soft synth) A synthesizer in software form.

WAV, Wave A computer audio file format for Windows that encodes sound without any data compression or data reduction. Its audio resolution can be 16-bit, 44.1 kHz, or higher.

Index